MATHEMATICAL SURVEYS AND MONOGRAPHS SERIES LIST

Volume

1. The problem of moments, J. A. Shohat and J. D. Tamarkin
2. The theory of rings, N. Jacobson
3. Geometry of polynomials, M. Marden
4. The theory of valuations, O. F. G. Schilling
5. The kernel function and conformal mapping, S. Bergman
6. Introduction to the theory of algebraic functions of one variable, C. C. Chevalley
7.1 The algebraic theory of semigroups, Volume I, A. H. Clifford and G. B. Preston
7.2 The algebraic theory of semigroups, Volume II, A. H. Clifford and G. B. Preston
8. Discontinuous groups and automorphic functions, J. Lehner
9. Linear approximation, Arthur Sard
10. An introduction to the analytic theory of numbers, R. Ayoub
11. Fixed points and topological degree in nonlinear analysis, J. Cronin
12. Uniform spaces, J. R. Isbell
13. Topics in operator theory, A. Brown, R. G. Douglas, C. Pearcy, D. Sarason, A. L. Shields; C. Pearcy, Editor
14. Geometric asymptotics, V. Guillemin and S. Sternberg
15. Vector measures, J. Diestel and J. J. Uhl, Jr.
16. Symplectic groups, O. Timothy O'Meara
17. Approximation by polynomials with integral coefficients, Le Baron O. Ferguson
18. Essentials of Brownian motion and diffusion, Frank B. Knight
19. Contributions to the theory of transcendental numbers, Gregory V. Chudnovsky
20. Partially ordered abelian groups with interpolation, Kenneth R. Goodearl
21. The Bieberbach conjecture: Proceedings of the symposium on the occasion of the proof, Albert Baernstein, David Drasin, Peter Duren, and Albert Marden, Editors
22. Noncommutative harmonic analysis, Michael E. Taylor
23. Introduction to various aspects of degree theory in Banach spaces, E. H. Rothe
24. Noetherian rings and their applications, Lance W. Small, Editor
25. Asymptotic behavior of dissipative systems, Jack K. Hale
26. Operator theory and arithmetic in H^∞, Hari Bercovici
27. Basic hypergeometric series and applications, Nathan J. Fine
28. Direct and inverse scattering on the lines, Richard Beals, Percy Deift, and Carlos Tomei
29. Amenability, Alan L. T. Paterson
30. The Markoff and Lagrange spectra, Thomas W. Cusick and Mary E. Flahive

MATHEMATICAL SURVEYS AND MONOGRAPHS SERIES LIST

Volume

31 **Representation theory and harmonic analysis on semisimple Lie groups,** Paul J. Sally, Jr. and David A. Vogan, Jr., Editors

32 **An introduction to CR structures,** Howard Jacobowitz

33 **Spectral theory and analytic geometry over non-Archimedean fields,** Vladimir G. Berkovich

34 **Inverse source problems,** Victor Isakov

35 **Algebraic geometry for scientists and engineers,** Shreeram S. Abhyankar

36 **The theory of subnormal operators,** John B. Conway

37 **Structural properties of polylogarithms,** Leonard Lewin, Editor

38 **Analysis of and on uniformly rectifiable sets,** Guy David and Stephen Semmes

Mathematical
Surveys and Monographs

Volume 38

Analysis of and on Uniformly Rectifiable Sets

Guy David
Stephen Semmes

American Mathematical Society
Providence, Rhode Island

1991 *Mathematics Subject Classification*. Primary 28A75; Secondary 42B20, 30C65, 30C85, 30G35, 49Q15.

Library of Congress Cataloging-in-Publication Data

David, Guy, 1957–
 Analysis of and on uniformly rectifiable sets / Guy David, Stephen Semmes.
  p. cm.—(Mathematical surveys and monographs, ISSN 0076-5376; v. 38)
 Includes bibliographical references and index.
 ISBN 0-8218-1537-7 (acid-free)
  1. Geometric measure theory. 2. Singular integrals. 3. Functions of complex variables.
I. Semmes, Stephen, 1962– . II. Title. III. Series: Mathematical surveys and monographs; no. 38
QA312.D27   1993
515'.42—dc20   93-36311
                                                                                    CIP

Copying and reprinting. Individual readers of this publication, and nonprofit libraries acting for them, are permitted to make fair use of the material, such as to copy a chapter for use in teaching or research. Permission is granted to quote brief passages from this publication in reviews, provided the customary acknowledgment of the source is given.

 Republication, systematic copying, or multiple reproduction of any material in this publication (including abstracts) is permitted only under license from the American Mathematical Society. Requests for such permission should be addressed to the Manager of Editorial Services, American Mathematical Society, P.O. Box 6248, Providence, Rhode Island 02940-6248. Requests can also be made by e-mail to reprint-permission@math.ams.org.

 The owner consents to copying beyond that permitted by Sections 107 or 108 of the U.S. Copyright Law, provided that a fee of $1.00 plus $.25 per page for each copy be paid directly to the Copyright Clearance Center, Inc., 222 Rosewood Drive, Danvers, Massachusetts 01923. When paying this fee please use the code 0076-5376/93 to refer to this publication. This consent does not extend to other kinds of copying, such as copying for general distribution, for advertising or promotional purposes, for creating new collective works, or for resale.

© Copyright 1993 by the American Mathematical Society. All rights reserved.
Printed in the United States of America
The American Mathematical Society retains all rights
except those granted to the United States Government.
⊚ The paper used in this book is acid-free and falls within the guidelines
established to ensure permanence and durability.
♻ Printed on recycled paper.
This publication was typeset using $\mathcal{A}_{\mathcal{M}}\mathcal{S}$-TEX,
the American Mathematical Society's TEX macro system.

10 9 8 7 6 5 4 3 2 1   97 96 95 94 93

Table of Contents

Preface   ix

Notation and Conventions   xi

PART I: Background Information and the Statements of the
        Main Results

Chapter 1. Reviews of Various Topics   3
  1.1 Review from geometric measure theory   3
  1.2 Review of some topics concerning singular integral operators
    and rectifiability   7
  1.3 Review of some aspects of Littlewood-Paley theory in
    connection with rectifiability   16
  1.4 Various characterizations of uniform rectifiability   21
  1.5 The weak geometric lemma and its relatives   26

Chapter 2. A Summary of the Main Results   31
  2.1 The results of Part II   31
  2.2 Bilateral approximation from a functorial point of view   37
  2.3 The results of Part III   42
  2.4 A rapid description of Part IV   50

Chapter 3. Dyadic Cubes and Corona Decompositions   53
  3.1 Cubes   53
  3.2 Corona decompositions   55
  3.3 Generalized corona decompositions   63

PART II: New Geometrical Conditions Related to Uniform
         Rectifiability

Chapter 1. One-Dimensional Sets   69
  1.1 The weak connectedness condition   69
  1.2 The weaker local symmetry condition $(d = 1)$ 77
  1.3 Weak constant density for one-dimensional sets   86
  1.4 The weak "two points on spheres" condition   93

Chapter 2. The Bilateral Weak Geometric Lemma and its Variants   97
   2.1 Introduction; the corona method   97
   2.2 Big projections in codimension 1   104
   2.3 Big projections in the higher codimension case   110
   2.4 The local convexity condition LCV   120
   2.5 The weaker local convexity condition WLCV   126
   2.6 Weak starlikeness   129
   2.7 Some questions about variants of the LCV and the LS   131

Chapter 3. The WHIP and Related Conditions   135
   3.1 The WHIP, the WTP, and uniform rectifiability   135
   3.2 The WHIP and weaker versions of the BWGL   138
   3.3 The weak exterior convexity condition and the GWEC   141
   3.4 The weak-no-mugs, weak-no-boxes, and weak-no-reels conditions   147
   3.5 The proof of Theorem 3.9 (part 1)   154
   3.6 Part 2 of the proof: The stopping-time argument   165

Chapter 4. Other Conditions in the Codimension 1 Case   183
   4.1 Introduction   183
   4.2 Labellings   187
   4.3 The derivation of Theorem 4.9 from Theorem 4.31   196

PART III: Applications

Chapter 1. Uniform Rectifiability and Singular Integral Operators   207
   1.1 Preliminaries   207
   1.2 Step one   208
   1.3 Step two   212
   1.4 An abstraction of §3   214

Chapter 2. Uniform Rectifiability and Square Function Estimates for the Cauchy Kernel   217
   2.1 Some general comments about square function estimates   217
   2.2 Uniform rectifiability implies the USFE when $d = 1$ 219
   2.3 From square function estimates to uniform rectifiability: Preliminary reductions and the plan of the proof   226
   2.4 The proof of Lemma 2.36   229
   2.5 A topological lemma   232
   2.6 The main step in the proof of Proposition 2.38   234
   2.7 The end of the proof of Proposition 2.38   244

Chapter 3. Square Function Estimates and Uniform Rectifiability in Higher Dimensions   249
   3.1 A brief review of Clifford analysis   249
   3.2 Clifford analysis and square function estimates   251

3.3 From square functions to uniform rectifiability: Preliminary
       reductions   252
   3.4 Cauchy flatness implies rectifiability   253
   3.5 The analogue of Proposition 2.59   256
   3.6 Cauchy flatness implies weak flatness   261
   3.7 Weak flatness implies exterior convexity   265
   3.8 Some remarks about the higher-codimension case   267

Chapter 4. Approximating Lipschitz Functions by Affine Functions   269
   4.1 The direct estimates   269
   4.2 The converse when $d = 1$ 279
   4.3 A more abstract version of the WALA   293

Chapter 5. The Weak Constant Density Condition   297
   5.1 Compactness will only get you so far   297
   5.2 The codimension 1 case, part 1   301
   5.3 A general lemma about Carleson packing conditions   305
   5.4 The codimension 1 case, part 2   306
   5.5 The weak dyadic density condition   307

PART IV: Direct Arguments for Some Stability Results

Chapter 1. Stability of Various Versions of the Geometric Lemma   313
   1.1 The statements   313
   1.2 A John-Nirenberg-Strömberg lemma for Carleson packing
       conditions   315
   1.3 Two lemmas on approximations of regular sets by d-planes   318
   1.4 The proof of the theorems   324

Chapter 2. Stability Properties of the Corona Decomposition   327
   2.1 Corona decompositions revisited   327
   2.2 Corona constructions and Lipschitz functions   328
   2.3 The statement of the main result   336
   2.4 Preliminaries   336
   2.5 The proof of Lemma 2.38   340

References   345

Table of Selected Notation   349

Table of Acronyms   351

Table of Theorems   353

Index   355

Preface

This monograph is concerned with quantitative notions of rectifiability of subsets of $\mathbf{R}^n$. It is not a priori obvious that there should be one "right" quantitative notion of rectifiability, but the results in [**DS2**] indicate that there is one that has many good features and connections with other questions; we shall refer to it as uniform rectifiability.

The motivations for the results presented here derive from a mixture of geometric measure theory and harmonic analysis. This mixture is slightly complicated because the two subjects interact in more than one way. For instance, we are interested in the analysis of functions that live on a d-dimensional set in $\mathbf{R}^n$ and in the behavior of certain linear operators that act on these functions (for instance, variants of the Cauchy integral operator), and these issues turn out to be related to the rectifiability properties of the set in question. However, we are also interested in the geometry of subsets of $\mathbf{R}^n$ for its own sake, and it happens that some of the ideas developed in harmonic analysis have significant counterparts in the geometric setting. It is also true that the problem of understanding what happens to rectifiability when you make it quantitative in a scale-invariant way is natural in its own right.

In order to help make manifest some of the ways in which these topics interact, we have included in Part I a fairly extensive review of the relevant background information (without proofs but with references to the literature).

We also give in Part I a summary of the main results of this monograph, as well as a review of some technical matters that will be used heavily. The proof of the main results are given in Part II and Part III, together with some variants that are not stated in the summary in Part I.

Part IV is somewhat separate from the rest. There are places in Part II where a proof can be completed by using the theorem in [**DS2**] to pass from one kind of geometric information to another (apparently stronger) kind. We provide an alternative to this in Part IV by giving new and more direct proofs of the needed consequences of [**DS2**]. These new proofs together with some of the results in Part II also provide new approaches to substantial portions of the main theorem in [**DS2**].

The authors wish to thank R. R. Coifman, Robert Hardt, and Peter Jones for helpful suggestions. The second author gratefully acknowledges support from the Marian and Speros Martel Foundation, the National Science Foundation, and the Alfred P. Sloan Foundation. Portions of this work were accomplished during visits of the second author to the Université de Paris-Sud and the Institut des Hautes Études Scientifiques.

Notation and Conventions

Generally speaking, we shall always be working with subsets of $\mathbf{R}^n$ that have Hausdorff dimension d, where d is an integer such that $0 < d < n$.

We shall denote by $H^d(A)$ the d-dimensional Hausdorff measure of A, the definition of which is recalled in §1.1 of Part I. We shall frequently write $|A|$ for $H^d(A)$. Similarly, when we write $\int_A f(x)\,dx$, "dx" should be assumed to be the restriction of $H^d(\cdot)$ to A, unless there is a statement to the contrary.

When $d = 1$, we often refer to $H^1(A)$ as the length of A.

$B(x, r)$ and $\overline{B}(x, r)$ will denote, respectively, the open and the closed balls with center x and radius r. Given a ball B, we write λB for the ball with the same center and λ times the radius.

By "Lipschitz" we always mean "Lipschitz of order 1". Thus if $X \subset \mathbf{R}^n$ and if f is a function defined on X, then f is Lipschitz if there is a constant $M \geq 0$ such that

$$|f(p) - f(q)| \leq M|p - q| \qquad \text{for all } p, q \in X.$$

The smallest M such that this holds is called the Lipschitz norm, or Lipschitz constant, of f.

When we speak of a d-dimensional Lipschitz graph, we mean a subset of $\mathbf{R}^n$ of the form $\{(x', x'') \in \mathbf{R}^d \times \mathbf{R}^{n-d} \simeq \mathbf{R}^n : x'' = A(x')\}$, where A is a Lipschitz mapping from $\mathbf{R}^d$ to $\mathbf{R}^{n-d}$, or a subset of $\mathbf{R}^n$ which is the image of such a set by a rotation. Thus Γ is a d-dimensional Lipschitz graph in $\mathbf{R}^n$ if there exists a d-plane $P \subset \mathbf{R}^n$, a $(n-d)$-plane $P^\perp$ orthogonal to P, and a Lipschitz mapping from P to $P^\perp$ such that Γ is the graph of A (i.e., the set $\{p + A(p), p \in P\}$).

We shall say that Γ is a Lipschitz graph with constant $\leq M$ if, in the representation above, we can take A to be Lipschitz with constant $\leq M$.

We shall follow the common practice of writing C for a positive real number which is likely to be somewhat large and whose precise value may change with each occurrence. Generally we make explicit the dependence of our constants on auxiliary parameters, or we state clearly our intention to ignore such dependence. We shall sometimes describe a constant as being "geometric" if it depends on particularly innocuous quantities, like the relevant dimensions, or the regularity constant of the set under consideration. (See Definition I.1.13.)

This monograph is composed of four parts, which are further divided into chapters, and then sections. Formulas, lemmas, etc., will be given labels like (2.7), where 2 refers to the second chapter of the current part, and 7 to the position of the formula in the whole chapter. When we need to refer to a statement from some other part, we shall use a more explicit labelling like Lemma II.3.1, which refers to the first lemma of Chapter 3 in Part II.

Part I

Background Information and the Statements of the Main Results

CHAPTER 1

Reviews of Various Topics

For the convenience of the reader, we have included in this monograph a fairly long description of our motivations and of relevant previous results. We have also tried to make it possible, and as painless as possible, for the reader to proceed directly to Chapter 2 of Part I, or at least skip the first three sections of Chapter 1.

1.1. Review from geometric measure theory.

General references for this section include [**Fe**], [**Fl1**], [**Ma1**], [**Ma3**], and [**Si**]. We shall largely follow [**Ma1**], especially Chapter 6.

Fix integers d and n, $0 < d < n$.

For the record, let us write down the definition of d-dimensional Hausdorff measure H^d that we shall use. For each $\delta > 0$ and $A \subset \mathbf{R}^n$, define

$$H^d_\delta(A) = \inf\left\{\sum_i (\operatorname{diam} E_i)^d\right\},$$

where the infimum is taken over all sequences E_i of subsets of $\mathbf{R}^n$ such that $A \subset \bigcup_i E_i$ and each E_i has diameter $\leq \delta$.

As usual, we note that $H^d_\delta(A)$ increases as δ decreases, and we set

(1.1) $$H^d(A) = \lim_{\delta \to 0} H^d_\delta(A) = \sup_{\delta > 0} H^d_\delta(A).$$

Let us record also the standard definition that $A \subset \mathbf{R}^n$ is H^d-measurable if

$$H^d(E) = H^d(E \cap A) + H^d(E \setminus A) \quad \text{for all} \ E \subset \mathbf{R}^n.$$

We shall be mostly concerned in this section with the notion of rectifiability. We say that a subset A of $\mathbf{R}^n$ is rectifiable (or, more precisely, d-rectifiable) if there exists a (countable!) sequence of Lipschitz mappings $f_j : \mathbf{R}^d \to \mathbf{R}^n$ such that A is covered by the union of their images except for a set of H^d-measure zero, i.e.,

$$H^d\left(A \setminus \left[\bigcup_j f_j(\mathbf{R}^d)\right]\right) = 0.$$

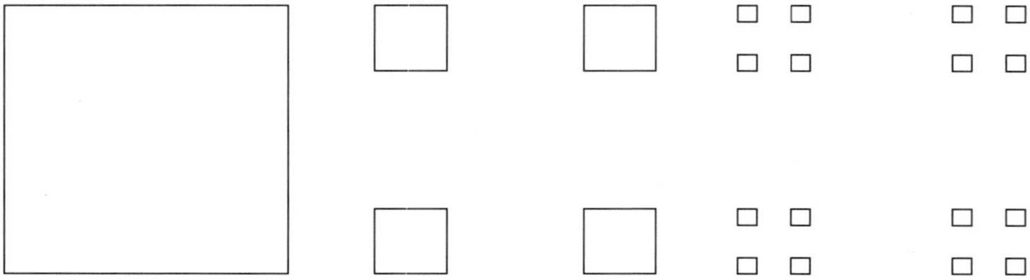

FIGURE 1.1

On the other hand, A is said to be totally unrectifiable if $H^d(A \cap B) = 0$ for every rectifiable set B in $\mathbf{R}^n$.

We should warn the reader that this terminology is slightly different from that of [**Fe**]; we are using the word "rectifiable" in places where "countably rectifiable" would be used in [**Fe**].

One of the good features of the notion of rectifiability is that it is very robust. For instance, A is rectifiable if and only if there is a countable sequence of d-dimensional Lipschitz graphs Γ_j such that $H^d\left(A \setminus \bigcup_j \Gamma_j\right) = 0$. We would also have an equivalent definition if we required instead that the Γ_j's be C^1-submanifolds.

Let us give an example of a one-dimensional totally unrectifiable set in $\mathbf{R}^2$. We define this set K by a Cantor-type construction. Let K_0 denote the unit square $[0, 1] \times [0, 1]$. Let K_1 denote the set that results by replacing K_0 with the four squares in the corners of K_0 that have sidelength $\frac{1}{4}$.

Repeat this procedure, so that K_j is the union of 4^j squares of sidelength 4^{-j} and K_{j+1} is obtained from K_j by replacing each of the squares composing K_j by four smaller squares in the corner, each having sidelength one-fourth of the sidelength of the parent. See Figure 1.1 for a picture of K_0, K_1, and K_2.

Thus we have $K_{j+1} \subset K_j$ for each j, and so $K = \bigcap_j K_j$ is a nonempty compact set in the plane. It can be shown that K has positive and finite one-dimensional Hausdorff measure and that K is totally unrectifiable. This set K has other interesting features. In particular, its analytic capacity is zero (see [**Ga1**]), and the Cauchy integral operator is unbounded on $L^2(K)$. This provides a nice illustration of the fact that rectifiability properties of a set E can be connected to more analytical properties of E, such as the boundedness of certain operators on $L^2(E)$. The set K is often referred to as the Garnett counterexample because of the result in [**Ga1**]. See §2 of this chapter for a few more details.

The dichotomy between rectifiable and totally unrectifiable sets is a rather clean one. One of the basic results is that if $H^d(A) < +\infty$, then there is a rectifiable set $B \subset A$ such that $B \backslash A$ is totally unrectifiable. This result is very useful when it comes to establishing various characterizations of rectifiability.

There is a wide variety of known geometric characterizations of rectifiability. We shall state three of them, which are given in terms of approximate tangent planes, densities, and projections. We first introduce some notation.

Given $A \subset \mathbf{R}^n$ and $a \in \mathbf{R}^n$, define the d-dimensional upper and lower densities of A at a by

$$(1.2) \qquad \Theta^*(A, a) = \limsup_{r \to 0} (2r)^{-d} H^d(A \cap B(a, r))$$

and

$$(1.3) \qquad \Theta_*(A, a) = \liminf_{r \to 0} (2r)^{-d} H^d(A \cap B(a, r)).$$

When these agree, we say that the (d-dimensional) density of A at a exists and denote it by $\Theta(A, a) = \Theta^*(A, a) = \Theta_*(A, a)$.

For any $A \subset \mathbf{R}^n$ and $a \in \mathbf{R}^n$, we say that a d-plane P that passes through a is the approximate tangent d-plane to A at a if $\Theta^*(A, a) > 0$ and if

$$(1.4) \qquad \limsup_{r \to 0} r^{-d} H^d(\{x \in A \cap B(a, r) : \operatorname{dist}(x, P) > s|x - a|\}) = 0$$

for all $s > 0$.

Clearly such a d-plane is unique if it exists.

THEOREM 1.5. *Suppose that $A \subset \mathbf{R}^n$ is H^d-measurable and that $H^d(A) < +\infty$. Then A is rectifiable if and only if there is an approximate tangent d-plane to A at a for almost every $a \in A$, and A is totally unrectifiable if and only if, for almost all $a \in A$, there does not exist an approximate tangent d-plane to A at a.*

(When we write "almost all $a \in A$", we mean almost all with respect to H^d.)

For this theorem, it is very important that we work with approximate tangent d-planes rather than the conventional notion of tangent d-plane. It is easy to build examples of sets that are rectifiable and have finite H^d-measure but for which there are no conventional tangent d-planes; we could, for instance, add to a set A a countable dense set, or do variants of this.

THEOREM 1.6. *Suppose that $A \subset \mathbf{R}^n$ is H^d-measurable and $H^d(A) < +\infty$. Then A is rectifiable if and only if $\Theta(A, a) = 1$ for almost all $a \in A$. On the other hand, A is totally unrectifiable if and only if $\Theta_*(A, a) < 1$ for almost all $a \in A$.*

In fact, David Preiss [P] has proved that A is rectifiable if $\Theta(A, a)$ merely exists for almost all $a \in A$. He has even proved that for each d

and n there exists a constant $\omega(n, d) > 0$ such that if $0 < \Theta^*(A, a) < [1 + \omega(n, d)]\Theta_*(A, a) < +\infty$ holds for almost all $a \in A$, then A must be rectifiable.

Let $G(n, d)$ denote the Grassmann manifold of d-planes in $\mathbf{R}^n$ that contain the origin. It is very easy to see that $G(n, d)$ is, in a natural way, a smooth manifold, so "almost all d-planes $P \in G(n, d)$" makes sense. Given $P \in G(n, d)$, let $\Pi_P : \mathbf{R}^n \to P$ denote the orthogonal projection onto P.

THEOREM 1.7. *Suppose that $A \subset \mathbf{R}^n$ is H^d-measurable and $H^d(A) < +\infty$. Then A is rectifiable if and only if $H^d(\Pi_P(B)) > 0$ for almost all $P \in G(n, d)$ whenever B is a H^d-measurable subset of A such that $H^d(B) > 0$. On the other hand, A is totally unrectifiable if and only if $H^d(\Pi_P(A)) = 0$ for almost all $P \in G(n, d)$.*

This theorem provides a very simple geometric criterion for a set to be rectifiable or at least to have a substantial rectifiable part. (Indeed, if we know that the mean value of $H^d(\Pi_P(A))$ over $P \in G(n, d)$ is, say, ≥ 1, then the rectifiable part of A should have measure ≥ 1, because the Π_P never increase Hausdorff measure.) One of the difficulties with quantitative rectifiability conditions is the absence (so far, anyway) of a suitable counterpart to this result.

The case when $d = 1$ is very special in geometric measure theory, in large part because of the role of connectedness. When $d = 1$, every compact connected set which has finite length is rectifiable. A proof of this can be found in [Fl1]. For us it will be helpful to know the following slightly more precise result.

THEOREM 1.8. *There is a constant $C = C(n)$ such that, whenever $A \subset \mathbf{R}^n$ is compact, connected and such that $H^1(A) < +\infty$, there is a positive number L and a Lipschitz function $f : [0, L] \to \mathbf{R}^n$ such that $A = f([0, L])$, $H^1(A) \leq L \leq CH^1(A)$, and $|f'(u)| = 1$ a.e. on $[0, L]$.*

Let us even sketch the proof of Theorem 1.8. The main point is the following elementary well-known result from graph theory.

LEMMA 1.9. *Let G be a connected graph, with only finitely many edges. Then there is a path that traverses each edge of G exactly twice (once in each direction).*

This is easily proved by induction on the number of edges.

To prove Theorem 1.8, we shall approximate A by a graph, apply Lemma 1.9, parameterize the path at constant speed, and then pass to the limit. (Actually, we shall also have to reparameterize by arclength at the end.)

Given $\rho > 0$, $\rho < \frac{1}{10} \operatorname{diam} A$, let X_ρ be a maximal subset of A with the property that $|x - y| \geq 2\rho$ whenever x, y are two distinct points of X_ρ. Thus X_ρ is finite for each $\rho > 0$. Let A_ρ be the set that you get by joining every pair of points x and $y \in X_\rho$ such that $|x - y| \leq 4\rho$ by a line segment. Then A_ρ can be associated to an abstract graph in the obvious way.

Let us check that A_ρ is connected. If it were not, then we could write $X_\rho = Y \cup Z$, where Y and Z are disjoint, nonempty, and satisfy $|y-z| > 4\rho$ for all $y \in Y$ and $z \in Z$. Then A itself would be disconnected because $A \cap (\bigcup_{y \in Y} \overline{B}(y, 2\rho))$ and $A \cap (\bigcup_{z \in Z} \overline{B}(z, 2\rho))$ would be nonempty disjoint closed subsets of A whose union contains A. This last is because X_ρ is maximal.

Let N_ρ be the number of points in X_ρ. We have $H^1(A_\rho) \leq C\rho N_\rho$. On the other hand, the connectedness of A implies that $H^1(A \cap B(x, \rho)) \geq \rho$ for each $x \in X$. This is because for each covering of $A \cap B(x, \rho)$ by sets E_i, the sets $\widetilde{E}_i = \{|x - t| : t \in E_i\}$ have smaller diameters and cover $[0, \rho)$. Since the $A \cap B(x, \rho)$ are disjoint, we get $\rho N_\rho \leq H^1(A)$, and therefore $H^1(A_\rho) \leq CH^1(A)$.

We now apply Lemma 1.9 to the graph associated to A_ρ. Taking a parameterization with constant speed of the path given by Lemma 1.9, we obtain a Lipschitz mapping $\gamma_\rho : [0, 1] \to A_\rho$ such that $\gamma_\rho([0, 1]) = A_\rho$ and $\|\gamma'_\rho\|_\infty \leq CH^1(A)$, where of course C does not depend on ρ.

By the Arzela-Ascoli theorem, we can find a sequence ρ_j such that ρ_j tends to 0 and the γ_{ρ_j} converge uniformly on $[0, 1]$. The limit is a function $\gamma : [0, 1] \to \mathbf{R}^n$, with Lipschitz norm $\leq CH^1(A)$. It is easy to check that $\gamma([0, 1]) = A$. The desired mapping f is obtained by reparameterizing γ by arclength. It is not hard to check that f has the required properties. This completes our sketch of the proof of Theorem 1.8.

1.2. Review of some topics concerning singular integral operators and rectifiability.

Some general references for this section include [**Ch**], [**D4**], [**Ma3**], and [**Mu**]. In particular, the reader may wish to consult [**Ch**], [**Ma3**], and [**Mu**] for information about the relationship between analytic capacity and the Cauchy integral operator.

Assume for the beginning that $d = 1$ and $n = 2$, and identify $\mathbf{R}^2$ with $\mathbf{C}$.

Roughly speaking, we would like to know the answer to the following question: under what conditions on an H^1-measurable set $E \subset \mathbf{C}$ is it true that

$$(1.10) \qquad Tf(z) = \text{p.v.} \int_E \frac{1}{z-w} f(w) \, dw$$

defines a bounded operator on $L^2(E)$? Here dw denotes the restriction of H^1 to E, and $L^2(E)$ is defined in terms of H^1 also.

In this formulation of the problem, the existence of the principal value is a slightly awkward issue. We shall avoid it in the following manner. For each

$\epsilon > 0$, define T_ϵ by

$$(1.11) \qquad T_\epsilon f(z) = \int_{\substack{w \in E \\ |z-w| > \epsilon}} \frac{1}{z-w} f(w)\, dw,$$

as a linear operator from $L^1(E)$ into the space of continuous functions on E.

It is reasonable to change the above question to the following: under what conditions on E is it true that T_ϵ defines a bounded linear operator on $L^2(E)$ for all $\epsilon > 0$, with operator norm bounded uniformly in ϵ?

Let us say that E is good for the Cauchy kernel if this is true. There is quite a bit known about the relationship between the geometry of E and the property of being good for the Cauchy kernel. Some of the main points are as follows.

If E is a line, then E is good for the Cauchy kernel. This is a classical result, and there are many proofs known. In this case, we have in particular that the T_ϵ are convolution operators, which is very helpful.

If E is, say, a bounded arc which is of class $C^{1,\alpha}$ for some $\alpha > 0$, then E is good for the Cauchy kernel. When E also does not cross itself, it is rather easy to reduce to the case of the line by standard methods. When E is not Jordan, you can cut it up into a finite number of Jordan pieces and deal with them separately. You just have to be a little careful when controlling the interaction between the pieces that cross each other.

If E is a Lipschitz graph, then E is also good for the Cauchy kernel. Unlike the $C^{1,\alpha}$ case, this does not reduce easily to the classical results for the line; new methods had to be developed. The fact that Lipschitz graphs are good for the Cauchy kernel was proved by Calderón [Ca] when the Lipschitz norm of the function being graphed is small and by Coifman, McIntosh and Meyer [CMM] in the general case. There are now several proofs of these results (see for instance [CJS], [D1], [D4], or [Mu]).

If E is connected, then E is good for the Cauchy kernel if and only if there is a constant $C > 0$ so that

$$(1.12) \qquad H^1(E \cap B(z, R)) \leq CR \quad \text{for all } z \in \mathbf{C} \text{ and } R > 0.$$

This follows from [D1].

It is not hard to prove that, even when E is not necessarily connected, the condition (1.12) is necessary for E to be good for the Cauchy kernel, at least if you make a minor auxiliary assumption like $H^1(E \cap B) < +\infty$ for all balls B. See Proposition 1.4 in Part II of [D4]. The converse is false: the Cantor set K described in §1.1 is a compact set for which (1.12) is true but which is not good for the Cauchy kernel. Indeed, Garnett [G1] proved that K has analytic capacity zero, which means that all bounded holomorphic functions on $\mathbf{C} \setminus K$ are constant, and it can be shown that if

1.2. SINGULAR INTEGRAL OPERATORS

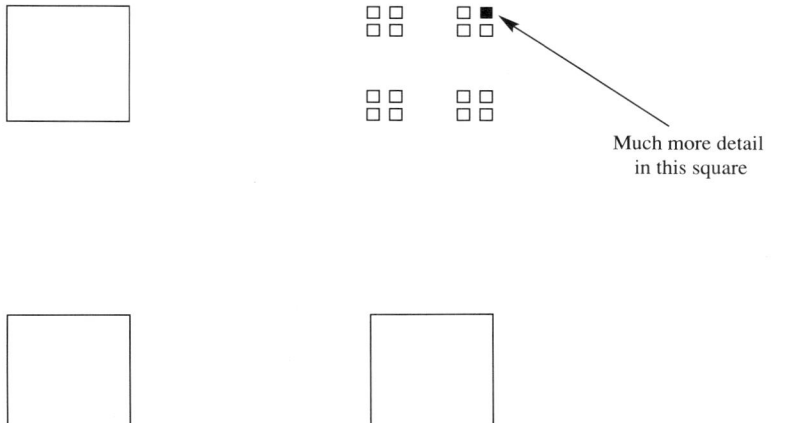

FIGURE 1.2. A suggestion for the construction of $\widetilde{K}$.

K were good for the Cauchy kernel, then there would be many nonconstant bounded holomorphic functions on $\mathbf{C} \setminus K$. (See [**G2**] and [**Mu**].)

These results suggest that there is a relationship between rectifiability and the property of being good for the Cauchy kernel. The naive guess that rectifiability together with (1.12) imply that E is good for the Cauchy integral is false, however, in essence because rectifiability is a qualitative notion, while being good for the Cauchy kernel is quantitative. More precisely, you can easily modify the construction of the Cantor set K from §1.1 in order to produce a compact subset $\widetilde{K}$ of $\mathbf{C}$ that still satisfies (1.12) but which is now rectifiable, and you can do this in such a way that $\widetilde{K}$ has pieces that are very well approximated by pieces of K, so that $\widetilde{K}$ is not good for the Cauchy kernel because K is not. See Figure 1.2.

In view of all of this it is natural to hope that the sets E which are good for the Cauchy kernel are characterized by some sort of quantitative rectifiability condition. This quantitative rectifiability condition should also be scale-invariant, because the property of being good for the Cauchy kernel is. There is a natural candidate for such a condition, but, before we state it, let us define the notion of (Ahlfors-) regularity, which is a quantitative and scale-invariant version of having Hausdorff dimension d. We state this definition in the case of general values of n and d for future reference.

DEFINITION 1.13. A subset E of $\mathbf{R}^n$ is said to be regular with dimension d if it is closed and if there is a constant $C_0 > 0$ so that

$$(1.14) \qquad C_0^{-1} R^d \leq H^d(E \cap B(x, R)) \leq C_0 R^d$$

for all $x \in E$ and $R > 0$.

The smallest constant C_0 will often be referred to as the regularity constant for E.

Our assumption that E be closed is not a big deal; if E is, say, H^d-measurable and satisfies (1.14), then its closure is regular. This is not hard to prove: the point is that (1.14) implies that, for all $x \in E$, $r > 0$, and $R > r$, $E \cap B(x, R)$ can be covered by $C(R/r)^d$ balls with radius r.

Our use of the word "regular" is slightly unfortunate, because this same term is used by some authors to mean something completely different (basically, rectifiability). We hope the reader will not be confused by this.

Simple examples of regular sets include d-planes and d-dimensional Lipschitz graphs. However, regular sets need not be even remotely like rectifiable sets. For example, the Cantor set K of §1.1 is a one-dimensional regular set in the plane, at least if you restrict (1.14) to R's in $(0, 1]$. (Of course, it is easy to modify K to get rid of this problem.) Unlike rectifiability, regularity also makes sense when d is not an integer, and there are plenty of examples of regular sets with noninteger dimension, such as Cantor sets or self-similar curves.

We shall be restricting ourselves, in this monograph, almost entirely to regular sets. This seems to be natural in the context of the sort of quantitative and scale-invariant phenomena that we are interested in. Let us also mention that if E is regular, then E has the structure of a "space of homogeneous type" (the notion introduced by Coifman and Weiss [CW]). This will allow us to use on E many of the standard real-variable techniques which are normally employed on $\mathbf{R}^n$.

Let us now come back to $d = 1$, $n = 2$, and sets that are good for the Cauchy kernel.

CONJECTURE 1.15. *If a one-dimensional regular set $E \subset \mathbf{C}$ is good for the Cauchy kernel, then there is a connected one-dimensional regular set $F \subset \mathbf{C}$ such that $E \subset F$.*

Notice that the converse to this is true, since connected one-dimensional regular sets in the plane are good for the Cauchy kernel, as we have pointed out earlier.

Although we are primarily concerned with quantitative issues here, we should at least mention that there are some natural qualitative versions of Conjecture 1.15, such as the following.

Suppose that E is an H^1-measurable set in $\mathbf{C}$ such that $0 < H^1(E) < +\infty$ and that $\Theta_*(E, z) > 0$ for H^1-almost all $z \in E$. Suppose also that

$$(1.16) \qquad \sup_{\epsilon > 0} \left| \int_{\substack{w \in E \\ |z-w| > \epsilon}} \frac{1}{z-w} \, dw \right| < +\infty$$

for H^1-almost all $z \in E$. Does this imply that E is rectifiable?

In order to make clear why the hypothesis of this question is a natural qualitative version of the property of being good for the Cauchy kernel, we should point out that if E is regular and good for the Cauchy kernel, then

$$(1.17) \qquad T_*f(z) = \sup_{\epsilon>0} |T_\epsilon f(z)| = \sup_{\epsilon>0} \left| \int_{\substack{w \in E \\ |w-z|>\epsilon}} \frac{1}{z-w} f(w)\, dw \right|$$

is a bounded sublinear operator on $L^2(E)$. This follows from standard Calderón-Zygmund theory (to wit, Cotlar's inequality). Also, the $T(1)$-Theorem implies that the maximal operator T_* can be controlled in terms of what it does to the function 1 (or, say, characteristic functions of balls) when E is regular. The precise statement is somewhat technical, and we shall not give it here; our point is only that the fact that we simply took the function $f(w) \equiv 1$ in (1.16) is not as important as one might think.

Mattila [Ma2] has a very nice result related to the question above. He proves that if μ is a finite Borel measure on $\mathbf{C}$ such that

$$(1.18) \qquad \liminf_{r \to 0} r^{-1} \mu(B(z,r)) > 0 \quad \text{for } \mu\text{-almost all } z \in \mathbf{C}$$

and if the limit

$$(1.19) \qquad \lim_{r \to 0} \int_{|w-z|>r} \frac{1}{z-w}\, d\mu(w)$$

exists and is finite for μ-almost all $z \in \mathbf{C}$, then there is a sequence of rectifiable curves Γ_j such that $\mu(\mathbf{C} \setminus [\bigcup_j \Gamma_j]) = 0$. Thus, if, in the question above, we replaced (1.16) by the (stronger) requirement that the limit of the truncated integral in (1.16) exists for H^1-almost all $z \in E$, then Mattila's theorem would imply that E is rectifiable. A version of this in higher dimensions has been given by Mattila and Preiss [MP]. (See also [Ma3].)

Now we want to describe some analogues of these issues for general choices of d and n. Notice that the preceding story was special not only because of the existence of complex analysis on $\mathbf{R}^2$ but also because $d=1$ is special, due to the role of connectedness.

As before, we want to relate the geometry of a set E to the behavior of some singular integral operators on E. We have to decide now which singular integral operators we want to allow. It will be convenient for us to give a name to a rather large class of kernels for these operators.

DEFINITION 1.20. We denote by $\mathscr{K}_d(\mathbf{R}^n)$ the set of smooth real-valued functions $K(x)$ on $\mathbf{R}^n \setminus \{0\}$ such that

$$(1.21) \qquad K \text{ is odd, i.e., } K(-x) = -K(x) \text{ for all } x,$$

and

$$(1.22) \qquad |x|^{d+j}|\nabla^j K(x)| \in L^\infty(\mathbf{R}^n \setminus \{0\}) \text{ for } j = 0, 1, 2, \ldots.$$

This is not the largest class of reasonable kernels that we could consider, but it is large enough for our purposes here.

Notice that $x_i/|x|^{d+1} \in \mathscr{K}_d(\mathbf{R}^n)$ for $1 \leq i \leq n$. This family of kernels provides the most obvious simple substitute for the Cauchy kernel in the general case. When $d = n - 1$, these kernels are the components of a generalization of the Cauchy kernel based on Clifford analysis. When $d = n - 1$, it is also true that the double-layer potential can be expressed in terms of the operators that correspond to the kernels $x_i/|x|^{d+1}$.

Another interesting subset of $\mathscr{K}_d(\mathbf{R}^n)$ is the class of all kernels $K(x)$ defined on $\mathbf{R}^n \setminus \{0\}$ which are smooth, odd, homogeneous of degree $-d$, and C^∞ (away from 0).

DEFINITION 1.23. Let us say that a set $E \subset \mathbf{R}^n$ which is measurable for H^d is good for all the kernels in $\mathscr{K}_d(\mathbf{R}^n)$ if for each $K \in \mathscr{K}_d(\mathbf{R}^n)$, the family of operators given by

$$(1.24) \qquad T_\epsilon f(x) = \int_{\substack{y \in E \\ |y-x| > \epsilon}} K(x-y) f(y) \, dy,$$

which are a priori defined only as operators from $L^1(E)$ into $C(E)$, actually determine bounded linear operators on $L^2(E)$ whose operator norms are bounded uniformly in ϵ.

Here, as well as later, dy denotes the restriction to E of H^d, and the spaces $L^1(E)$ and $L^2(E)$ are defined in terms of dy.

Notice that we do not require any sort of uniform estimate in terms of K on the norms of the T_ϵ.

We would like to have a geometrical characterization of the sets E that are good for all the kernels in $\mathscr{K}_d(\mathbf{R}^n)$. We would also like to know whether we get the same class of sets if we replace $\mathscr{K}_d(\mathbf{R}^n)$ by the smaller class composed of its elements that are homogeneous of degree $-d$, say, or even by the n kernels $x_i/|x|^{d+1}$.

We shall restrict our attention to d-dimensional regular sets E, as we indicated before. It is again true that if E is good for all kernels in $\mathscr{K}_d(\mathbf{R}^n)$, or even just the kernels $x_i/|x|^{d+1}$, $1 \leq i \leq n$, and if E satisfies some reasonable a priori assumption like $H^d(E \cap B) < +\infty$ for all balls B, then there is a $C > 0$ so that

$$H^d(E \cap B(x, R)) \leq C R^d \quad \text{for all } x \in E \text{ and } R > 0.$$

(See Proposition 1.4 in Part III of [**D4**].) Thus the second half of the regularity condition (1.14) is necessary for E to be good for all kernels in $\mathscr{K}_d(\mathbf{R}^n)$.

The assumption that E be regular allows us in particular to have access to Calderón-Zygmund theory, which implies for instance that if the T_ϵ, $\epsilon > 0$, form a uniformly bounded family of operators on $L^2(E)$ for some $K \in \mathcal{K}_d(\mathbf{R}^n)$, then

(1.25) $\quad T_*f(x) = \sup_{\epsilon > 0} | T_\epsilon f(x) | = \sup_{\epsilon > 0} \left| \int_{\substack{y \in E \\ |y-x| > \epsilon}} K(x-y) f(y) \, dy \right|$

defines a sublinear operator which is bounded on $L^2(E)$ and also on $L^p(E)$ for all p, $1 < p < +\infty$.

The known sufficient conditions for a set E to be good for all $K \in \mathcal{K}_d(\mathbf{R}^n)$ are largely the same as for the Cauchy kernel. When E is a d-plane, this is true for classical reasons; see, for example, [St]. When E is a $C^{1,\alpha}$ submanifold (for some $\alpha > 0$) that is nice at infinity, then E is still good for all kernels in $\mathcal{K}_d(\mathbf{R}^n)$. This is also pretty classical and not hard to deduce from the result for d-planes.

If E is a d-dimensional Lipschitz graph (see the Notation page for a definition), then E is good for all kernels in $\mathcal{K}_d(\mathbf{R}^n)$. For kernels that are homogeneous of degree $-d$, this is an immediate consequence of the theorem of Coifman, McIntosh, and Meyer [CMM], obtained by applying the rotation method. For general K's you need an extra argument, like the one in [CDM], to deduce from [CMM] that Lipschitz graphs are good for all $K \in \mathcal{K}_d(\mathbf{R}^n)$. An alternate proof is sketched in Example 6.7 of Part II of [D4]; the theorem in [CS] could also be used.

When $d = 1$ it is true, as for the Cauchy kernel, that if E is regular and connected, then E is good for all the kernels in $\mathcal{K}_1(\mathbf{R}^n)$. This follows from [D1]. When $d > 1$, this is certainly not true, and there is not any such simple condition that plays the same role.

We shall actually give later some necessary and sufficient conditions for a regular set E to be good for all the kernels in $\mathcal{K}_d(\mathbf{R}^n)$, but let us first describe more clearly what we are looking for.

As for the Cauchy kernel, we would like to find a geometric description of the sets E that are good for all the kernels in $\mathcal{K}_d(\mathbf{R}^n)$ (or some interesting subclass of $\mathcal{K}_d(\mathbf{R}^n)$). We know that all Lipschitz graphs are good, so smoothness is not an issue, and we know examples of regular sets (variants of the set K in §1.1) that are not good, so we suspect that the right sort of conditions are quantitative rectifiability conditions. The conditions that we are seeking should allow sets that are much more complicated than Lipschitz graphs, in the sense that our sets should be allowed to cross themselves or have reasonably many handles, and so on. (This is because we know classes of good sets that may exhibit such behavior.) On the other hand, these conditions should not allow the sets to be too scattered (like Cantor sets).

Here is a first candidate for such a condition.

DEFINITION 1.26. We say that E has BPLG (big pieces of Lipschitz graphs) if E is regular and if there exist constants $C_1 > 0$ and $\theta > 0$ so that, for each $x \in E$ and $R > 0$, there is a d-dimensional Lipschitz graph Γ with constant $\leq C_1$ (see the Notation page for the definition) such that

(1.27) $$H^d(E \cap \Gamma \cap B(x, R)) \geq \theta R^d.$$

In many ways, this is the sort of geometric condition on E that we want. It implies rectifiability (this is not hard to prove), but it is not implied by rectifiability because θ and C_1 are not allowed to depend on x and R. (It is not difficult to produce the appropriate counterexamples. Sets like the $\widetilde{K}$ alluded to at the beginning of this section and in Figure 1.2 would work.)

BPLG is a sufficient condition for E to be good for all $K \in \mathscr{K}_d(\mathbf{R}^n)$, because of the following result.

PROPOSITION 1.28. *Let $K \in \mathscr{K}_d(\mathbf{R}^n)$ be given, and let $E \subset \mathbf{R}^n$ be a regular set. Then E is good for K (i.e., $\{T_\epsilon\}_{\epsilon>0}$ is a (uniformly) bounded family of bounded operators on $L^2(E)$, or equivalently T_* is bounded on $L^2(E)$) as soon as there exist constants $\eta > 0$, $C_0 > 0$, and $M \geq 0$ such that the following holds: For each ball B centered on E, there is a regular set $F = F_B$ such that:*

(1.29) $H^d(E \cap F \cap B) \geq \eta H^d(E \cap B)$;

(1.30) *F satisfies the regularity condition (1.14) with constant C_0;*

(1.31) *if $\widetilde{T}_\epsilon$ is defined by*

$$\widetilde{T}_\epsilon f(x) = \int_{\substack{y \in F \\ |y-x| > \epsilon}} K(x-y) f(y) \, dy$$

for $x \in f$, then $\widetilde{T}_\epsilon$ is a bounded operator on $L^2(F)$, with norm $\leq M$.

Notice that we require uniform bounds on the regularity constants for the F_B, and in (1.31) we require uniform bounds on the norms of the $\widetilde{T}_\epsilon$ with respect to both ϵ and B.

Proposition 1.28 says that if E has big pieces of (uniformly) regular sets which are already known to be (uniformly) good for a given kernel K, then E is itself good for that kernel. Thus, for instance, if E has BPLG, then E is good for all kernels in $\mathscr{K}_d(\mathbf{R}^n)$, because we know that Lipschitz graphs are good for all $K \in \mathscr{K}_d(\mathbf{R}^n)$. (To be precise, we also use the fact that, for a given K, we get an estimate on the operator defined by K on the Lipschitz graph that depends only on the Lipschitz constant.)

See [D4] for a proof of Proposition 1.28, where it is given a slightly different formulation as Proposition 3.2 in Part III. See also [D1].

BPLG is a reasonably general condition that implies that E is good for all kernels in $\mathscr{K}_d(\mathbf{R}^n)$. When $d = 1$, all connected regular sets have BPLG. (This is not very difficult to prove; see [D1], [D4].) In higher dimensions too there are some nice conditions on E which are known to imply that E has BPLG. A survey of the relevant literature and related topics can be found in Part III of [D4].

However, the BPLG condition has two main weaknesses. The first is that it is not always very easy to check. The second one is that it is not at all clear that it is the right condition. In other words, although BPLG is a reasonably nice looking scale-invariant quantitative rectifiability condition that implies estimates for many singular integral operators, it is not clear that there are no other conditions that would be more general and perhaps simpler to check as well. It turns out that this concern is quite justified, as we shall soon see. For the moment let us proceed somewhat naively.

DEFINITION 1.32. A set $E \subset \mathbf{R}^n$ has $(BP)^2 LG$ (big pieces of BPLG) if E is regular and if there exist constants, C_1, θ, and $\alpha > 0$ so that, if B is any ball centered on E, then there is a regular set $F \subset \mathbf{R}^n$ such that

$$H^d(E \cap F \cap B) \geq \alpha H^d(E \cap B),$$

F is regular with constant $\leq C_1$, and F satisfies the BPLG condition with constants θ and C_1.

Because of Proposition 1.28 (applied twice), we know that if E has $(BP)^2 LG$, then it is good for all $K \in \mathscr{K}_d(\mathbf{R}^n)$. Thus $(BP)^2 LG$ is another scale-invariant quantitative rectifiability condition, which is apparently weaker than BPLG and has the same consequences in terms of singular integral operators. In fact, we can define the conditions $(BP)^j LG$ for all $j \geq 1$, and these are apparently weaker and weaker conditions that all imply that E is good for all kernels.

A more concrete alternative to BPLG is given by the following.

DEFINITION 1.33. A set $E \subset \mathbf{R}^n$ has BPBI (big pieces of bilipschitz images of $\mathbf{R}^d$) if it is regular, and if there are constants $\theta > 0$ and $C_1 > 0$ such that, for each ball B centered on E, we can find a compact set $A \subset \mathbf{R}^d$ and a mapping $\rho : A \to \mathbf{R}^n$ such that ρ is bilipschitz with constant $\leq C_1$, i.e.,

$$(1.34) \qquad C_1^{-1}|x-y| \leq |\rho(x) - \rho(y)| \leq C_1|x-y| \qquad \text{for all } x, y \in A,$$

and

$$(1.35) \qquad H^d(E \cap \rho(A) \cap B) \geq \theta H^d(E \cap B).$$

This is another scale-invariant quantitative rectifiability condition which is implied by BPLG and which implies estimates for the singular integral operators with kernels in $\mathscr{K}_d(\mathbf{R}^n)$. This last is a consequence of Proposition 1.28 and the following three facts. The first is that if A is a subset of $\mathbf{R}^d$ and if $\rho : A \to \mathbf{R}^n$ is bilipschitz, then ρ admits an extension $\hat{\rho} : \mathbf{R}^d \to \mathbf{R}^m$, where m is a sufficiently large integer, which is also bilipschitz. You can in fact take $m = \max(n, 2d+1)$ (and perhaps even smaller than that), but it is a lot easier to construct a $\hat{\rho}$ if you allow m to depend also on the bilipschitz constant of ρ, which is all we need here. (See §17 of [**DS2**].) The second fact is that a bilipschitz image of $\mathbf{R}^d$ into $\mathbf{R}^m$ is good for all kernels in $\mathscr{K}_d(\mathbf{R}^m)$. This is known, and the same references as for the case of Lipschitz graphs

([**CDM**], Example 6.7 in Part II of [**D4**], and [**CS**]) still apply. The third fact is that if $\mathbf{R}^n$ is identified with an n-plane through the origin in $\mathbf{R}^m$, then every element of $\mathscr{K}_d(\mathbf{R}^n)$ can be extended to an element of $\mathscr{K}_d(\mathbf{R}^m)$. This is easy to check.

There is an amusing relationship between BPBI and BPLG: if $n \geq 2d + 1$ and if E has BPBI, then E has $(\text{BP})^2\text{LG}$.

The condition $n \geq 2d + 1$ is not such a big deal, because we can always increase the ambient dimension if we so desire. The fact that BPBI implies $(\text{BP})^2\text{LG}$ when $n \geq 2d + 1$ follows from the bilipschitz extension result that we mentioned above and the fact that a bilipschitz image of $\mathbf{R}^d$ inside $\mathbf{R}^n$ always has BPLG. (See [**D3**] and then [**J2**].)

We hope that the reader will find it pleasant to learn that if E is a regular set in $\mathbf{R}^n$, then E is good for all kernels in $\mathscr{K}_d(\mathbf{R}^n)$ if and only if E has BPBI. This and some related results were proved in [**DS2**] and will be recapitulated in §1.4. From this and Proposition 1.28 we deduce that all the $(\text{BP})^j\text{LG}$ conditions, $j \geq 1$, imply BPBI. Thus, if $n \geq 2d + 1$, all the conditions $(\text{BP})^j\text{LG}$, $j \geq 2$, and BPBI are equivalent. (A more direct proof of this can be given using the results of Chapter 2 of Part IV.)

It was recently discovered by Tomasz Hrycak that BPBI does not imply BPLG. He observed that the "venetian blinds" construction on p. 88 of [**Fl2**] can be used to produce a counterexample. Peter Jones had previously a more complicated method for producing a weaker result.

Let us also mention, without waiting for §1.4, that when $d = 1$, if E is a regular set that has BPBI, then there is a connected regular set F such that $E \subset F$. Thus Conjecture 1.5 is true if you replace the hypothesis that E is good for the Cauchy kernel by the stronger requirement that E be good for all kernels in $\mathscr{K}_1(\mathbf{R}^n)$. (Here the value of n no longer matters.)

Because of the special status that the BPBI condition has, we shall distinguish it with a special name. We say that E is uniformly rectifiable if E has BPBI. We shall encounter other characterizations of uniform rectifiability in §1.4, and we shall obtain some new characterizations in Parts II and III.

1.3. Review of some aspects of Littlewood-Paley theory in connection with rectifiability.

There is an obvious analogy between rectifiability properties of sets and differentiability properties of functions, and it is natural to try to apply techniques from the study of the second to the study of the first. This idea is of course very old, but it has only been in the last few years—at least as far as we know—that the techniques of Littlewood-Paley theory have been applied to the geometrical setting, beginning with the seminal paper [**J1**] of Peter Jones. In this section we shall review some results from Littlewood-Paley theory in the context of the differentiability properties of functions and also some of the related works of Jones in [**J1**], [**J3**]. We shall also touch upon other aspects of Littlewood-Paley theory that will be relevant in the geometrical setting.

Let f be a real-valued function on the real line. For the purposes of exposition it will be convenient for us to look at the differentiability of f in terms of knowing when $f' \in L^2$.

PROPOSITION 1.36. *If $f : \mathbf{R} \to \mathbf{R}$ lies in $L^2(\mathbf{R})$, say, then the following are equivalent:*

(a) *f is locally absolutely continuous and $f' \in L^2$;*
(b) $\sup\limits_{t>0} \int_{\mathbf{R}} t^{-2} |f(x+t) - f(x)|^2 \, dx < +\infty$;
(c) $\int_0^\infty \int_{\mathbf{R}} t^{-2} |f(x+t) + f(x-t) - 2f(x)|^2 \, dx \frac{dt}{t} < +\infty$.

For a proof of this see Chapter 5 of [St], especially pages 139 and 140. In Chapter 8 of [St] there are also some pointwise versions of this result.

The key point about this is that the differentiability of a function f is reflected both in the boundedness of its first difference quotient $\frac{f(x+t)-f(x)}{t}$ and in quadratic estimates on the second difference quotient

$$(1.37) \qquad \frac{f(x+t) + f(x-t) - 2f(x)}{t}.$$

In other words, the second differences somehow involve enough cancellations so that there is some significant orthogonality around, which is not true of the first differences.

Let us sketch the proof of the equivalence of (a) and (c) to convince any skeptical reader of the existence of this orthogonality. We use Plancherel's theorem to compute on the Fourier transform side:

$$\int_0^\infty \int_{\mathbf{R}} t^{-2} |f(x+t) + f(x-t) - 2f(x)|^2 \, dx \frac{dt}{t}$$
$$= c \int_0^\infty \int_{\mathbf{R}} t^{-2} | \hat{f}(\xi) \{ e^{i\xi t} + e^{-i\xi t} - 2 \} |^2 \frac{d\xi \, dt}{t},$$

where c is some constant (involving π). By integrating in t first and making the change of variables $t \to |\xi|^{-1} t$, we get that this is equal to

$$c \left\{ \int_{\mathbf{R}} |\xi|^2 | \hat{f}(\xi) |^2 \, d\xi \right\} \left\{ \int_0^\infty t^{-3} | e^{it} + e^{-it} - 2 |^2 \, dt \right\}.$$

It is easy to see that the integral in t is positive and finite, while the integral in ξ is just $\int_{\mathbf{R}} |f'(x)|^2 \, dx$. Note that the integral in t would have diverged if we had used only the first difference instead of the second.

Proposition 1.36 is a simple example of a general phenomenon. Given a function f, there are many quantities that one can form that somehow measure the oscillations of f and in terms of which one can characterize certain types of behavior of f, for example, its differentiability properties. These characterizations frequently involve quadratic expressions that encode some sort of underlying orthogonality. In our case we have used (1.37) as our method of measuring the oscillations of f.

We are interested in being able to implement such ideas in the geometrical setting. Thus we want to find quantities that measure the "oscillations" of a set, and we want to use them to characterize certain kinds of behaviors of the set related to rectifiability. There are natural counterparts to (1.37) in the geometrical setting, which measure the extent to which a given set is symmetric about a given point, and we shall study some of them later. For the moment, let us look at a variant of (1.37), in the case of functions, and its geometrical version.

One way to view (1.37) is as a measurement of the deviation of f from being affine. Instead of using second differences to do this indirectly, we can simply do this directly.

Given a function $f : \mathbf{R}^d \to \mathbf{R}$ and $1 \leq q < \infty$, set

$$(1.38) \qquad \gamma_q(x, t) = \inf_a t^{-1} \left\{ t^{-d} \int_{B(x,t)} |f(y) - a(y)|^q dy \right\}^{1/q}$$

for all $x \in \mathbf{R}^d$ and $t > 0$, where the infimum is taken over all affine functions a. Also let

$$(1.39) \qquad \gamma_\infty(x, t) = \inf_a \left\{ t^{-1} \|f - a\|_{L^\infty(B(x,t))} \right\}$$

be the natural extension to $q = +\infty$.

THEOREM 1.40. *Let f be a locally integrable function on $\mathbf{R}^d$. Suppose that $1 \leq q \leq \infty$ when $d = 1$ and that $1 \leq q < \frac{2d}{d-2}$ when $d \geq 2$. Then the distributional gradient of f lies in $L^2(\mathbf{R}^d)$ if and only if*

$$(1.41) \qquad \int_0^\infty \int_{\mathbf{R}^d} \gamma_q(x, t)^2 \, dx \, \frac{dt}{t}$$

is finite. Also, (1.41) is comparable to $\|\nabla f\|_{L^2}^2$.

See [Do], especially Theorem 6.

The advantage of this over the previous story is that the γ_q's have a simpler geometrical counterpart; instead of measuring the extent to which a function differs from being affine, we can measure the extent to which a set differs from being a d-plane. Before doing this, we should record the appropriate version of Theorem 1.40 for Lipschitz functions.

THEOREM 1.42. *Suppose that $f : \mathbf{R}^d \to \mathbf{R}$ is Lipschitz and that $1 \leq q \leq \infty$ when $d = 1$ and $1 \leq q < \frac{2d}{d-2}$ when $d \geq 2$. Then*

$$(1.43) \qquad \int_0^R \int_{B(x,R)} \gamma_q(y, t)^2 \, dy \, \frac{dt}{t} \leq C \|f\|_{\text{Lip}}^2 R^d$$

for all $x \in \mathbf{R}^d$ and $R > 0$. Here $C = C(q, d)$ does not depend on f, x, or R.

Theorem 1.42 is easy to obtain from Theorem 1.40 by performing the appropriate localization. The converse to Theorem 1.42 is not true (the estimate

(1.43) does not imply that f is Lipschitz). In order to get an equivalence, you would have to replace the requirement that f be Lipschitz with the condition that its distributional gradient lie in BMO. (We shall not need this fact.)

The conclusion of Theorem 1.42 provides an example of how Carleson measures can arise. Recall that a measure λ on $\mathbf{R}^{d+1}_+$ is called a Carleson measure if there is a $C > 0$ such that

$$(1.44) \qquad |\lambda|(B(x, R) \times (0, R]) \leq CR^d$$

for all $x \in \mathbf{R}^d$ and $R > 0$. The smallest such C is called the Carleson norm (or Carleson constant) of λ. Thus Theorem 1.42 says that $\gamma_q(y, t)^2 dy \frac{dt}{t}$ is a Carleson measure on $\mathbf{R}^{d+1}_+$ when f is Lipschitz and q lies in a certain range, with Carleson norm dominated by the square of the Lipschitz norm of f.

Carleson measure conditions—in various guises—will play a fundamental role in this monograph. Roughly speaking, Carleson measures live on a $(d+1)$-dimensional space but behave as if they were living on a d-dimensional set, at least from the perspective of the boundary.

Now let us consider geometrical versions of the γ_q's and Theorem 1.42. Let E be a d-dimensional regular set in $\mathbf{R}^n$, and define, for $1 \leq q < +\infty$, $x \in E$, and $t > 0$,

$$(1.45) \qquad \beta_q(x, t) = \inf_P \left\{ t^{-d} \int_{E \cap B(x,t)} \left[t^{-1} \operatorname{dist}(y, P) \right]^q dy \right\}^{1/q}.$$

Here the infimum is taken over all affine d-planes contained in $\mathbf{R}^n$. Similarly, let

$$(1.46) \qquad \beta(x, t) = \beta_\infty(x, t) = \inf_P \left\{ t^{-1} \sup_{y \in E \cap B(x,t)} \operatorname{dist}(y, P) \right\}$$

be the natural extension to $q = +\infty$.

This is the most obvious geometrical version of (1.38); we measure the deviation of a set from a d-plane rather than the deviation of a function from an affine function. In the case where E is a Lipschitz graph, it is easy to relate the two quantities.

Notice that, since E is regular and $x \in E$, the expression in braces in (1.45) is essentially the L^q-mean of $t^{-1} \operatorname{dist}(y, P)$ over $B(x, t)$. In particular, $\beta_q(x, t) \leq C$.

In [**J3**], Peter Jones proved the following very striking result.

THEOREM 1.47. *Let E be a one-dimensional regular set in $\mathbf{R}^n$. Then the following are equivalent:*

(1.48) $\beta_\infty(x, t)^2 \frac{dx\, dt}{t}$ *is a Carleson measure on $E \times \mathbf{R}_+$;*

(1.49) *there is a one-dimensional regular set F in $\mathbf{R}^n$ which is connected and which contains E.*

The definition of a Carleson measure on $E \times \mathbf{R}_+$ is essentially the same as on $\mathbf{R}_+^{d+1}$. We give it directly in the general case for future reference.

DEFINITION 1.50. Let E be a d-dimensional regular set in $\mathbf{R}^n$, and let λ be a nonnegative Borel measure on $E \times \mathbf{R}_+$. Then λ is a Carleson measure if there is a constant $C \geq 0$ such that

$$\lambda(B(x, R) \times (0, R]) \leq CR^d$$

for all $x \in E$ and all $R > 0$.

The smallest constant C such that (1.51) holds is called the Carleson constant (or norm) of λ.

Jones proved in fact quite a bit more than what is stated in Theorem 1.47. He formulated the definition of $\beta(x, t)$ and the condition (1.48) in a slightly different manner, in such a way so as to be able to drop the requirement that E be regular. He also proved, when $n = 2$, an un-uniformized version of Theorem 1.47 in which E is assumed to be compact, say, (1.49) is replaced by the requirement that E be contained in a compact connected set of finite length, and (1.48) is replaced by an appropriate quadratic estimate on the (modified) $\beta(x, t)$'s. When $n > 2$, his arguments only gave one-half of the un-uniformized result, but the other half has since been supplied by Okikiolu [O].

The condition (1.48) had appeared earlier in [J1], where Jones observed that it is satisfied by one-dimensional Lipschitz graphs and applied this to certain issues in complex analysis, such as the L^2-boundedness of the Cauchy integral operator.

The statement and proof of Theorem 1.47 rely heavily on the assumption that $d = 1$. There is a version of this theorem when $d > 1$, which was proved in [DS2] and will be recalled in the next section. Basically (1.49) should be replaced by the requirement of uniform rectifiability, while the $\beta_\infty(x, t)$ in (1.48) should be replaced by $\beta_q(x, t)$ for q in the appropriate range.

Notice, incidentally, that the functions $\beta_q(x, t)$, like the $\gamma_q(x, t)$, are measurable. This is not hard to check, and we leave the details to the reader. It is also easy to see that this could not possibly be a serious issue; using inequalities like

$$(1.51) \quad \beta_q(y, s) \leq C\beta_q(x, t) \quad \text{when } |x - y| < \tfrac{t}{2} \text{ and } \tfrac{t}{10} \leq s < \tfrac{t}{2},$$

it is easy to replace (1.48) by an equivalent discretized condition with integrals replaced by sums and measurability issues sent to oblivion.

There is another type of Littlewood-Paley result that we want to record in this section. We shall be considering geometrical counterparts of this later.

PROPOSITION 1.52. *Let ψ be a C^1 function over $\mathbf{R}^d$ such that*

$$|\psi(x)| + |\nabla \psi(x)| \leq C(1 + |x|)^{-d-1} \quad \text{for all } x \in \mathbf{R}^d$$

and $\int_{\mathbf{R}^d} \psi(x)\, dx = 0$. Define ψ_t, $t > 0$, by $\psi_t(x) = t^{-d}\psi\left(\frac{x}{t}\right)$. Then

$$\text{(1.53)} \qquad \int_0^\infty \int_{\mathbf{R}^d} |\psi_t * f(x)|^2\, dx\, \frac{dt}{t} \leq C \int_{\mathbf{R}^d} |f(x)|^2 dx$$

for all $f \in L^2(\mathbf{R}^d)$, and

$$\text{(1.54)} \qquad |\psi_t * f(x)|^2 dx\, \frac{dt}{t} \text{ is a Carleson measure on } \mathbf{R}^{d+1}_+$$

for all $f \in L^\infty(\mathbf{R}^d)$.

This is not hard to prove, using Plancherel for (1.53) and then a simple localization for (1.54). See [Jé].

There are versions of Proposition 1.52 for functions whose distributional gradient lies in L^2 or L^∞. For example, if you assume also that

$$\int_{\mathbf{R}^d} \psi(x) x_j\, dx = 0 \quad \text{for } j = 1, \ldots, d,$$

then

$$\text{(1.55)} \qquad \int_0^\infty \int_{\mathbf{R}^d} |t^{-1}(\psi_t * f)(x)|^2\, dx\, \frac{dt}{t} \leq C \int_{\mathbf{R}^d} |\nabla f(x)|^2\, dx$$

whenever $\nabla f \in L^2(\mathbf{R}^d)$.

A particularly important example of Proposition 1.52 occurs when we take $d = 1$ and $\psi(x) = (x + i)^{-2}$. In that case we can write

$$(\psi_y * f)(x) = \int_{\mathbf{R}} \frac{y}{(z-u)^2}\, f(u)\, du,$$

where $z = x + iy$. This is of course $-y$ times $F'(z)$, where

$$F(z) = \int_{\mathbf{R}} \frac{1}{z-u}\, f(u)\, du$$

is the Cauchy integral of f. In this case (1.53) gives

$$\text{(1.56)} \qquad \iint_{\mathbf{C}\setminus\mathbf{R}} |yF'(z)|^2 y^{-1}\, dz \leq C \int_{\mathbf{R}} |f(u)|^2\, du,$$

where the dz on the left represents two-dimensional Lebesgue measure. Actually, (1.53) only gives (1.56) with the domain of integration $\mathbf{C}\setminus\mathbf{R}$ on the left side replaced by the upper half-plane. However, the part that comes from the lower half-plane can be controlled similarly, by taking $\psi(x) = (x - i)^{-2}$.

1.4. Various characterizations of uniform rectifiability.

This section is mostly a review of the results in [DS2]. Let us list a first series of equivalent conditions.

THEOREM 1.57. *Let E be a d-dimensional regular set in $\mathbf{R}^n$. Then the conditions (1.58)–(1.62) are equivalent.*

(1.58) $\qquad E$ is good for all kernels K in $\mathcal{K}_d(\mathbf{R}^n)$.

This terminology was explained in §1.2 (see Definitions 1.20 and 1.23).

(1.59) $\qquad \beta_1(x,t)^2 \frac{dx\,dt}{t}$ is a Carleson measure on $E \times \mathbf{R}_+$.

See (1.45) for the definition of β_1 and Definition 1.50 concerning Carleson measures.

If $\beta_1(x,t)$ in (1.59) is replaced by $\beta_q(x,t)$, then we still get an equivalent condition as long as $1 \leq q < \frac{2d}{d-2}$ when $d \geq 2$ and $1 \leq q \leq +\infty$ when $d = 1$. Of course (1.59) becomes stronger, a priori, as you increase q.

Following Jones [J1], we shall often refer to condition (1.59), or any of its variants with $\beta_q(x,t)$ and q in the appropriate range, as "the geometric lemma".

(1.60) $\qquad E$ has big pieces of Lipschitz images of $\mathbf{R}^d$ (BPLI).

This means that there exist $\theta, M > 0$ so that, for each $x \in E$ and $R > 0$, there is a Lipschitz mapping ρ from the ball $B_d(0,R)$ in $\mathbf{R}^d$ into $\mathbf{R}^n$ such that ρ has Lipschitz norm $\leq M$ and

$$H^d(E \cap B(x,R) \cap \rho(B_d(0,R))) \geq \theta R^d.$$

Notice that if E has BPBI (see Definition 1.33), then E has BPLI. This follows immediately from the Whitney extension theorem, which allows you to extend to $\mathbf{R}^d$ a Lipschitz mapping defined on a subset of $\mathbf{R}^d$. The converse (the fact that BPLI implies BPBI) is also true. To prove this, one shows that, if ρ is a Lipschitz function as in the definition of BPLI, it is possible to find a reasonably large subset A of $B_d(0,R)$ such that $\rho(A) \subset E \cap B(x,R)$ and ρ is bilipschitz on A. This follows from the main result of [J2].

Actually, the BPLI condition even implies the following (a priori) stronger condition.

(1.61) E has very big pieces of bilipschitz images of $\mathbf{R}^d$ inside $\mathbf{R}^{n^*}$, where
$\qquad n^* = \max(n, 2d+1)$.

This means that for each $\epsilon > 0$ there is an $M > 0$ so that, for each $x \in E$ and $R > 0$, there is a mapping $\rho : \mathbf{R}^d \to \mathbf{R}^{n^*}$ which is bilipschitz with constant M and satisfies

$$H^d(E \cap B(x,R) \setminus \rho(\mathbf{R}^d)) \leq \epsilon R^d.$$

(Recall that "ρ is bilipschitz with constant M" means that $M^{-1}|u-v| \leq |\rho(u) - \rho(v)| \leq M|u-v|$ for all $u, v \in \mathbf{R}^d$.)

We shall abbreviate (1.61) in the usual fashion by saying that E has VBPBI. In simpler terms, this condition says that, for each ball B centered on E, you can cover nearly all of $E \cap B$ by a bilipschitz image of $\mathbf{R}^d$ (and this with uniform bounds).

The proof that BPBI implies VBPBI which is given in [DS2] is rather roundabout, but, as suggested to us by Brian White, there is a more direct geometrical proof. The main point is to go from "big pieces" to "very big pieces" via an iterative construction. This has to be done with care, because of the problem of the interactions between the new pieces that are being added. To complete the proof, one then has to extend bilipschitz mappings defined on a subset of $\mathbf{R}^d$; this is where the integer n^* shows up, because the authors do not know how to make the extension when there is not enough room in the ambient space.

(1.62) *There is an A_1-weight ω on $\mathbf{R}^d$ and an ω-regular mapping from $\mathbf{R}^d$ into $\mathbf{R}^{n+1}$ whose image contains E.*

The reader should consult [DS2] for a decoding of the jargon. Basically, (1.62) means that E is contained in a set which admits a fairly nice parametrization.

When $n \geq 2d$ we are allowed to replace the $\mathbf{R}^{n+1}$ in (1.62) by $\mathbf{R}^n$.

When $d = 1$, (1.62) is easily seen to be equivalent to the following:

(1.63)   *There is a regular curve $\Gamma \subset \mathbf{R}^n$ such that $E \subset \Gamma$.*

We call a closed set in $\mathbf{R}^n$ a regular curve if it is the image of $\mathbf{R}$ by a Lipschitz mapping $z : \mathbf{R} \to \mathbf{R}^n$ with the property that

(1.64) $|\{x \in \mathbf{R} : z(x) \in B(y, R)\}| \leq CR$

for all $x \in \mathbf{R}^n$, $R > 0$, and some constant C.

It is easy to see that regular curves are connected one-dimensional regular sets. The converse is not exactly true. (It is an amusing exercise to check that the union of the two coordinate axes in $\mathbf{R}^2$ does not have a nice enough parameterization.) However, one can check (directly) that every connected one-dimensional regular set in $\mathbf{R}^n$ is contained in some regular curve. The proof is a minor modification of the proof of Theorem 1.8 that was given in §1.1. Thus the condition (1.63) is essentially the same as the geometric condition in Conjecture 1.15 in §1.2.

Coming back to (1.62), ω-regular mappings are the appropriate generalization to higher dimensions of Lipschitz mappings from $\mathbf{R}$ to $\mathbf{R}^n$ that satisfy (1.64). The presence of the A_1 weight ω stems from the increased difficulty in parameterizing sets with dimension larger than 1; when $d = 1$ we are able to simply take $\omega \equiv 1$. It may be that we could take $\omega \equiv 1$ when $d > 1$, but at present we need the extra flexibility afforded by general A_1 weights. The fact that, in condition (1.62), we were not able to find an ω-regular mapping from $\mathbf{R}^d$ to $\mathbf{R}^n$ (when $n < 2d$) is again related to extension problems.

We hope that the reader is now convinced of the interest of attaching a special name to all the equivalent properties of Theorem 1.57. Note that the BPBI condition of Definition 1.33 is clearly stronger than (1.60) and weaker than (1.61), so that the following definition is consistent with the provisional definition of uniform rectifiability given at the end of §1.2.

DEFINITION 1.65. We say that the set $E \subset \mathbf{R}^n$ is uniformly rectifiable if it is regular and satisfies the equivalent conditions (1.58)–(1.62) of Theorem 1.57.

Let us rapidly discuss some of the nice features, and also the shortcomings, of Theorem 1.57.

First, the various conditions defining uniform rectifiability are not always very easy to check. In particular, we do not have any characterization of uniform rectifiability in terms of projections that is as simple and useful as Theorem 1.7 is in the context of (ordinary) rectifiability. Nonetheless, some rather good criteria for uniform rectifiability are known; see Part III of [**D4**] for an overview.

Although Theorem 1.57 provides a reasonable characterization of uniform rectifiability in terms of the behavior of singular integral operators on E, it also leaves room for improvement. In particular we still do not know whether a one-dimensional regular set in the plane must be uniformly rectifiable if it is good for the Cauchy kernel (Conjecture 1.15). We shall obtain in Part III a partial result in this direction, which will also have suitable versions for all d and n. See Theorems 2.32 and 2.33.

Theorem 1.57 provides an appropriate generalization of Theorem 1.47 (the uniform version of Jones's travelling salesman theorem) to dimensions $d > 1$, and it has the additional pleasing feature of permitting some flexibility with the choice of q (this was new already for $d = 1$). Unfortunately, the method of proof in [**DS2**] relies heavily on the assumption that E is regular. It does not give, even when $d = 1$, the sort of nonuniform results that were obtained in [**J3**] and [**O**] in the case when E is not assumed to be regular. Of course, it is not clear that there are nice higher-dimensional versions of the travelling salesman theorem in the nonregular case. Even when $d = 1$, it is not clear that there should be reasonable results in the spirit of [**J3**] and [**O**], in terms of versions of the $\beta_q(x, t)$ with $q < +\infty$, for sets that are not regular.

To conclude this section, we quote from [**DS2**] other characterizations of uniform rectifiability which are connected to Littlewood-Paley theory.

THEOREM 1.66. *Let E be a d-dimensional regular set in $\mathbf{R}^n$. Then each of the two conditions* (1.67) *and* (1.68) *below is equivalent to uniform rectifiability of E.*

1.4. UNIFORM RECTIFIABILITY

(1.67)

For each compactly supported, C^∞ odd function ψ on $\mathbf{R}^n$, there is a $C > 0$ so that

$$\sum_{k=-\infty}^{\infty} \int_E \left| \int_E \psi_k(x-y) f(y) dy \right|^2 dx \leq C \int_E |f(x)|^2 dx$$

for all $f \in L^2(E)$;

(1.68)

For each compactly supported, C^∞, odd function ψ on $\mathbf{R}^n$, we have that

$$\sum_{k=-\infty}^{\infty} \left| \int_E \psi_k(x-y) dy \right|^2 dx \, d\delta_{2^k}(t)$$

is a Carleson measure on $E \times \mathbf{R}_+$.

The notations for this theorem are as follows. We set $\psi_k(x) = 2^{-kd} \psi(2^{-k} x)$ (notice the d-dimensional normalization, even though the ambient space is n-dimensional); the restriction to E of d-dimensional Hausdorff measure is still denoted by dx or dy; $d\delta_{2^k}$ is the Dirac mass at 2^k. Finally, Carleson measures have been defined in §1.3, Definition 1.50.

Notice that (1.67) is very similar to (1.53). The main differences are that $\mathbf{R}^d$ has been replaced by E and the integral in t has been replaced by a sum over k. (This last difference is not at all major.) Thus (1.67) says, roughly speaking, that one has the same sort of square function estimates on E as on $\mathbf{R}^d$.

Similarly, (1.68) should be compared with (1.54). Once again we are replacing the integral in t by a discrete sum on k, and we are integrating on E instead of $\mathbf{R}^d$. This time we are studying E more directly; the restriction to E of Hausdorff measure plays in (1.68) the same role as $f(y)dy$ in (1.54).

In other words, (1.67) concerns some version of Littlewood-Paley theory on E, while (1.68) can be viewed as a Littlewood-Paley condition applied to E itself.

An important difference between Theorem 1.66 and standard results of Littlewood-Paley theory is that, for Theorem 1.66, all functions ψ (or at least a very large number of functions ψ) are needed. This is of course quite different from what is usually done in Littlewood-Paley theory, where a single nondegenerate function ψ is chosen once and for all and everything is computed in terms of convolutions with the ψ_k. This is related to the fact that we do not have any formula to reconstruct E from the values of the $\int_E \psi_k(x-y) dy$, $x \in E$.

One of the reasons why (1.68), for instance, measures the extent to which E is rectifiable is the following. Since the most notable property of ψ is its oddness, and we make sure to look at $\int_E \psi_k(x-y) dy$ only for points $x \in E$,

(1.68) can be seen as a measure of the symmetry of E. In particular, these integrals vanish when E is a d-plane. Notice that, if we had measured the size of $\int_E \psi_k(x-y)\,dy$ on the whole space $\mathbf{R}^n$, we would have measured the regularity of $dx = H^d\big|_E$, seen as a measure on $\mathbf{R}^n$, with the help of Littlewood-Paley theory (which is not the point and does not work so well).

Concerning the proof of Theorem 1.66, notice that the fact that (1.67) implies (1.68) is essentially trivial. The converse is also rather easy and is obtained by a standard variation of the $T(1)$ argument. It is also pretty easy to show that (1.58) (the fact that E is good for all kernels in $\mathcal{K}_d(\mathbf{R}^n)$) implies (1.67). The fact that (1.68) implies uniform rectifiability is much harder, and is one of the main points of [**DS2**].

The characterizations of uniform rectifiability by (1.59) (the geometric lemma) and (1.68) are natural geometric analogues of classical results of Littlewood-Paley type, as we have discussed. However, one of the main points of this monograph is that they are also quite misleading. We shall encounter other characterizations of uniform rectifiability that deal with similar quantities but which involve much weaker estimates. In particular, the "Littlewood-Paley condition" (1.68) can be weakened substantially (while still characterizing uniform rectifiability).

These new characterizations are very natural geometrically, but they do not have any reasonable counterpart at the level of functions and traditional harmonic analysis.

1.5. The weak geometric lemma and its relatives.

One of the unfortunate aspects of the higher-dimensional version of Theorem 1.47 (the uniform travelling salesman theorem) provided by Theorem 1.57 is that, in the definition of the geometric lemma (1.59), we cannot use $\beta_q(x,t)$ with $q = +\infty$. This is unfortunate because the β_∞'s are more geometric and easier with which to work.

The weak geometric lemma (WGL) defined below is a quite simple condition which is defined in terms of β_∞. It is a necessary, but not sufficient, condition for uniform rectifiability.

As will be the case with many of the rectifiability conditions discussed in this monograph, the WGL is most conveniently stated in terms of Carleson sets.

DEFINITION 1.69. Let E be a d-dimensional regular set in $\mathbf{R}^n$. A Carleson set is a measurable subset A of $E \times \mathbf{R}_+$ such that $\chi_A(x,t)\,dx\frac{dt}{t}$ is a Carleson measure on $E \times \mathbf{R}_+$ (where, as usual, dx denotes the restriction to E of H^d).

By Definition 1.50, this means that there is a $C \geq 0$ such that

$$(1.70) \qquad \int_0^R \int_{E \cap B(w,R)} \chi_A(x,t)\,dx\,\frac{dt}{t} \leq CR^d$$

for all $w \in E$ and $R > 0$.

1.5. THE WEAK GEOMETRIC LEMMA

Note that (1.70) is a nicely invariant way of saying that A is a fairly small set and even behaves as though it were d-dimensional from the perspective of $E \times \{0\}$. Simple examples of Carleson sets are

$$A = \{(x, t) : x \in E \quad \text{and} \quad t_0 \leq t \leq 10^6 t_0\}$$

(where t_0 is any given positive number), or

$$A = \{(x, t) \in E \times \mathbf{R}_+ : |x - x_0| \leq 10^6 t\}$$

(where x_0 is any given point of E).

The notion of Carleson set makes sense for nonmeasurable sets as well if you work with outer measures. In most cases there will be inequalities like (1.51) available to render this point irrelevant.

DEFINITION 1.71. Let E be a d-dimensional regular set. Then E satisfies the weak geometric lemma (WGL) if for each $\epsilon > 0$ the set

(1.72) $$\{(x, t) \in E \times \mathbf{R}_+ : \beta_\infty(x, t) > \epsilon\}$$

is a Carleson set.

Recall that β_∞ is defined in (1.46). In our definition of the WGL, it is important that the Carleson constant for the set (1.72) be allowed to depend on ϵ as wildly as it wants. This is one of the reasons why the WGL, and conditions like it, will be so easy to use. The WGL is a quantified way of saying that, for each precision level $\epsilon > 0$, it is true for almost all balls $B(x, t)$ centered on E that $E \cap B(x, t)$ stays ϵt-close to some d-plane $P = P(x, t)$.

We should also mention that, in applications, the fact that (1.72) is a Carleson set will be used for only one value of ϵ (but of course this value might be very small and might depend on the application we have in mind).

The name "weak geometric lemma" stems from the practice of referring to conditions like (1.48) and (1.59) by saying that E satisfies a geometric lemma. It is obvious, by Tchebytchev, that (1.48) implies the WGL. The fact that (1.59) implies the WGL is also obvious once you have checked that

(1.73) $$\beta_\infty(x, t) \leq C(q) \beta_q(x, 2t)^{q/d+q}$$

for all $x \in E$, $t > 0$, and $0 < q < +\infty$. Here $C(q)$ depends on q and the regularity constant for E not but on x or t.

The inequality (1.73) is so simple that we provide a proof here for the convenience of the reader. Let P be a d-plane that realizes the infimum in the definition of $\beta_q(x, 2t)$ (see (1.45)), let z be a point of $E \cap B(x, t)$ that maximizes the distance to P, and let $D = \mathrm{dist}(z, P)$. If $D \leq t$, we have

$$\begin{aligned}
\beta_q(x, 2t)^q &= (2t)^{-d-q} \int_{E \cap B(x, 2t)} \mathrm{dist}(y, P)^q \, dy \\
&\geq (2t)^{-d-q} \int_{E \cap B(z, \frac{D}{2})} \mathrm{dist}(y, P)^q \, dy \\
&\geq (2t)^{-d-q} \left| E \cap B\left(z, \frac{D}{2}\right) \right| \left(\frac{D}{2}\right)^q \\
&\geq C^{-1} t^{-d-q} D^{d+q},
\end{aligned}$$

so $\beta_\infty(x, t) \leq \frac{D}{t} \leq C\beta_q(x, 2t)^{q/d+q}$, as promised. When $D \geq t$, we also have $\beta_q(x, 2t) \geq C^{-1}$, and there is nothing to prove. This proves (1.73).

Of course, the (WGL) is much weaker than (1.48) or (1.59), and it is not difficult to find examples of regular sets that satisfy the WGL but not the geometric lemma (1.59) and which are not even rectifiable. See §20 of [**DS2**].

Regular sets that satisfy the WGL can be seen, in the spirit of §1.3, as an analogue of the class of functions on $\mathbf{R}^d$ that lie in the closure, inside the Zygmund class Z, of the set of functions whose distributional gradient lies in BMO. (The Zygmund class can be defined as the set of functions f such that $\gamma_\infty(x, t)$ is bounded uniformly in x and t, where $\gamma_\infty(x, t)$ is as in (1.39).) It is well known that there exist functions which lie in the closure in Z of the class of smooth functions but which do not have locally integrable first derivatives. The counterexamples in [**DS2**] are geometric analogues of these functions.

Although the WGL does not by itself imply uniform rectifiability, it is useful in an auxiliary role. We are now going to give an example of this which will be used later. We need a definition first.

DEFINITION 1.74. A d-dimensional regular set E is said to have big projections if there is a $\theta > 0$ so that, for each $x \in E$ and $R > 0$, there is a d-plane P such that

$$(1.75) \qquad |\Pi_P(B(x, R) \cap E)| \geq \theta R^d,$$

where Π_P denotes the orthogonal projection onto P.

THEOREM 1.76. *If E is a regular set that has big projections and satisfies the WGL, then E has BPLG.*

Recall that BPLG is defined in §1.2 (Definition 1.26) and is a sufficient condition for uniform rectifiability.

The converse to Theorem 1.76 is also true. One can derive the fact that BPLG implies the WGL from [**DS2**], but a more direct proof (of a more general result, even) is given in Chapter 1 of Part IV. (A direct proof was also found by Peter Jones.) The fact that BPLG implies big projections is pretty trivial (since a piece of a Lipschitz graph has a projection of roughly the same size).

See [**DS3**] for a proof of Theorem 1.76. The main point is to realize that the techniques of [**J2**] are applicable.

It is not true that E must be uniformly rectifiable, or even rectifiable, as soon as it has big projections. For example, the Cantor set K described in §1.1 has big projections (as long as you restrict yourself to $R < 1$, of course). It is not clear whether the following stronger condition implies uniform rectifiability.

Let $G(n, d)$ be the Grassmann manifold of all d-planes in $\mathbf{R}^n$ that contain the origin, and choose a reasonable probability measure on $G(n, d)$. Let us say that the regular set E has "plenty of big projections" if there exists a $\theta > 0$ such that, for all $x \in E$ and $R > 0$, there is a measurable subset $\mathscr{P}$ of $G(n, d)$ such that $|\mathscr{P}| \geq \theta$ and such that (1.75) holds for every $P \in \mathscr{P}$.

If E has plenty of big projections, then it is rectifiable (this uses Theorem 1.7), but we do not know whether E is necessarily uniformly rectifiable.

To conclude this section, we record a couple of criteria for the WGL to hold.

The first one is given in terms of quantities that measure the extent to which E is symmetric about all its elements.

Let E be a d-dimensional regular set in $\mathbf{R}^n$. Given $x \in E$ and $t > 0$, set

$$(1.77) \qquad sy(x, t) = \sup_{y, z \in E \cap B(x, t)} \left[t^{-1} \operatorname{dist}(2y - z, E) \right]$$

and, more generally,

$$(1.78) \qquad sy_q(x, t) = \left\{ t^{-2d} \int_{E \cap B(x, t)} \int_{E \cap B(x, t)} \left[t^{-1} \operatorname{dist}(2y - z, E) \right]^q dy\, dz \right\}^{1/q}$$

for $1 \leq q < +\infty$.

DEFINITION 1.79. Let E be a regular set of dimension d in $\mathbf{R}^n$. We say that E satisfies the local symmetry condition (LS) if, for all $\epsilon > 0$,

$$(1.80) \qquad \{(x, t) \in E \times \mathbf{R}_+ : sy(x, t) > \epsilon\}$$

is a Carleson set.

See Definition 1.69 for Carleson sets. Notice that, as in the definition of the WGL, we do not require any specific control on how the Carleson norm for (1.80) varies with ϵ.

PROPOSITION 1.81. *If E is regular and satisfies LS, then E also satisfies the WGL.*

This is proved in §5 of [**DS2**]. The proof is reasonably simple and self-contained.

In many ways the $sy_q(x, t)$ behave like the $\beta_q(x, t)$. For example, if E is a d-dimensional regular subset of $\mathbf{R}^n$, and if $1 \leq q \leq +\infty$ when $d = 1$, or $1 \leq q < \frac{2d}{d-2}$ when $d \geq 2$, then

(1.82) $sy_q(x, t)^2 dx \, \frac{dt}{t}$ is a Carleson measure on $E \times \mathbf{R}_+$

if and only if E is uniformly rectifiable. See §19 of [**DS2**] for a proof.

We also have that

(1.83) $C^{-1} sy_q(x, t) \leq sy(x, t) \leq C \, sy_q(x, 2t)^{q/2d+q}$

for all $0 < q < +\infty$, where C is a constant that depends on the regularity constant for E and q. This is proved exactly like (1.73). This implies that we could replace the $sy(x, t)$ in Definition 1.79 by $sy_q(x, t)$ for any q and still get an equivalent definition of LS.

Like the $\beta_q(x, t)$, the $sy_q(x, t)$ have a natural counterpart for the analysis of functions. Given $f : \mathbf{R}^d \to \mathbf{R}$, consider

(1.84) $\left\{ t^{-2d} \int_{B(x,t)} \int_{B(x,t)} \left[t^{-1} \, | f(2y - z) + f(z) - 2f(y) | \right]^q dy \, dz \right\}^{1/q}$.

This is just an L^q-average of the second difference of f already considered in §1.3 (see (1.37)), and the versions of Theorems 1.40 and 1.42 with $\gamma_q(x, t)$ replaced by (1.84) are still true.

Notice that (1.84) is easily controlled by $\gamma_q(x, 2t)$, where $\gamma_q(x, t)$ is as in (1.38). However, the geometric analogue of this is false; the $sy_q(x, t)$ are not controlled by the $\beta_q(x, 2t)$. This is connected to the fact that, as we shall see later, the local symmetry condition alone implies uniform rectifiability (whereas the WGL does not).

The other criterion for the WGL that we want to record here is the following weakened version of the Littlewood-Paley-type condition (1.68):

for each compactly supported, odd, C^∞ function ψ on $\mathbf{R}^n$ and for each $\epsilon > 0$, we have that

(1.85) $\left\{ (x, t) \in E \times \mathbf{R}_+ : \left| \int_E \psi_k(x - y) dy \right| > \epsilon \right.$

$\left. \text{for the integer } k \text{ such that } 2^k \leq t < 2^{k+1} \right\}$

is a Carleson set.

It is clear that (1.85) is a weaker condition than (1.68). We shall refer to it as the weak Littlewood-Paley condition.

LEMMA 1.86. *If E is a d-dimensional regular set that satisfies (1.85), then E also satisfies LS.*

This is fairly easy to prove. See §4 of [**DS2**].

There are other conditions similar to the local symmetry condition that can play essentially the same role. We shall describe some of these in the next chapter. (See, in particular, the local convexity condition LCV of Definition 2.7.)

CHAPTER 2

A Summary of the Main Results

2.1. The results of Part II.
One of the principal themes of [DS2] was that many results concerning the analysis of functions have counterparts for the analysis of the geometry of sets. The essential nonlinearities in the geometric setting, however, render many of the (basically linear) methods from harmonic analysis difficult or almost impossible to apply. (See [DS1] for additional comments along these lines.)

One of the principal themes of this monograph is that the behavior of a set can be studied in many interesting ways which do not have obvious counterparts in the world of functions. In particular, if E is a closed set and $B(x, t)$ is a ball centered on E, there are ways other than $\beta(x, t)$ (defined by (1.46)) to measure how close $E \cap B(x, t)$ is to a d-plane. For instance, we could measure the distance from $E \cap B(x, t)$ to d-planes in terms of Hausdorff distance. We shall see that some methods for measuring the rectifiability of a set that are not directly analogous to techniques for analyzing functions are in fact much more efficient for capturing relevant geometric information about E.

A simple-minded difference between sets and functions is, for instance, that you might want to measure how disconnected a set is (or how many holes or gaps it has). This can be relevant for the study of rectifiability, but we do not know of any satisfactory analogue in the context of functions.

Here is another example of a way in which sets should be viewed differently from functions. Consider the Grassmann manifold $G(n, d)$ of all d-planes in $\mathbf{R}^n$ that contain the origin. Its counterpart for functions is the set of linear mappings from $\mathbf{R}^d$ to $\mathbf{R}^{n-d}$. Traditionally, one uses the fact that this is a vector space (which is not true for $G(n, d)$!), while we shall, in effect, use the fact that $G(n, d)$ is compact when equipped with its natural topology.

Let us now describe some of the new geometrical characterizations of uniform rectifiability that will be studied in Part II.

In what follows, E will always denote a d-dimensional regular set in $\mathbf{R}^n$ (see Definition 1.13).

Our first condition, the "bilateral weak geometric lemma", is a version of the WGL (see Definition 1.71) where the distance from E to P inside a

given ball $B(x, t)$ takes into account not only the distance from points of $E \cap B(x, t)$ to P (as in the definition of $\beta(x, t)$) but also the distance from points of $P \cap B(x, t)$ to E. More precisely, set
(2.1)
$$b\beta(x, t) = \inf_{P} \left\{ \sup_{y \in E \cap B(x,t)} t^{-1} \text{dist}(y, P) + \sup_{z \in P \cap B(x,t)} t^{-1} \text{dist}(z, E) \right\},$$

where the infimum is taken over all d-planes P in $\mathbf{R}^n$. [The letter b in $b\beta$ stands for "bilateral".]

DEFINITION 2.2. A d-dimensional regular set E in $\mathbf{R}^n$ is said to satisfy the bilateral weak geometric lemma (BWGL) if, for each $\epsilon > 0$,

(2.3) $$\{(x, t) \in E \times \mathbf{R}_+ : b\beta(x, t) > \epsilon\}$$

is a Carleson set.

For the definition of Carleson sets, see Definition 1.69 and (1.70). As with the WGL and LS, it is important that we do not require any control on the way the Carleson constants of (2.3) depend on ϵ. Also, we should mention that in applications (such as Theorem 2.4), the fact that (2.3) is a Carleson set is only used for a single value of ϵ (but which might be very small and depend on the given situation).

THEOREM 2.4. *A regular set satisfies the BWGL if and only if it is uniformly rectifiable.*

Recall from Definition 1.65 that uniformly rectifiable sets are sets that satisfy the equivalent conditions of Theorem 1.57. For instance, when $d = 1$, E is uniformly rectifiable if and only if it is contained in some regular curve (see the discussion around (1.63)).

The fact that uniform rectifiability implies the BWGL was already known and not very difficult. To prove this, one first checks that if E is good for all kernels in $\mathcal{K}_d(\mathbf{R}^n)$ (condition (1.58)), then the Littlewood-Paley condition (1.68) holds. (See Theorem 1.66.) Next, (1.68) implies the apparently much weaker condition (1.85), which also implies LS easily. (See Lemma 1.86, and also Definition 1.79 for the definition of LS.) The fact that the BWGL is implied by uniform rectifiability is then a consequence of the following.

PROPOSITION 2.5. *If a regular set E satisfies LS, then it satisfies the BWGL.*

The details of the proof that uniform rectifiability implies the BWGL can be found in [DS2]. In particular, Proposition 2.5 is essentially the same as Proposition 5.5 in [DS2].

Presumably one can give a more direct proof of this fact in the spirit of Chapter 1 of Part IV, but in this regard the BWGL is more complicated than the WGL.

The fact that the BWGL is enough to imply uniform rectifiability is the new part of Theorem 2.4 and was quite a surprise to us. Here are two reasons why. First, when one goes from uniform rectifiability to the BWGL, one seems to lose a lot of information when passing from the Littlewood-Paley estimate (1.68) to its weak version (1.85). This is apparently confirmed by the large amount of additional work needed in [**DS2**] to go from the WGL (actually, the BWGL) to the uniform rectifiability conditions, again using (1.68). The second reason is that we know that the WGL is not enough to imply uniform rectifiability, while the geometric lemma (1.59) is necessary and sufficient. Actually, we even know that (1.59), with $\beta_q(x,t)$ replaced by an appropriate bilateral version $b\beta_q(x,t)$, is also necessary and sufficient. Thus the BWGL implies conditions which appear to be much stronger than itself.

The proof of the fact that the BWGL implies uniform rectifiability will be given in Chapter 2 of Part II, along with some variants. Let us mention one of these.

Given a regular set E in $\mathbf{R}^n$, define a function cv on $E \times \mathbf{R}_+$ by

$$(2.6) \qquad cv(x,t) = \sup_{y,z \in E \cap B(x,t)} \left[t^{-1} \operatorname{dist}\left(\frac{y+z}{2}, E\right)\right].$$

The function cv, which measures the extent to which $E \cap B(x,t)$ is convex, is a close relative of the function sy used to define LS. It can also be interpreted as an analogue for sets of a supremum of second difference quotients of a function.

DEFINITION 2.7. A d-dimensional regular set E is said to satisfy the local convexity condition (in short, LCV) if, for each $\epsilon > 0$,

$$(2.8) \qquad \{(x,t) \in E \times \mathbf{R}_+ : cv(x,t) > \epsilon\}$$

is a Carleson set.

The following result will be proved in Part II, Chapter 2.

PROPOSITION 2.9. *If the d-dimensional regular set E satisfies the LCV, then it satisfies the BWGL.*

Notice that the BWGL trivially implies both LS and LCV, so we have the following consequence of Theorem 2.4.

COROLLARY 2.10. *For d-dimensional regular sets in $\mathbf{R}^n$, uniform rectifiability is equivalent to each of the BWGL, LS, and LCV conditions.*

It is amusing to note that LS is thus equivalent to the (apparently much stronger) condition that $sy_q(x,t)^2 \frac{dxdt}{t}$ be a Carleson measure on $E \times \mathbf{R}_+$ when q is in the appropriate range (see (1.82)). Similarly, one could also show that LCV is equivalent to conditions like

$$(2.11) \qquad cv_q(x,t)^2 \frac{dxdt}{t} \text{ is a Carleson measure},$$

where cv_q is the appropriate generalization of cv and q is in the same range as usual. Of course, the analogous statements in the context of functions are wrong.

The advantage of conditions like the BWGL, LS, or LCV over, say, the geometric lemma is that they capture what appears to be an important piece of information about E, namely, the absence of holes or gaps in E. (This is the main difference between the bilateral weak geometric lemma and the WGL.) Another important advantage is that in practice it is much easier to work with the assumption that something is a Carleson set than the more traditional quadratic Carleson measure conditions.

For one-dimensional sets, the connection between rectifiability properties of E and the absence of too many holes or gaps in E is quite clear. If you want to show that E is contained in a single connected set with finite length, you should somehow control the sum of the diameters of the gaps between the various pieces composing E, because these gaps will have to be filled using additional pieces of curve. The special role of the notion of connectedness allows us to do even better than Theorem 2.4 in dimension $d=1$.

DEFINITION 2.12. Let E be a one-dimensional regular set in $\mathbf{R}^n$. We shall say that E satisfies the weak connectedness condition (WCC) (or sometimes that E is weakly connected) if there exist $r \in (0, 1)$ and $M \geq 1$ such that $E \times \mathbf{R}^+ \setminus \mathscr{G}(r, M)$ is a Carleson set, where

(2.13)
$$\begin{aligned}\mathscr{G}(r, M) = \{(x, t) \in E \times \mathbf{R}^+ : \text{ for all points } u, v \in E \cap B(x, t) \\ \text{such that } |u-v| \geq \tfrac{t}{10}, \text{ there exists a chain of points} \\ y_0 = u, \ y_1, y_2, \ldots, y_N = v, \\ \text{with } y_i \in E \cap B(x, Mt) \\ \text{and } |y_{i+1} - y_i| \leq r|u-v| \text{ for } i=0, \ldots, N-1\}.\end{aligned}$$

We owe the present formulation of this definition to an observation of P. Jones (our previous definition was unnecessarily restrictive).

The integer N in (2.13) may depend on x, t, u, v, but it is easy to show that you can control the best values of N in terms of M and the regularity constant for E.

The choice of the constant $\frac{1}{10}$ in (2.13) is not really relevant, and even the constant r could be fixed arbitrarily without really altering the content of the definition. (More details will be given in Part II, Chapter 1.) Also, the nervous reader might be concerned about the issue of measurability of $\mathscr{G}(r, M)$. It can be shown that $\mathscr{G}(r, M)$ is measurable, but this does not really matter. You can define the notion of Carleson sets for nonmeasurable sets using outer measures, and anyway we shall in fact be working with an equivalent reformulation of Definition 2.12 which is discretized (and therefore insensitive to measurability issues).

It is clear that the BWGL implies the WCC, and so the following theorem is stronger than Theorem 2.4 (when $d=1$).

THEOREM 2.14. *For one-dimensional regular sets, uniform rectifiability is equivalent to the WCC.*

The fact that uniform rectifiability implies the WCC follows from the "if" part of Theorem 2.4 and was already known. For the converse, we shall actually show that

(2.15)

If the one-dimensional regular set E satisfies the WCC, then there is a connected regular set of dimension 1 in $\mathbf{R}^{n+1}$ that contains E.

The proof of (2.15) is surprisingly easy. The basic idea is that the total additional length needed to connect the various pieces of E is controlled in terms of $\mathscr{G}(r, M)$. (See Part II, §1.1 for more details.) Notice that once (2.15) is established, showing that E is contained in some regular curve in $\mathbf{R}^{n+1}$ only requires a little additional work (see the remark after (1.63) and (1.64)). We could even change the curve away from E to make it fit inside $\mathbf{R}^n$ (but the proof is not very pleasant).

Our proof of (2.15) will give a new, and much simpler, proof of a large part of the main result in [**DS2**] when $d = 1$. Namely, the fact that the condition (1.58) of Theorem 1.57 (which says that E is good for all the kernels in $\mathscr{K}_1(\mathbf{R}^n)$) implies that there is a connected regular set that contains E can be proved as follows. First, (1.58) implies the Littlewood-Paley estimate (1.68) and hence its weak form (1.85); then (1.85) implies the local symmetry condition LS (see Lemma 1.86), which implies the BWGL (see Proposition 2.5), and thus the WCC and the conclusion of (2.15). The first steps (up to the BWGL) are the same as in [**DS2**], but the proof of (2.15) allows us to shunt the most painful part of the proof (the fact that the Littlewood-Paley estimate and the BWGL imply that E has a "corona decomposition").

The WCC is more satisfying as a criterion for uniform rectifiability than the BWGL or the LS, because it is much less rigid and, in particular, much less dependent on the special geometry of $\mathbf{R}^n$. For instance, it is easy to modify slightly the definition of the WCC to make it clearly invariant under bilipschitz mappings. One way would be to define $\mathscr{G}(r, M, \tau)$ the same way as $\mathscr{G}(r, M)$, but with $\frac{1}{10}$ replaced by τ, and then demand that, for all r and τ small enough, there be an M such that the complement of $\mathscr{G}(r, M, \tau)$ in $E \times \mathbf{R}_+$ is a Carleson set. It is easy to check that the new definition would be equivalent to Definition 2.12. Of course we know that uniform rectifiability is invariant under bilipschitz mappings because (1.60), (1.61), and (1.62) are, and it would have been nicer if the BWGL were more apparently invariant under bilipschitz mappings.

Chapter 1 of Part II contains a proof of (2.15), along with some other characterizations of uniform rectifiability for one-dimensional regular sets (all of them proved using Theorem 2.14).

Although the WCC and the BWGL are nice criteria for uniform rectifiability, there are situations in which they do not seem to apply. The condition

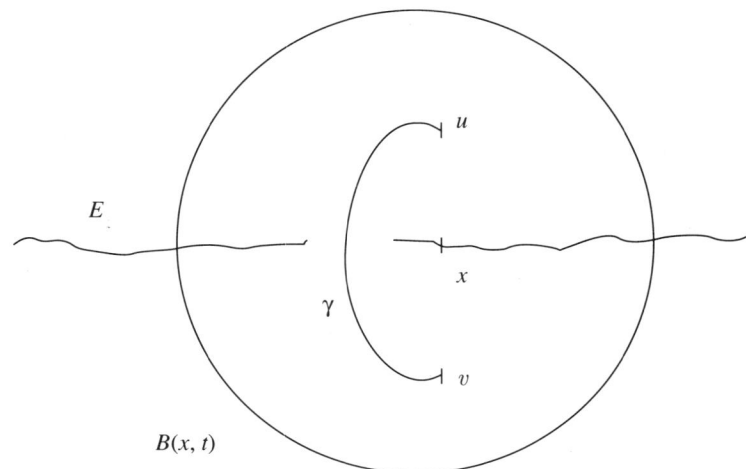

FIGURE 2.1. This (x, t) is not in $\mathscr{G}_e(\epsilon)$.

we discuss next will work some of the times that the WCC or the BWGL do not.

The next definition concerns a regular set E of codimension 1, and it is a way to ask that, inside most of the balls $B(x, t)$, each of the connected components of the complement of E be approximately convex. This can be seen as a more subtle cousin of the BWGL, because the complement of a d-plane in $\mathbf{R}^{d+1}$ is indeed composed of two convex components.

For a d-dimensional regular set E in $\mathbf{R}^{d+1}$ and $\epsilon > 0$ (small), let $\mathscr{G}_e(\epsilon)$ be the set of $(x, t) \in E \times \mathbf{R}_+$ such that, whenever u and v are two points of $B(x, t) \setminus E$ that can be joined by a path $\gamma \subset \mathbf{R}^{d+1} \setminus E$ which satisfies

(2.16) $\qquad \gamma \subset B(x, t) \quad \text{and} \quad \operatorname{dist}(\gamma, E) \geq \epsilon t,$

the line segment from u to v does not touch E.

DEFINITION 2.17. The d-dimensional regular set $E \subset \mathbf{R}^{d+1}$ is said to satisfy the weak exterior convexity condition (WEC) if $E \times \mathbf{R}_+ \setminus \mathscr{G}_e(\epsilon)$ is a Carleson set for each $\epsilon > 0$.

Notice that the WEC is still another condition that prevents E from having too many holes (see Figure 2.1). In the same way as one can see the BWGL as a property of approximation of E by d-planes (for each $\epsilon > 0$, we require that for most balls $B(x, t)$ the Hausdorff distance between the set $E \cap B(x, t)$ and the intersection of $B(x, t)$ with some d-plane be less than ϵt), one can also interpret the WEC as a condition of approximation of E by sets F whose complementary components are all convex. This point of view will be taken up in the next section.

THEOREM 2.18. *For a d-dimensional regular set E in $\mathbf{R}^{d+1}$, the WEC is equivalent to uniform rectifiability.*

Of course, the fact that uniform rectifiability implies the WEC follows from the "if" part of Theorem 2.4 and the easy observation that the BWGL implies the WEC.

The converse is proved in Chapter 3 of Part II. Actually, the main result of that chapter is the equivalence between uniform rectifiability and the combination of the WHIP and the WTP. The precise statement of these two conditions is a little complicated, so we refer to §3.1 of Part II for their definition. The main interest of these conditions is that they can easily be deduced from a variety of other more geometrically significant conditions, such as the WEC, the GWEC, the OUWGL, or the WNM. The GWEC (generalized weak exterior convexity condition) is a higher-codimension version of the WEC; the OUWGL (other unilateral weak geometric lemma) is a condition similar to the BWGL, but where one only demands, for each ϵ and most balls $B(x, t)$, that there exist a d-plane P containing x such that all points of $P \cap B(x, t)$ are within ϵt of E. For precise definitions of these conditions and a few others, we refer the reader to §§3.2–3.4 of Part II.

In the last chapter of Part II we shall provide yet another condition on a regular set E of codimension 1 that is equivalent to uniform rectifiability. We do not describe this condition here but only say that it is satisfied as soon as E has the property of bilateral approximation by Lipschitz graphs (or even by bilipschitz images of $\mathbf{R}^d$ in $\mathbf{R}^{d+1}$). These last are defined exactly like the BWGL but with d-planes replaced by Lipschitz graphs with uniformly bounded constants or by bilipschitz images of $\mathbf{R}^d$ in $\mathbf{R}^{d+1}$ with uniformly bounded bilipschitz constants (see the next section and, of course, Chapter 4 of Part II for more details).

The advantage of these bilateral approximation properties over the BWGL is that they are much more flexible. In particular, the second one is clearly invariant under bilipschitz mappings of $\mathbf{R}^{d+1}$, which is always a good thing.

Although we expect that there are versions of at least some of the results of Chapter II.4 when the codimension is larger than 1, we do not have any proofs.

2.2. Bilateral approximation from a functorial point of view.

Many of the conditions that were discussed in the previous section can be seen as conditions of bilateral approximation of E by some classes of sets. To make this more precise, we need a few definitions.

Given two subsets A_1, A_2 of $\mathbf{R}^n$, we shall need to measure the distance between the two sets in, say, the closed unit ball. The first thing that comes to mind is the Hausdorff distance between the two sets $A_i \cap \overline{B}(0, 1)$, $i = 1, 2$, but it will be more convenient to take a version of this which has a smoother behavior across the boundary of $B(0, 1)$. We set

$$(2.19) \quad D(A_1, A_2) = \sup_{x \in \overline{A}_1 \cap \overline{B}(0,1)} \mathrm{dist}(x, A_2) + \sup_{y \in \overline{A}_2 \cap \overline{B}(0,1)} \mathrm{dist}(y, A_1).$$

Notice that $D(A_1, A_2)$ only depends on the sets $A_i \cap \overline{B}(0, 3)$, at least if

$A_1 \cup A_2$ intersects $B(0, 1)$. Also, $D(A_1, A_2)$ is not affected if you replace A_1 by its closure. The fact that D does not satisfy the triangle inequality will not disturb us.

For each set $E \subset \mathbf{R}^n$, each point $x \in E$, and each radius $t > 0$, let

(2.20) $$E_{x,t} = t^{-1}(E - x)$$

be the image of E by the translation and dilation that maps $B(x, t)$ to the unit ball.

We are now ready to associate, to each class $\mathscr{E}$ of closed subsets of $\mathbf{R}^n$, a property of bilateral approximation which has the same structure as the BWGL.

DEFINITION 2.21. Let $\mathscr{E}$ be a class of closed sets in $\mathbf{R}^n$. We denote by Approx($\mathscr{E}$) the class of d-dimensional regular sets E such that, for each $\epsilon > 0$ small enough, the complement in $E \times \mathbf{R}_+$ of the set

(2.22) $\{(x, t) \in E \times \mathbf{R}_+ :$ there is a set $A \in \mathscr{E}$ such that $D(A, E_{x,t}) \leq \epsilon\}$

is a Carleson set.

Notice that the class Approx($\mathscr{E}$) does not depend on what the sets of $\mathscr{E}$ do outside of $\overline{B}(0, 2)$ and, in fact, only depends on the intersections of the sets of $\mathscr{E}$ with a small neighborhood of the closed unit ball. Also, our assumption that all the sets of $\mathscr{E}$ are closed is not a serious restriction, because taking the closure of all sets of A would not change (2.22). There are other restrictions on the class $\mathscr{E}$ which could be made without loss of generality. For instance, we might as well assume that the class $\mathscr{E}$ is closed with respect to the Hausdorff distance between sets, because adding Hausdorff limits of sets of $\mathscr{E}$ does not change (2.22) either. On the other hand, it is not always convenient to choose classes $\mathscr{E}$ that contain only regular sets. (However, if $\mathscr{E}$ is closed for the Hausdorff distance and if you do not want Approx($\mathscr{E}$) to be empty, you should make sure that $\mathscr{E}$ contains enough sets that satisfy the regularity condition for balls contained in $B(0, 1)$.)

If we take $\mathscr{E}$ to be the class of d-planes, then Approx($\mathscr{E}$) is the class of regular sets that satisfy the BWGL.

If $d = 1$, we can choose $\mathscr{E}$ to be the class of all compact connected subsets of the closed unit ball. Then Approx($\mathscr{E}$) is a class which is a priori a little smaller than the class of one-dimensional regular sets that satisfy the WCC. [Of course, these classes turn out to be the same a posteriori.]

Now let $n = d + 1$, and take $\mathscr{E} = \mathscr{E}_0$ to be the class of all closed subsets A of $\mathbf{R}^n$ such that every connected component of $\mathbf{R}^d \setminus A$ is convex.

LEMMA 2.23. *If $\mathscr{E}_0$ is as above, then Approx($\mathscr{E}_0$) is the class of regular sets which satisfy the WEC (see Definition 2.17).*

Let us first check that if $E \in \text{Approx}(\mathscr{E}_0)$, then it satisfies the WEC. It is enough to check that $(x, t) \in \mathscr{G}_e(\epsilon)$ as soon as there is a set $A \in \mathscr{E}_0$ such that $D(A, E_{x,2t}) \leq \frac{\epsilon}{10}$, say. To prove that $(x, t) \in \mathscr{G}_e(\epsilon)$, we consider two points

u, $v \in B(x, t) \setminus E$, and we suppose that there is a path $\gamma \subset B(x, t) \setminus E$ that joins them and which satisfies $\mathrm{dist}(\gamma, E) \geq \epsilon t$. We want to check that the line segment $[u, v]$ avoids E.

Let $\widetilde{A} = x + 2tA$. Since $D(A, E_{x, 2t}) \leq \frac{\epsilon}{10}$, we get that

$$(2.24) \qquad \sup_{y \in E \cap B(x, 2t)} \mathrm{dist}(y, \widetilde{A}) + \sup_{y \in \widetilde{A} \cap B(x, 2t)} \mathrm{dist}(y, E) \leq \frac{\epsilon t}{5}.$$

Thus γ stays at distance $\geq \frac{4\epsilon t}{5}$ from $\widetilde{A}$, and so u and v lie in the same connected component of the complement of $\widetilde{A}$. Call this component $\mathcal{O}$. We also have that u' and $v' \in \mathcal{O}$, provided that $|u - u'| \leq \frac{\epsilon}{2}$ and $|v - v'| \leq \frac{\epsilon}{2}$. Since $\mathcal{O}$ is convex by definition of A and $\widetilde{A}$, the line segment $[u', v']$ does not meet $\widetilde{A}$, and since this holds for all choices of u', v', we see that $[u, v]$ stays at distance $\geq \frac{\epsilon}{2}$ from $\widetilde{A}$. Using (2.24) again, we deduce that $[u, v]$ does not meet E, which proves the first half of the lemma.

We shall never use the other half, but let us sketch its proof anyway. This time we suppose that $(x, 2t) \in \mathcal{G}_e(\epsilon)$, where $\epsilon > 0$ is small, and we shall prove that there exists a set $A \in \mathcal{E}_0$ such that

$$(2.25) \qquad \sup_{y \in E \cap \overline{B}(x, t)} \mathrm{dist}(y, A) + \sup_{y \in A \cap \overline{B}(x, t)} \mathrm{dist}(y, E) \leq (2d + 10)\epsilon t.$$

From here Lemma 2.23 will follow.

To construct the set A, we proceed as follows. First let $\mathcal{U} = \{w \in B(x, 2t) \setminus E : \mathrm{dist}(w, E) > 2\epsilon t\}$, and let $\mathcal{U}_i$, $i \in I$, be all the connected components of $\mathcal{U}$. We let $\mathcal{V}_i = \{w \in \mathcal{U}_i \cap \overline{B}(x, t) : \mathrm{dist}(w, E) > (2d+9)\epsilon t\}$, and we take $\mathcal{W}_i$ to be the convex hull of $\mathcal{V}_i$. Finally let $\mathcal{W} = \bigcup_{i \in I} \mathcal{W}_i$ and $A = \mathbf{R}^{d+1} \setminus \mathcal{W}$. The first thing we shall check is that $\mathcal{W}_i$ is still contained in $\mathcal{U}_i$. Since the $\mathcal{U}_i$ are disjoint, it will follow that the $\mathcal{W}_i$ are actually the components of the complement of A and, therefore, that $A \in \mathcal{E}_0$.

To prove that $\mathcal{W}_i \subset \mathcal{U}_i$ we shall use the following.

LEMMA 2.26. *If a point x in $\mathbf{R}^n$ lies in the convex hull of $F \subseteq \mathbf{R}^n$, then it lies in the convex hull of some subset of F which contains at most $n + 1$ elements.*

This is very well known. See p. 73 of [**R**] for a proof.

Let us now show that $\mathcal{W}_i \subseteq \mathcal{U}_i$. Consider first elements of $\mathcal{W}_i$ which are a convex combination of two points in $\mathcal{V}_i$. Let u, v be two points of $\mathcal{V}_i$. Notice that if u', v' are such that $|u' - u| \leq (2d + 7)\epsilon t$ and $|v' - v| \leq (2d + 7)\epsilon t$, then u' and v' are still in $\mathcal{U}_i$, and therefore they can be joined by a path $\gamma \subset \mathcal{U}_i$. We can use our assumption that $(x, 2t) \in \mathcal{G}_e(\epsilon)$ to deduce that the segment $[u', v']$ does not meet E. Since this holds for all choices u', v', we see that $[u, v]$ stays at distance $> (2d + 7)\epsilon t$ from E. Hence convex combinations of ≤ 2 points of $\mathcal{V}_i$ still lie in $\mathcal{U}_i$ and remain at distance $> (2d + 7)\epsilon t$ from E. Iterating this argument, we find out that all convex combinations of $\leq (d + 2)$ points of $\mathcal{V}_i$ still lie in $\mathcal{U}_i$ (and even at distance $> 7\epsilon t$ from E). Hence $\mathscr{W}_i \subset \mathscr{U}_i$.

The estimate (2.25) is now very easy to check. First notice that $E \subset A$ because $\mathscr{W}$ does not meet E. Hence the first term in (2.25) is zero. For the second term, notice that all the points of $B(x, t)$ that are at distance $\geq (2d + 10)\epsilon t$ from E lie in one of the $\mathscr{V}_i$'s. This proves (2.25) and hence Lemma 2.23.

Let us now come back to the WEC. The functorial definition in terms of Approx($\mathscr{E}_0$) given by Lemma 2.23 probably provides a better understanding of the WEC than the original definition. The WEC is a priori a much weaker condition than the BWGL, because there are many regular sets in $\mathscr{E}_0$ besides hyperplanes. For instance, finite unions of hyperplanes are elements of $\mathscr{E}_0$, and so the WEC is also a priori weaker than the property characterized by the class Approx($\mathscr{E}_1$), where $\mathscr{E}_1$ is the class of finite unions of d-planes.

Notice that Approx($\mathscr{E}_1$) still makes sense in higher codimensions. A regular set of dimension d in $\mathbf{R}^n$ will be said to satisfy the BAFUP (property of bilateral approximation by finite unions of d-planes) if it lies in Approx($\mathscr{E}_1$).

When $d < n - 1$, we cannot say that the BAFUP implies the WEC (since the latter no longer makes sense), but it is still true that it implies uniform rectifiability. Actually, there is a generalization of the WEC (called the GWEC) which works in all codimensions and is implied by the BAFUP. Even the weaker condition OUWGL (other unilateral weak geometric lemma)—where one only demands that for every $\epsilon > 0$ and most of the balls $B(x, t)$ there exists a d-plane P containing x such that every point of $P \cap B(x, t)$ is within ϵt of E—implies uniform rectifiability. See §3.2 in Part II for details.

Although the WEC is a little less rigid than the BWGL, it is still not obvious from its definition that it is invariant under bilipschitz mappings of $\mathbf{R}^{d+1}$. A better condition in this regard is the property of bilateral approximation by bilipschitz images of $\mathbf{R}^d$ inside $\mathbf{R}^{d+1}$ (BABI) already alluded to at the end of the previous section. This condition corresponds to the union of the classes Approx($\mathscr{E}(M)$), $M \in (1, \infty)$, where $\mathscr{E}(M)$ consists of all images of $\mathbf{R}^d$ by M-bilipschitz mappings of $\mathbf{R}^d$ into $\mathbf{R}^{d+1}$.

Before we leave this section we would like to discuss briefly qualitative versions of Approx($\mathscr{E}$). There are a number of different ways to replace the condition in Definition 2.21 by a nonquantitative version, but in the nice examples they are often equivalent. Let us describe one of them.

Denote by $\mathscr{B}(\epsilon)$ the complement of the set in (2.22), i.e.,

(2.27) $\qquad \mathscr{B}(\epsilon) = \{(x, t) \in E \times \mathbf{R}_+ : D(A, E_{x,t}) \geq \epsilon \text{ for all } A \in \mathscr{E}\}$.

A simple condition on E would be that

(2.28)
> for each $\epsilon > 0$ and almost all $x \in E$, there are only finitely many $j \in \mathbf{Z}_+$ such that $(x, 2^{-j}) \in \mathscr{B}(\epsilon)$.

Let us explain why this can reasonably be described as a qualitative version of the definition of Approx($\mathscr{E}$). Let us assume that the discretization does not create any problem (for instance, assume that if $(x, t) \in \mathscr{B}(\epsilon)$, then $(x, u) \in \mathscr{B}\left(\frac{\epsilon}{100}\right)$ for all u such that $t \leq u \leq 2t$). The Carleson set condition in Definition 2.21 can then be interpreted as follows. For each ball $B(x, R)$ centered on E and each $y \in E \cap B(x, R)$, count the number of $j \in \mathbf{Z}$ such that $2^{-j} \leq R$ and $(y, 2^{-j}) \in \mathscr{B}(\epsilon)$. The mean value over $E \cap B(x, R)$ of this number should be $\leq C$. In (2.28), we only demand that this number be finite for almost every $y \in E$.

For each closed subset B of $\mathbf{R}^{n+1}$, set

(2.29) $$\mathrm{Dist}(B, \mathscr{E}) = \inf\{D(B, A) : A \in \mathscr{E}\}.$$

Notice that $\mathrm{Dist}(B, \mathscr{E})$ does not come from a real metric, but this will not create any problem because our restricted distance function $D(A, B)$ between sets (defined by (2.19)) is controlled by the Hausdorff distance. Let us be more precise. The Hausdorff distance between two sets A and B is given by

(2.30) $$\mathscr{H}(A, B) = \sup_{x \in A} \mathrm{dist}(x, B) + \sup_{x \in B} \mathrm{dist}(x, A).$$

Unlike our function D, the distance $\mathscr{H}$ satisfies the triangle inequality. Moreover, it is well known and easy to prove that if you restrict $\mathscr{H}$ to the set $\mathscr{F}$ of compact subsets of $\overline{B}(0, 2)$, say, you get a metric for which $\mathscr{F}$ is a compact Hausdorff space. This will be useful for us because convergence in $\mathscr{F}$ implies convergence for D. (Notice that $D(A, B) \leq \mathscr{H}(A, B)$ trivially.)

There is a variant of this that is better for our purposes. Given $l > 0$ define $\mathscr{H}_l(A, B)$ as in (2.30), except that we add the restriction $|x| \leq l$ to the suprema. With these functions we can define a topology on the space of all closed subsets of $\mathbf{R}^n$, whereby $A_j \to A$ if and only if $\mathscr{H}_l(A_j, A) \to 0$ for every l.

Coming back to our condition (2.28), we can (and shall) assume that $\mathscr{E}$ is closed with respect to the topology just described. Indeed, if we replace $\mathscr{E}$ by its closure, then we do not change the sets $\mathscr{B}(\epsilon)$, the condition (2.28), or (2.29).

The condition (2.28) is equivalent to saying that, for almost all $x \in E$,

(2.31) $$\lim_{j \to +\infty} \mathrm{Dist}(E_{x, 2^{-j}}, \mathscr{E}) = 0.$$

This condition has a reformulation which has the same structure as conditions about weak tangents that arise frequently in geometric measure theory and the study of minimal surfaces, harmonic maps, etc. Let us be more precise.

For each $x \in E$, consider the set $\mathscr{L}_0(E, x)$ of closed subsets A of $\mathbf{R}^n$ with the property that $D(E_{x,2^{-j_m}}, A)$ tends to 0 for some sequence j_m of integers that tends to $+\infty$. Of course, what really matters is the set $\mathscr{L}(E, x)$ of intersections $A \cap \overline{B}(0, 1)$, where $A \in \mathscr{L}_0(E, x)$. This set is interesting in particular because it is never empty. [The collection $\mathscr{L}$ of intersections $E_{x,2^{-j}} \cap \overline{B}(0, 2)$, $j \in \mathbf{Z}$, is relatively compact for the Hausdorff metric, and all the intersections of $\overline{B}(0, 1)$ with limits of sequences in $\mathscr{L}$ are in $\mathscr{L}(E, x)$.] With our usual assumption that E be regular, we even get that the sets which are elements of $\mathscr{L}(E, x)$ satisfy the regularity condition (1.14), restricted to balls that are contained in $B(0, 1)$.

It is easy to check that our condition (2.28) is equivalent to saying that for almost every $x \in E$, $\mathscr{L}(E, x)$ is contained in the set of intersections $\overline{B}(0, 1) \cap A$, $A \in \mathscr{E}$ (at least if you add the minor condition that $A \in \mathscr{E}$ if and only if $2A \in \mathscr{E}$).

A difference with what is traditionally done in geometric measure theory is that we have defined our weakly tangent sets $\mathscr{L}(E, x)$ as limits of sets $E_{x,2^{-j}}$ for a variant of Hausdorff distance rather than defining weakly tangent measures by taking limits of measures on the $E_{x,2^{-j}}$. (See for instance [**P**], especially the introduction.) Our approach is reasonable because we have conditions, such as the regularity assumption on E, that prevent the occurrence of many tiny pieces of E that would float around and eventually create artificially large sets in the Hausdorff limit. Of course, such conditions are not always natural in other contexts.

The bottom line of this discussion is that one can view conditions like the BWGL, the WEC, or other conditions of bilateral approximation as being quantitative versions of conditions which require that all the weak tangents at almost all points in a given regular set lie in some prescribed class. In some cases the corresponding qualitative conditions have been studied before in connection with rectifiability properties of sets. This is true of the BWGL, for instance. (See [**Ma3**], especially Chapter 16.) In other cases they may be new and interesting in their own right.

2.3. The results of Part III.

The main results of Part III deal mostly with characterizations of uniform rectifiability in terms of various properties of the analysis of functions on E or of certain linear operators defined on E. There is one exception, which is a relative of a theorem of Preiss on densities [**P**]. All the results in Part III rely heavily on Part II.

As indicated in §1.2, we do not know whether a 1-dimensional regular set in the complex plane that is good for the Cauchy kernel must be uniformly rectifiable. We have the following partial result, though.

THEOREM 2.32. *Let E be a regular set of dimension* 1 *in $\mathbf{R}^2$. If E is good for the Cauchy kernel and satisfies the weak geometric lemma (WGL), then E is uniformly rectifiable.*

2.3. THE RESULTS OF PART III

Recall that E is said to be good for the Cauchy kernel if the operators T_ϵ defined by (1.11) are bounded on $L^2(E)$ with uniform bounds with respect to ϵ. For the WGL, see Definition 1.71 (and maybe (1.46) and Definition 1.50).

The converse to Theorem 2.32 is true and well known; namely, if E is uniformly rectifiable (i.e., if it is contained in a regular curve), then it is good for the Cauchy kernel and also satisfies the WGL. What is not known yet is whether being good for the Cauchy kernel implies the WGL.

Theorem 2.32 is an improvement of a theorem of Fang [**Fg**]. Fang proved that if E is a one-dimensional regular set which is good for the Cauchy kernel and if $\sup\{\beta_\infty(x, t) : (x, t) \in E \times \mathbf{R}_+\}$ is sufficiently small, then E is uniformly rectifiable. It is more natural to assume only that E satisfies the WGL instead of requiring that the supremum of the $\beta_\infty(x, t)$'s be small, because the WGL is a necessary condition for E to be uniformly rectifiable.

One can probably adapt Fang's method to allow the WGL to replace the assumption that the $\beta_\infty(x, t)$'s are small, but the proof of Theorem 2.32 given in Part III has the advantage of not relying so heavily on the notion of connectedness and even allowing the following generalization to higher dimensions and codimensions.

THEOREM 2.33. *Let E be a d-dimensional regular set in $\mathbf{R}^n$. Suppose that E satisfies the WGL and that there exists a constant $C > 0$ such that*

$$(2.34) \quad \int_{x \in E} \left| \int_{y \in E \setminus B(x, \epsilon)} \frac{x - y}{|x - y|^{d+1}} f(y) \, dy \right|^2 dx \leq C \int_E |f(x)|^2 \, dx$$

for all $f \in L^2(E)$ and $\epsilon > 0$.

Then E is uniformly rectifiable.

Just as with Theorem 2.33, the converse is true and already known, since uniform rectifiability implies the WGL and the fact that E is good for all kernels in $\mathscr{K}_d(\mathbf{R}^n)$. Theorem 2.33 will be proved in Chapter 1 of Part III. More precisely, we shall prove that the WGL and the estimate (2.34) together imply the BWGL, so that Theorem 2.4 can be applied.

Let us consider another condition related to complex analysis, which turns out also to be equivalent to uniform rectifiability. We first define this condition in the simplest case.

DEFINITION 2.35. Let E be a one-dimensional regular set in $\mathbf{R}^2 \approx \mathbf{C}$. To each $f \in L^2$ we associate a function F (the Cauchy integral of f) defined on $\mathbf{C} \setminus E$ by

$$(2.36) \quad F(z) = \int_E \frac{1}{z - w} f(w) \, dw.$$

We say that E satisfies the usual square function estimates for the Cauchy kernel (USFE) if there exists a constant $C > 0$ so that

$$(2.37) \quad \iint_{\mathbf{C}} |F'(z)|^2 \operatorname{dist}(z, E) \, dz \leq C \int_E |f|^2$$

for all $f \in L^2(E)$.

Here dz denotes the Lebesgue measure on $\mathbf{R}^2$. Although $F'(z)$ is only defined on $\mathbf{C} \setminus E$, the left side of (2.37) still makes sense because E, being a one-dimensional regular set, has vanishing planar measure.

The USFE has a natural generalization to arbitrary dimensions d (but codimension 1). Let $N(x)$ denote the fundamental solution for the Laplacian on $\mathbf{R}^{d+1}$. Thus $N(x) = c(d)/|x|^{d-1}$ when $d > 1$. Let $\nabla^2 N$ be its second gradient (as a matrix-valued C^∞ function on $\mathbf{R}^{d+1} \setminus \{0\}$ which is homogeneous of degree $-d-1$).

DEFINITION 2.38. Let E be a d-dimensional regular set in $\mathbf{R}^{d+1}$. We say that E satisfies the USFE if there exists a constant $C > 0$ so that

$$(2.39) \qquad \int_{\mathbf{R}^{d+1} \setminus E} |G(x)|^2 \, \mathrm{dist}(x, E) \, dx \leq C \int_E |f|^2$$

for all $f \in L^2(E)$, where we set

$$(2.40) \qquad G(x) = \int_E \nabla^2 N(x - y) f(y) \, dy.$$

The two definitions clearly coincide when $d = 1$. It turns out that the USFE is easier to relate to uniform rectifiability than the boundedness on $L^2(E)$ of the Cauchy integral operator.

THEOREM 2.41. *Let E be a d-dimensional regular set in $\mathbf{R}^{d+1}$. Then E is uniformly rectifiable if and only if it satisfies the USFE (usual square function estimates for the Cauchy kernel).*

The "only if" part of this theorem is basically known. Since it does not seem to be stated in the literature in quite this manner, we shall sketch a proof for the convenience of the reader in §2.2 of Part III. To prove the "if" part, we shall show that the USFE implies the WEC and then apply Theorem 2.18. Actually, we shall even prove that the WEC follows from the following apparently weaker version of the USFE.

For each d-dimensional regular set $E \subset \mathbf{R}^{d+1}$, we define a function $e(z)$ on $\mathbf{R}^{d+1} \setminus E$ by

$$(2.42) \qquad e(z) = \mathrm{dist}(z, E) \left| \int_E \nabla^2 N(z - y) \, dy \right|.$$

(When $d = 1$, $e(z)$ is simply $\mathrm{dist}(z, E) \left| \int_E dw/(z-w)^2 \right|$.) Thus $e(z)$ is obtained by taking the absolute value of the function G of (2.40) associated to $f \equiv 1$ and then renormalizing. (Notice that $e(z) \leq C$, because E is regular.)

DEFINITION 2.43. Let E be a regular set of dimension d in $\mathbf{R}^{d+1}$. We say that E satisfies the weaker usual square function estimates for the Cauchy kernel (WUSFE) if, for each $\epsilon > 0$, the set $\{z \in \mathbf{R}^{d+1} \setminus E : e(z) > \epsilon\}$ is a

Carleson set. This means that there exists a constant $C(\epsilon) > 0$ such that

$$(2.44) \qquad \int_{B(x,R)\setminus E} \chi_{\{e(z)>\epsilon\}}(z) \operatorname{dist}(z, E)^{-1} dz \leq C(\epsilon) R^d$$

for all $x \in E$ and $R > 0$, where dz denotes the Lebesgue measure on $\mathbf{R}^{d+1}$.

The fact that the USFE implies the WUSFE is easy (the details are given in Chapter 2 of Part III). The relationship between the WUSFE and the USFE is essentially the same as that between the weak version (1.85) of the Littlewood-Paley condition (1.68) and (1.68) itself.

THEOREM 2.45. *If the regular set E satisfies the WUSFE, then it satisfies the WEC.*

This will be proved in Chapter 2 of Part III for one-dimensional regular sets in the complex plane and in Chapter 3 of Part III in the general (codimension 1) case. Theorem 2.41 will then follow.

An amusing consequence of Theorem 2.41 is that if E is a one-dimensional regular set which satisfies the USFE, then it is good for the Cauchy kernel. We do not know how to prove this directly, i.e., without going through a substantial amount of geometry. There are classical methods for passing back and forth between the L^2-boundedness of certain linear operators and associated square function estimates (see, for instance, Chapter 4 of [St]), but it is not clear that anything similar will work in this situation (with sets E that would be much more general than Lipschitz graphs).

Our next result concerns the behavior of Lipschitz functions defined on E. Given a Lipschitz function $f : E \to \mathbf{R}$ and a constant $K > 0$, define a function $\gamma^{(K)}$ on $E \times \mathbf{R}_+$ by

$$(2.46) \qquad \gamma^{(K)}(x, t) = t^{-1} \inf_a \left\{ \sup_{y \in E \cap B(x,t)} |f(y) - a(y)| \right\}$$

where the infimum is taken over all affine functions $a : \mathbf{R}^n \to \mathbf{R}$ such that $|\nabla a| \leq K$.

DEFINITION 2.47. The d-dimensional regular set E in $\mathbf{R}^n$ is said to satisfy the property of weak approximation of Lipschitz functions on E by affine functions (WALA) if there is a $K \geq 1$ such that, for all Lipschitz functions $f : E \to \mathbf{R}$ with norm ≤ 1 and each $\alpha > 0$, the set

$$(2.48) \qquad \left\{ (x, t) \in E \times \mathbf{R}_+ : \gamma^{(K)}(x, t) > \alpha \right\}$$

is a Carleson set.

Let us say a few words about the role of the constant K. When we defined the numbers $\gamma_\infty(x, t)$ for a function f defined on $\mathbf{R}^d$ (see (1.39)), we did not impose any restriction on the Lipschitz constants of the approximating functions a. However, it is very easy to check that any function a that realizes the infimum in (1.39) must be Lipschitz with norm ≤ 2 if f is Lipschitz with norm ≤ 1. Thus the $\gamma_\infty(x, t)$ of (1.39) is the same as the new $\gamma^{(2)}(x, t)$ for a 1-Lipschitz function f defined on the whole space. In particular, if we take $d = n$ and $E = \mathbf{R}^n$, then E satisfies the WALA because of Theorem 1.42. (See also Remark IV.2.28.)

In the general case, we cannot expect to have any control on the Lipschitz norm of the approximating affine functions a. Even if $E = \mathbf{R}^d \subset \mathbf{R}^n$ with $d < n$, the function $f \equiv 0$ is perfectly approximated by all affine functions that vanish on E (which can have arbitrarily large norms). We introduced the additional restriction $|\nabla a| \leq K$ to avoid cases where a 1-Lipschitz function would be approximated by affine functions with extremely large Lipschitz norms.

THEOREM 2.49. *Let E be a regular set of dimension d in $\mathbf{R}^n$. If E is uniformly rectifiable, then it satisfies the WALA (see Definition 2.47). Conversely, if $d = 1$ and if E satisfies the WALA, then E is uniformly rectifiable.*

The fact that uniform rectifiability implies the WALA is pretty easy to prove, using the existence of corona decompositions for uniformly rectifiable sets. This fact was observed in [DS3], and we shall review it in some detail in Chapter 4 of Part III. We shall even prove that if E is uniformly rectifiable, then the analogue of Theorem 1.42 (with Lipschitz functions defined on E and $\gamma_q(x, t)$ replaced by appropriate $\gamma_q^{(K)}(x, t)$) also holds on E, and in fact any $K > 1$ will work. See Chapter 4 of Part III for more details.

For the converse result, we shall show that the WALA implies one of the conditions described in Chapter 3 of Part II (the weak no-box condition). This condition implies uniform rectifiability when $d = 1$ by Proposition II.3.45 but unfortunately not in general. It could very well be true that the WALA implies uniform rectifiability also when $d > 1$, but we do not know how to prove that.

We shall in fact be able to do much better for the second part of Theorem 2.49 than what we already stated. We shall only need the weaker version of the WALA which corresponds to taking $K = \infty$, i.e., we do not need to restrict ourselves to affine functions with bounded gradient. (See Proposition III.4.35.) We do not even need to restrict ourselves to affine functions. In §III.4.3 we define a generalization of the WALA (called the GWALA), in which the space of affine functions is replaced by any finite-dimensional vector space of functions, and this space is also allowed to depend on (x, t). (See Definition III.4.91.) This generalization of the WALA has the attractive feature that it is much less dependent on the special structure of the ambient space $\mathbf{R}^n$. However, our proof of the fact that the GWALA implies uniform rectifiability when $d = 1$ (see Theorem III.4.94) does rely on the special structure of $\mathbf{R}^n$, through results of Chapter II.3.

See [DS3] and [S3] for other results that are related to Theorem 2.49. In particular, [DS3] contains a result where the Carleson condition on the set (2.48) is used to derive interesting information about Lipschitz mappings on E.

The next result that we want to discuss (Theorem 2.56) is a cousin of the theorem of D. Preiss concerning densities that we stated just after Theorem 1.6. We need to cover some preliminaries first.

DEFINITION 2.50. Let μ be a nonnegative Borel measure on $\mathbf{R}^n$. We say that μ is regular (of dimension d) if there is a $C_0 > 0$ such that

$$(2.51) \qquad C_0^{-1} R^d \leq \mu(B(x, R)) \leq C_0 R^d$$

for all $x \in \operatorname{supp} \mu$ and $R > 0$.

It is easy to show that if μ is a regular measure, then $E = \operatorname{supp} \mu$ is a d-dimensional regular set and μ is comparable in size to the restriction of d-dimensional Hausdorff measure to E (i.e., $C^{-1} dx \leq d\mu \leq C dx$).

Our motivation comes from the following "direct" result.

THEOREM 2.52. *Let E be a uniformly rectifiable d-dimensional subset of $\mathbf{R}^n$, and let μ be a d-dimensional regular measure such that $\operatorname{supp} \mu = E$. Then the complement in $E \times \mathbf{R}_+$ of the set*
$$(2.53)$$
$\{(x, t) \in E \times \mathbf{R}_+ : \text{ there exists a positive number } \delta = \delta(x, t) \text{ such that}$

$$|\mu(B(y, s)) - \delta s^d| \leq \epsilon t^d \quad \text{whenever } y \in E \cap B(x, t) \text{ and } 0 < s \leq t\}$$

is a Carleson set for each $\epsilon > 0$.

This is basically known (see §6 of [DS2]). The condition that (2.53) be a Carleson set is quite similar to the weak Littlewood-Paley condition (1.85), and it is actually reasonably easy to pass from the latter to the former. The fact that $d\mu$ might be different from dx does not play any significant role in the proofs.

We shall state a converse to Theorem 2.52 that works in some dimensions, but let us first define a condition that is a little less restrictive than the conclusion of Theorem 2.52.

If E is a d-dimensional regular set in $\mathbf{R}^n$ and C_0, ϵ are two positive constants, we let $\mathscr{G}_d(C_0, \epsilon)$ be the set of all $(x, t) \in E \times \mathbf{R}_+$ for which there exists a measure $\mu = \mu_{x,t}$ which satisfies $\operatorname{supp} \mu = E$, the regularity condition (2.51) with the constant C_0, and

$$(2.54) \qquad |\mu(B(y, s)) - s^d| \leq \epsilon t^d$$

for all $y \in E \cap B(x, t)$ and all $0 < s \leq t$. (We do not need a constant δ as in (2.53) any longer, because we can always divide $\mu_{x,t}$ by a constant.)

DEFINITION 2.55. The d-dimensional regular set E in $\mathbf{R}^n$ is said to satisfy the weak constant density condition (WCD) if there exists a $C_0 > 0$ such that the complement in $E \times \mathbf{R}_+$ of the set $\mathscr{G}_d(C_0, \epsilon)$ defined above is a Carleson set for every $\epsilon > 0$.

THEOREM 2.56. *Let E be a d-dimensional regular set in $\mathbf{R}^n$. Suppose that $d = 1$, 2, or $n - 1$ and that E satisfies the WCD. Then E is uniformly rectifiable.*

The proof of this theorem will be given in Chapter 5 of Part III. It uses a compactness argument to show that if E satisfies the WCD, then it satisfies a property of bilateral approximation by sets which support measures with (truly) constant density. Previously known properties of these sets allow us to apply directly the results of Chapter II.2 when $d = 1$ or 2. The $d = n - 1$ case is more tricky, because the approximating sets can have singularities, and we have to show that these singularities are far away most of the time.

When $d = 1$, the geometry is much simpler, and it is possible to give a more direct proof (see Chapter 1 of Part II).

The restrictions on the dimension in the statement of Theorem 2.56 come from the fact that less is known about the measures with constant density in most dimensions. We expect, though, that the general result is true.

One of the main differences between Theorem 2.52 and Preiss's theorem on densities is that the former concerns a quantitative form of rectifiability. Another difference is that the WCD should not be viewed as simply a quantitative version of the requirement that the density exists; it is a priori much weaker than that, because even the condition that the complement of the set (2.53) be a Carleson set does not imply directly that the density exists almost everywhere (the constant $\delta(x, t)$ might change with t, even when (x, t) stays in (2.53)).

An amusing consequence of Theorem 2.56 is that, for a regular subset of $\mathbf{R}^n$ of dimension $d = 1$, 2, or $(n-1)$, uniform rectifiability is also equivalent to the condition of being good for all kernels in the slightly different class $\widetilde{\mathscr{K}}_d(\mathbf{R}^n)$. Here $\widetilde{\mathscr{K}}_d(\mathbf{R}^n)$ denotes the class of radial kernels $K(x)$, defined and C^∞ on $\mathbf{R}^n \setminus \{0\}$, with the following size and cancellation properties:

(2.57) $$|x|^{d+l}|\nabla^l K(x)| \in L^\infty \quad \text{for all } \ell \geq 0$$

and

(2.58) $$\left| \int_{r_1 \leq |x| \leq r_2} K(x)|x|^{d-n} dx \right| \leq C \quad \text{for all } r_1, r_2 > 0,$$

where C does not depend on r_1 or r_2. (The size condition is the same as for $\mathscr{K}_d(\mathbf{R}^n)$, but we had to replace oddness by a cancellation property that makes sense for radial kernels.)

The fact that uniform rectifiability implies goodness for all kernels of $\widetilde{\mathscr{K}}_d(\mathbf{R}^n)$ is true in all dimensions, and has the same proof as for $\mathscr{K}_d(\mathbf{R}^n)$. (This condition is satisfied by Lipschitz graphs and is stable under the big pieces functor.) The converse follows from Theorem 2.56 because E must satisfy the WCD if it is good for all kernels in $\widetilde{\mathscr{K}}_d(\mathbf{R}^n)$. The proof of this last fact is essentially the same as the proof that if E is good for all kernels in $\mathscr{K}_d(\mathbf{R}^n)$, then it satisfies the weak Littlewood-Paley condition (1.85) and the local symmetry condition. (See [**DS2**], especially §§3–6.)

There is a variation of this story about radial kernels that works out nicely and which we shall describe now. Similar results were observed by Mattila and Preiss [**MP**], and their work plays a crucial role in what follows.

Let $\mathscr{K}'_d(\mathbf{R}^n)$ denote the set of functions on $\mathbf{R}^n \setminus \{0\}$ of the form $x_j K(x)$, where $K(x)$ is C^∞, radial, and satisfies

$$|x|^{d+l+1} |\nabla^l K(x)| \in L^\infty \quad \text{for all } l \geq 0.$$

Thus $\mathscr{K}'_d(\mathbf{R}^n) \subseteq \mathscr{K}_d(\mathbf{R}^n)$. (See Definition 1.20.)

THEOREM 2.59. *Let E be a d-dimensional regular set in $\mathbf{R}^n$. Then E is good for all kernels in $\mathscr{K}'_d(\mathbf{R}^n)$ if and only if E is uniformly rectifiable.*

Of course, we already knew that "if" part, since $\mathscr{K}'_d(\mathbf{R}^n) \subseteq \mathscr{K}_d(\mathbf{R}^n)$, so it is the converse about which we really care. Let us sketch the argument.

Fix d and n, $0 < d < n$, and let E be a d-dimensional regular set in $\mathbf{R}^n$ which satisfies

(2.60) $\qquad E$ is good for all kernels in $\mathscr{K}'_d(\mathbf{R}^n)$.

Let $\theta(x)$ be any C^∞ radial function on $\mathbf{R}^n$ with compact support. Given $\epsilon > 0$, consider

(2.61)
$$\left\{ (x, t) \in E \times \mathbf{R}_+ : \left| \int_E 2^{-j(d+1)} (x-y) \theta(2^{-j}(x-y)) \, dy \right| > \epsilon \right.$$
$$\left. \text{for the integer } j \text{ such that } 2^j \leq t < 2^{j+1} \right\}.$$

Then (2.60) implies that (2.61) is a Carleson set for all $\epsilon > 0$ and all $\theta(x)$ as above. This can be proved using the same sort of arguments as in §§3 and 4 of [**DS2**]. In other words, the Carleson condition on (2.61) is to (2.60) as (1.85) is to (1.58).

Following Mattila [**Ma2**], let us say that a nonnegative locally finite Borel measure μ on $\mathbf{R}^n$ is symmetric if

$$\int (x-a) \phi(x-a) \, d\mu(x) = 0$$

for all $a \in \operatorname{supp} \mu$ and all continuous radial functions ϕ. This is not exactly the same as the definition in [**Ma2**], but it is easily seen to be equivalent. The point is that our Carleson condition on (2.61) implies that $H^d \big|_E$ behaves like it is almost a symmetric measure at most locations and scales. It is not too hard to formulate precisely a condition which expresses this property and which is analogous to the conclusion of Theorem 2.52. There is also a reasonable version of Definition 2.55 in this context.

A fairly simple compactness argument can then be used to show that E must lie in $\operatorname{Approx}(\mathscr{S})$, where $\mathscr{S}$ is the class of closed sets in $\mathbf{R}^n$ of the form $\operatorname{supp} \mu$, where μ is a d-dimensional regular measure on $\mathbf{R}^n$ which is also symmetric.

It turns out that $\mathscr{S}$ is nothing but the set of d-planes in $\mathbf{R}^n$. This follows from [**Ma2**] when $d = 1$ and $n = 2$ and from [**MP**] in the general case. (See also Chapter 20 in [**Ma3**].) Altogether, then, we get that (2.60) implies that E satisfies the BWGL and, hence, is uniformly rectifiable, by Theorem 2.4.

2.4. A rapid description of Part IV.

The main goal of Part IV is to give direct proofs of some stability results. The first result concerns stability of various types of geometric lemmas under the "big pieces functor". The following will be proved in Chapter IV.1.

Let E be a d-dimensional regular set in $\mathbf{R}^n$. Suppose that E has big pieces of elements of some class Σ. (This is defined the same way as BPLG in Definition 1.26, but with Lipschitz graphs replaced by elements of Σ.) Assume that the elements of Σ are regular (with uniform bounds) and that they satisfy uniformly some version of the geometric lemma (for instance, (1.59), or one of its variants with 1 replaced by q, or the WGL). Then E satisfies the same version of the geometric lemma.

In some cases the given version of the geometric lemma (e.g., (1.59)) is equivalent to uniform rectifiability. Then the stability under the big pieces functor can be derived from the corresponding result for uniformly rectifiable sets. This last can be obtained, for instance, from the characterization of uniform rectifiability in terms of the L^2 boundedness of the singular integral operators with kernels in $\mathscr{K}_d(\mathbf{R}^n)$, together with Proposition 1.28. The proof given in Part IV is of course more direct than this and also more general.

Our second result concerns corona decompositions. We refer to the next section for a precise definition, but let us give a very vague idea of what we mean by "E has a corona decomposition". For each $\epsilon > 0$, there should be a partition of $E \times \mathbf{R}_+$ into "stopping-time regions" $\mathscr{R}_i$ with the main properties that there should not be too many regions (this will be measured using some Carleson packing condition), and also that, for each region $\mathscr{R}_i$, there exists a Lipschitz graph Γ_i with constant $\leq \epsilon$ which approximates E very well, with errors around a point $x \in E$ of the size of the smallest t's such that $(x, t) \in \mathscr{R}_i$.

Corona decompositions are a very useful notion. For instance, one of the key ingredients in the proof of the equivalence of various characterizations of uniform rectifiability given in [**DS2**] was the fact that uniformly rectifiable sets have corona decompositions.

We shall give in Chapter 2 of Part IV a direct proof of the fact that the existence of corona decompositions is stable under the action of the big pieces functor. This is again not a new result, because we know that some characterizations of uniform rectifiability are invariant under taking big pieces and also that uniform rectifiability is equivalent to the exisitence of a corona decomposition. The proof given in Part IV will have the advantage of being much more direct and of allowing more natural proofs of some of our results.

2.4. A RAPID DESCRIPTION OF PART IV

Let us give an example. To prove that the BWGL (see Definition 2.2) implies uniform rectifiability, we first prove that it implies some sort of generalized corona decomposition, where ϵ-Lipschitz graphs are replaced by regular sets that have BPLG. Once we know that every regular set that has BPLG has a (true) corona decomposition, we can deduce rather easily the existence of a corona decomposition for E. (See the next chapter for more details about this.) The fact that BPLG implies the existence of a corona decomposition can be proved using the stability result mentioned above, together with the fact that Lipschitz graphs (with constants that are not necessarily small) have corona decompositions. See Chapter 2 of Part IV.

CHAPTER 3

Dyadic Cubes and Corona Decompositions

In this chapter we shall review some background information of a more technical nature, which will be used extensively throughout the rest of this monograph.

3.1. Cubes.
Let E be a d-dimensional regular set in $\mathbf{R}^n$. It is possible to construct a family of subsets of E that behave in much the same way as do the dyadic cubes in $\mathbf{R}^n$. More precisely, one can construct a family Δ_j, $j \in \mathbf{Z}$, of measurable subsets of E with the following properties:

(3.1) each Δ_j is a partition of E, i.e., $E = \bigcup_{Q \in \Delta_j} Q$ and $Q \cap Q' = \varnothing$

whenever $Q, Q' \in \Delta_j$ and $Q \neq Q'$;

(3.2) if $Q \in \Delta_j$ and $Q' \in \Delta_k$ for some $k \geq j$, then either $Q \subset Q'$ or $Q \cap Q' = \varnothing$;

(3.3) for all $j \in \mathbf{Z}$ and all $Q \in \Delta_j$, we have that $C^{-1} 2^j \leq \operatorname{diam} Q \leq C 2^j$ and $C^{-1} 2^{jd} \leq |Q| \leq C 2^{jd}$;

(3.4) for all $j \in \mathbf{Z}$ and $Q \in \Delta_j$, Q has a small boundary, in the sense that
$$|\{x \in Q : \operatorname{dist}(x, E \setminus Q) \leq \tau 2^j\}|$$
$$+ |\{x \in E \setminus Q : \operatorname{dist}(x, Q) \leq \tau 2^j\}| \leq C \tau^{1/C} 2^{jd}$$
for all $0 < \tau < 1$.

Here C is a constant that depends only on d, n, and the regularity constant for E. See Appendix 1 of [**D4**] for a proof of the existence of such a family of partitions of E.

In the rest of this monograph we shall always assume that for each regular set E a family Δ_j, $j \in \mathbf{Z}$, of partitions with the above properties has been chosen once and for all. [Of course, the precise choice of the Δ_j will not really matter.] Let $\Delta = \bigcup_{j \in \mathbb{Z}} \Delta_j$. We shall refer to the elements of Δ as cubes or even dyadic cubes. If $Q \in \Delta_j$, the unique cube $Q' \in \Delta_{j+1}$ such that $Q \subset Q'$ will be called the parent of Q and the cubes $R \in \Delta_{j-1}$ such that $R \subset Q$ will be called its children. It might happen that a cube Q has only one child (itself), but, because of (3.3), Q never has more than $2^d C^2$ children.

Property (3.4) is sort of special. For many purposes it is not needed, and it is easier to construct a family of cubes that satisfy (3.1)–(3.3) without worrying about (3.4). However, there are some occasions in which it is very useful. Condition (3.4) can be interpreted as a quantified way of demanding that each cube Q have a "boundary" (from the point of view of E) with dimension $< d$. The consequence of (3.4) that will be used most often here is the existence of a point of Q which is not too close to $E \setminus Q$. Let us state this as a lemma.

LEMMA 3.5. *There is a constant $C_1 > 0$, depending only on d, n, and the regularity constant for E, such that we can associate, to each cube $Q \in \Delta$, a "center" $c(Q) \in Q$ which satisfies*

(3.6) $$\operatorname{dist}(c(Q), E \setminus Q) \geq C_1^{-1} \operatorname{diam} Q.$$

A convenient bit of notation that we shall use frequently is

(3.7) $$\lambda Q = \{x \in E : \operatorname{dist}(x, Q) \leq (\lambda - 1) \operatorname{diam} Q\},$$

where $Q \in \Delta$ and $\lambda > 1$.

The set Δ of all dyadic cubes on E provides a convenient discrete model for $E \times \mathbf{R}_+$. Most of our conditions involving $E \times \mathbf{R}_+$ can easily be stated in terms of dyadic cubes as well. For example, given $Q \in \Delta$, we can define $\beta(Q)$ by

(3.8) $$\beta(Q) = (\operatorname{diam} Q)^{-1} \inf_P \left\{ \sup_{y \in 2Q} \operatorname{dist}(y, P) \right\},$$

where the infimum is taken over all d-planes in $\mathbf{R}^n$. This is essentially the same as $\beta_\infty(x, t)$ (see (1.46)), except that it is defined on Δ instead of $E \times \mathbf{R}_+$. It is easy to reformulate the WGL in terms of the $\beta(Q)$'s instead of the $\beta_\infty(x, t)$'s. Before doing this, let us state a definition.

DEFINITION 3.9. Let $\mathscr{A}$ be a subset of Δ. We say that $\mathscr{A}$ satisfies a Carleson packing condition if there is a $C \geq 0$ so that

$$\sum_{\substack{Q \in \mathscr{A} \\ Q \subset R}} |Q| \leq C |R|$$

for all $R \in \Delta$.

This is the analogue for subsets of Δ of the notion of Carleson sets (see Definition 1.69) for subsets of $E \times \mathbf{R}_+$.

LEMMA 3.10. *E satisfies the WGL (Definition 1.71) if and only if $\{Q \in \Delta : \beta(Q) > \epsilon\}$ satisfies a Carleson packing condition for all $\epsilon > 0$.*

This is basically trivial. It follows from the observation that for each $A > 1$ there is a $C > 0$ so that

(3.11) $\quad \beta_\infty(x, t) \leq C\beta(Q) \quad$ when $\quad x \in Q$ and $A^{-1} \operatorname{diam} Q \leq t \leq \operatorname{diam} Q$

and

(3.12) $\quad \beta(Q) \leq C\beta_\infty(x, t) \quad$ when $\quad x \in Q$ and $2 \operatorname{diam} Q \leq t \leq A \operatorname{diam} Q$.

3.2. Corona decompositions.

The following definition will be quite useful.

DEFINITION 3.13. Let E be a d-dimensional regular set in $\mathbf{R}^n$. Assume, as usual, that a family Δ of dyadic cubes with properties (3.1)–(3.4) has been chosen. A coronization of E is a triple $(\mathscr{B}, \mathscr{G}, \mathscr{F})$, where $\mathscr{B}$ and $\mathscr{G}$ are two subsets of Δ (the "bad cubes" and the "good cubes") and $\mathscr{F}$ is a family of subsets S, $S \in \mathscr{F}$, of $\mathscr{G}$, which satisfy the following five conditions.

(3.14) $\quad\quad\quad\quad \Delta = \mathscr{B} \cup \mathscr{G} \quad$ and $\quad \mathscr{B} \cap \mathscr{G} = \varnothing$.

(3.15) $\mathscr{B}$ satisfies a Carleson packing condition, i.e., there is a $C > 0$ so that $\sum_{Q \in \mathscr{B}, Q \subset R} |Q| \leq C|R| \quad$ for all $R \in \Delta$.

(3.16) $\quad\quad \mathscr{G}$ is the disjoint union of the S, $S \in \mathscr{F}$.

(3.17) Each $S \in \mathscr{F}$ is coherent. This means that each $S \in \mathscr{F}$ has a (unique) maximal element $Q(S)$ which contains all other elements of S as subsets, that $Q' \in S$ as soon as $Q' \in \Delta$ satisfies $Q \subset Q' \subset Q(S)$ for some $Q \in S$, and that if $Q \in S$ then either all of the children of Q lie in S or none of them do.

(3.18) The maximal cubes $Q(S)$, $S \in \mathscr{F}$, satisfy a Carleson packing condition (there is a $C > 0$ such that $\sum_{S \in \mathscr{F}, Q(S) \subset R} |Q(S)| \leq C|R|$ for all $R \in \Delta$).

We shall often refer to the sets S, $S \in \mathscr{F}$, as "stopping-time regions". This is because coronizations are frequently constructed using a stopping-time argument (where the stopping-time regions are maximal coherent regions where something nice happens). The maximal cube $Q(S)$ of a stopping-time region S will often be called the top cube of S.

Notice that

(∗) if $S \subseteq \Delta$ is coherent, then every $x \in Q(S)$ is either contained in a minimal cube of S or in arbitrarily small cubes in S.

This is easy to check.

REMARK. We call a subset S of Δ semicoherent if it has the first two features mentioned in (3.17), i.e., if there is a (unique) maximal top cube $Q(S)$ in S which contains all the other cubes in S and if $Q \in S$, $Q' \in \Delta$, and $Q \subseteq Q' \subseteq Q(S)$ imply that $Q' \in S$. Semicoherent families of cubes are fairly nice, but they can be rather badly behaved in terms of their minimal cubes. For instance, (∗) is extremely false if "coherent" is replaced by "semicoherent". This is not a serious problem, but it is a nuisance that we prefer to avoid.

One reason that the notion of semicoherence deserves a special name is that there has been some confusion in the literature between the notions of coherence and semicoherence, even to the point where some authors have actually claimed (at least implicitly) that the semicoherent version of (∗) is true. This information is likely to surprise the reader, given the simplicity of the counterexamples to (∗) in the semicoherent case, but nonetheless one can find this mistake in [S4] and [DS2].

Fortunately, the mathematical consequences of this mistake are pretty minor. If, for instance, the definition of "coherent" given in (2.5) of [DS2] is replaced with (3.17), then everything else in [DS2] still works, with only minor adjustments. (The changes needed in the proof of Lemma 7.1 in [DS2] require a bit of thought, but not much. See Lemma IV.2.35.) That is also true for [S4], except for the fact that no term like "coherent" was ever formally introduced. The main point is that anything which satisfies the weaker version of the definition of a coronization in which (3.17) is replaced by semicoherence can always be modified slightly to get a true coronization.

Let us be more precise. Suppose that $(\mathcal{B}, \mathcal{G}, \mathcal{F})$ satisfies all the requirements of a coronization but with (3.17) replaced by semicoherence. Then there is a different partition $\mathcal{F}'$ of $\mathcal{G}$ such that $(\mathcal{B}, \mathcal{G}, \mathcal{F}')$ is a true coronization, and each $S' \in \mathcal{F}'$ is contained in some $S \in \mathcal{F}$. Indeed, let $(\mathcal{B}, \mathcal{G}, \mathcal{F})$ be given, and fix $S \in \mathcal{F}$. Call a cube $Q \in S$ a troublemaker if $Q \neq Q(S)$ and if the parent of Q (which must be an element of S) has another child which is not an element of S. It is not difficult to produce a partition $\mathcal{F}'(S)$ of S by semicoherent subsets of Δ with the property that the top cube of any $S' \in \mathcal{F}'(S)$ is either $Q(S)$ or a troublemaker and so that no element of any $S' \in \mathcal{F}'(S)$ except its top cube is a troublemaker. [The elements of $\mathcal{F}'(S)$ can be generated by a simple stopping-time argument: start with $Q(S)$ or a troublemaker, and go down until you get to minimal cubes of S or parents of troublemakers.] Each $S' \in \mathcal{F}'(S)$ is actually coherent; this is not hard to check, using the easy observation that if $Q_1, Q_2 \in S$ have the same parent and if Q_1 is a troublemaker, then so is Q_2. Let $\mathcal{F}'$ be the union of the $\mathscr{F}'(S)$'s. To show that $(\mathscr{B}, \mathscr{G}, \mathscr{F}')$ is a coronization we need only check that the top cubes of the elements of $\mathscr{F}'$ satisfy a Carleson packing condition. This is not hard to do, using the simple fact that the collection of all troublemakers (for all $S \in \mathscr{F}$) satisfies a Carleson packing condition. [Each troublemaker has a sibling which lies in $\mathscr{B}$ or which is the top cube for some element of $\mathscr{F}$.]

The preceding observation provides a convenient tool for producing coronizations. Sometimes it is easier at first to work with semicoherent stopping-time regions and then worry about coherence afterward.

We are now ready to state the definition of a corona decomposition. Very roughly, the idea is that the regular set E has a corona decomposition if, for each $\eta > 0$, there is a coronization of E such that, for each stopping-time region $S \in \mathscr{F}$, one can find a Lipschitz graph with constant $\leq \eta$ which is a good approximation of E at the scale of the cubes of S. Let us make this more precise.

DEFINITION 3.19. Let E be a d-dimensional regular set in $\mathbf{R}^n$. We say that E admits a corona decomposition if, for each $\eta > 0$ and $\theta > 0$, one can find a coronization $(\mathscr{B}, \mathscr{G}, \mathscr{F})$ of E such that

(3.20) for each $S \in \mathscr{F}$, there exists a d-dimensional Lipschitz graph

$\Gamma = \Gamma(S)$ with constant $\leq \eta$ such that $\operatorname{dist}(x, \Gamma(S)) \leq \theta \operatorname{diam} Q$

whenever $x \in 2Q$ and $Q \in S$.

This condition is a little technical, but it is extremely useful, in part because it is very versatile. It should be viewed as another quantitative rectifiability condition, and in fact it is equivalent to uniform rectifiability. This is one of the main points of [**DS2**].

In the definition of a corona decomposition, it is important that we do not demand any control on the way the Carleson packing constants for the coronization $(\mathscr{B}, \mathscr{G}, \mathscr{F})$ depend on η or θ. Nonetheless, the existence of a corona decomposition (and even the existence of $(\mathscr{B}, \mathscr{G}, \mathscr{F})$ for a single value of η and θ) contains more information than one might think. For instance, it is easy to verify that, for almost every $x \in E$, all sufficiently small cubes Q such that $x \in 2Q$ belong to a single $S \in \mathscr{F}$. [This is because Q is a top cube whenever it belongs to a different stopping-time region than its parent and because the Carleson packing condition implies that a single point cannot usually be contained in too many top cubes.] This implies in particular that the point x lies in the corresponding Lipschitz graph $\Gamma(S)$. Thus the existence of a corona decomposition of E implies immediately that E is rectifiable.

There are a number of places where something like a corona decomposition has been used. Of course, the ideas go back to Carleson's corona construction, although that construction dealt with the analysis of functions rather than the geometry of sets. In the geometric setting, something like a corona decomposition was used in [**GJ**], [**J1**], [**J3**], [**S4**], and [**DS2**]. In

[GJ] and [J1], [J3] one actually partitioned the complement of the given set into nicer sets rather than decomposing the upper half-space associated to E into regions that correspond to places where E is well-behaved, as we are (essentially) doing here and as was done in [S4] and [DS2].

The reason why the notion of a corona decomposition is so useful is that it organizes geometric information well, both from the points of view of deriving consequences from it and verifying it. See [DS2] and [S4] for examples of how one can derive geometrical and analytical consequences from the existence of a corona decomposition. As to the task of verifying its existence, the definition of a corona decomposition has several nice features. It is designed to be amenable to stopping-time arguments. The fact that we can throw away cubes into $\mathscr{B}$ with no other restriction than the Carleson packing condition, without even caring about the precise value of the Carleson packing constants, is extremely convenient. It is also helpful to be able to localize in terms of scale as well as location. We shall see in Part II, Chapters 2–4, a few examples of proofs of uniform rectifiability that go through corona decompositions.

For the rest of this chapter we are going to discuss some variations of the definition of a corona decomposition that will be useful later. In this section we begin with some particularly simple variations.

We should perhaps first point out that the notion of existence of a corona decomposition does not depend on the choice of Δ. That is, if you replace Δ with another family Δ' of subsets of E that still satisfy (3.1), (3.2), and (3.3), then E admits a corona decomposition with respect to Δ if and only if it admits a corona decomposition with respect to Δ'. This is not hard to prove, but since we shall not need this fact and the proof is not exactly entertaining, we shall omit it. (It is much easier once you have Lemma 3.26.)

LEMMA 3.21. *Suppose that E satisfies the a priori weaker version of Definition 3.19 in which you ask only that, for each $\eta > 0$, there exist a coronization $(\mathscr{B}, \mathscr{G}, \mathscr{F})$ of E such that (3.20) holds for some value of θ. Then E has a corona decomposition.*

This is a rather routine verification, but it will be worthwhile to give the proof in some detail, because this will give us an opportunity to prove a few lemmas that will be used again later.

LEMMA 3.22. *Let $\mathscr{A}$ be a subset of Δ that satisfies a Carleson packing condition. Then there is a coronization $(\mathscr{B}, \mathscr{G}, \mathscr{F})$ of E such that $\mathscr{A} \subset \mathscr{B}$.*

Let us first make the simple observation that there always exists a coronization of E. Indeed, fix any point p in E, let $\mathscr{B}_0$ be the collection of cubes Q in Δ such that $p \in 2Q$, and set $\mathscr{G}_0 = \Delta \setminus \mathscr{B}_0$. Thus every element of $\mathscr{G}_0$ is contained in a maximal element of $\mathscr{G}_0$. If we take $\mathscr{F}_0$ to be the family of subsets of $\mathscr{G}_0$ of the form $\{Q : Q \subseteq M\}$, where M is a maximal element of $\mathscr{G}_0$, then $\mathscr{F}_0$ is a partition of $\mathscr{G}_0$, and it is not difficult to verify that $(\mathscr{B}_0, \mathscr{G}_0, \mathscr{F}_0)$ is a coronization of E.

Now let $\mathscr{A}$ be a given subset of Δ which satisfies a Carleson packing condition. To prove the lemma, we are going to show that we can produce a coronization $(\mathscr{B}, \mathscr{G}, \mathscr{F})$ of E such that $\mathscr{B} = \mathscr{A} \cup \mathscr{B}_0$ and $\mathscr{G} = \mathscr{G}_0 \setminus \mathscr{A}$. For this the remark after Definition 3.13 will come in handy.

Fix $S_0 \in \mathscr{F}_0$ for the moment. Suppose that R is a cube in $S_0 \cap \mathscr{G}$ whose parent does not lie in $S_0 \cap \mathscr{G}$. It is easy to generate a maximal semicoherent subset of $S_0 \cap \mathscr{G}$ with R as its top cube; you simply go down until you have to stop. Let $\mathscr{F}_1(S_0)$ denote the family of all semicoherent subsets of $S_0 \cap \mathscr{G}$ which are produced in this way. It is not hard to show that $\mathscr{F}_1(S_0)$ is a partition of $S_0 \cap \mathscr{G}$.

Set $\mathscr{F}_1 = \bigcup \{\mathscr{F}_1(S_0) : S_0 \in \mathscr{F}_0\}$. Then $\mathscr{F}_1$ is a partition of $\mathscr{G}$ by semicoherent subsets, and the top cube of each $S_1 \in \mathscr{F}_1$ is either the child of an element of $\mathscr{B}$ or the top cube of some $S_0 \in \mathscr{F}_0$. This implies that $(\mathscr{B}, \mathscr{G}, \mathscr{F}_1)$ satisfies all the requirements of coronization, except that the elements of $\mathscr{F}_1$ are merely semicoherent (and not coherent). However, we can modify $(\mathscr{B}, \mathscr{G}, \mathscr{F}_1)$ to get a true coronization $(\mathscr{B}, \mathscr{G}, \mathscr{F})$, as in the remark after Definition 3.13. This proves Lemma 3.22.

We continue in the general direction of the proof of Lemma 3.21 with a definition.

DEFINITION 3.23. Given a (large) constant $A > 1$, we say that two cubes Q_1, Q_2 are A-close if

(3.24) $$A^{-1} \operatorname{diam} Q_1 \leq \operatorname{diam} Q_2 \leq A \operatorname{diam} Q_1$$

and

(3.25) $$\operatorname{dist}(Q_1, Q_2) \leq A(\operatorname{diam} Q_1 + \operatorname{diam} Q_2).$$

LEMMA 3.26. *Let $(\mathscr{B}, \mathscr{G}, \mathscr{F})$ be a coronization of E, and let $A > 1$ be given. Then there exists a coronization $(\mathscr{B}', \mathscr{G}', \mathscr{F}')$ of E such that, for each $S' \in \mathscr{F}'$, there exists a region $S \in \mathscr{F}$ that contains all the cubes $Q \in \Delta$ that are A-close to any cube $Q' \in S'$.*

To prove Lemma 3.26, consider the set $\mathscr{A}$ of all cubes $Q \in \Delta$ that are A-close to a cube of $\mathscr{B}$ or that lie in one of the stopping-time regions $S \in \mathscr{F}$ but are A-close to a cube that belongs to a different region $\widetilde{S}$. If we can show that $\mathscr{A}$ satisfies a Carleson packing condition, then Lemma 3.26 will follow from Lemma 3.22 (or, more directly, the second part of its proof).

The set of cubes that are A-close to a cube of $\mathscr{B}$ is taken care of by the following.

LEMMA 3.27. *If $\mathscr{B}$ is a set of cubes that satisfies a Carleson packing condition, the set*

$$\mathscr{B}_A = \{Q \in \Delta : Q \text{ is } A\text{-close to some } Q' \in \mathscr{B}\}$$

also satisfies a Carleson packing condition.

We omit the easy proof.

To prove Lemma 3.26, we still have to check that
(3.28)
$$\mathscr{A}_0 = \{Q \in \mathscr{G} : \text{there is a cube } Q' \in \mathscr{G} \text{ that is } A\text{-close to } Q$$
but which belongs to a different stopping-time region than $Q\}$

satisfies a Carleson packing condition. The proof is in the same spirit as the (omitted) proof of Lemma 3.27 but a little more complicated.

Let us first state a helpful auxiliary fact.

LEMMA 3.29. *Let $(\mathscr{B}, \mathscr{G}, \mathscr{F})$ be a coronization of E, and let $K > 1$ be given. Set*
$$\mathscr{C} = \bigcup_{S \in \mathscr{F}} \{Q \in S : \text{dist}(Q, E \setminus Q(S)) \le K \operatorname{diam} Q\}.$$

Then $\mathscr{C}$ satisfies a Carleson packing condition.

Let us assume this for the moment and use it to prove the required Carleson packing condition for $\mathscr{A}_0$.

Let $Q \in \mathscr{A}_0$ be given. Thus $Q \in S$ for some $S \in \mathscr{F}$, and there is a $Q' \in \mathscr{G}$ such that $Q' \notin S$ and Q' is A-close to Q. If Q' is disjoint from $Q(S)$, then $Q \in \mathscr{C}$, as long as K is large enough. If $Q' \supseteq Q(S)$, then $\operatorname{diam} Q \ge A^{-1} \operatorname{diam} Q(S)$, and again we have $Q \in \mathscr{C}$ if K is large enough. Now suppose that $Q' \subseteq Q(S)$, and let us show that this implies that $Q' \in \mathscr{C}$ if K is large enough. Let S' be the stopping-time region that contains Q'. Then $Q(S') \subseteq Q(S)$ (since otherwise $Q' \subseteq Q(S) \subseteq Q(S')$ and hence $Q(S) \in S'$, a contradiction). Hence $Q \not\subseteq Q(S')$ (for the same reason), so either $Q \supseteq Q(S')$ or $Q \cap Q(S') = \varnothing$. Each of these two possibilities implies that $Q' \in \mathscr{C}$, as before.

Altogether we get that if $Q \in \mathscr{A}_0$, then either $Q \in \mathscr{C}$ or Q is A-close to an element of $\mathscr{C}$. Therefore, the desired Carleson packing condition for $\mathscr{A}_0$ follows from Lemmas 3.29 and 3.27. This proves Lemma 3.26.

Now let us check Lemma 3.29. Given $Q_0 \in \Delta$ set
$$\mathscr{C}(Q_0) = \{Q \in \Delta : Q \subseteq Q_0, \text{dist}(Q, E \setminus Q_0) \le K \operatorname{diam} Q\}.$$

LEMMA 3.30. *For each $Q_0 \in \Delta$ we have*
$$\sum_{Q \in \mathscr{C}(Q_0)} |Q| \le C|Q_0|$$

for some C (which depends on K).

This follows easily from (3.4).

Let us come back to $\mathscr{C}$ itself. By definitions,
$$\mathscr{C} = \bigcup_{S \in \mathscr{F}} [\mathscr{C}(Q(S)) \cap S].$$

Let $R \in \Delta$ be given, and consider

$$\sum_{\substack{Q \in \mathscr{C} \\ Q \subseteq R}} |Q| = \sum_{S \in \mathscr{F}} \left(\sum_{\substack{Q \in \mathscr{C}(Q(S)) \cap S \\ Q \subseteq R}} |Q| \right).$$

If $S \in \mathscr{F}$ makes a nontrivial contribution to the right-hand side, then either $Q(S) \subseteq R$ or $R \in S$ (since S is coherent). For the first case we have that

$$\sum_{\substack{S \in \mathscr{F} \\ Q(S) \subseteq R}} \sum_{Q \in \mathscr{C}(Q(S))} |Q| \leq C \sum_{\substack{S \in \mathscr{F} \\ Q(S) \subseteq R}} |Q(S)| \leq C|R|$$

by Lemma 3.30 and the Carleson packing condition on the top cubes $Q(S)$. On the other hand, there can be at most one $S \in \mathscr{F}$ such that $R \in S$. Call it S_0 if it exists. (Otherwise, we are already finished.) Then $R \subseteq Q(S_0)$, $\{Q \in \mathscr{C}(Q(S_0)) : Q \subseteq R\} \subseteq \mathscr{C}(R)$, and hence

$$\sum_{\substack{Q \in \mathscr{C}(Q(S_0)) \\ Q \subseteq R}} |Q| \leq \sum_{Q \in \mathscr{C}(R)} |Q| \leq C|R|.$$

This proves Lemma 3.29.

Coming back to Lemma 3.21, we see that if $(\mathscr{B}, \mathscr{G}, \mathscr{F})$ is a coronization of E such that (3.20) holds with a given value of θ (say, $\theta = L$) and if we choose A large enough, then the new coronization of E given by Lemma 3.26 will satisfy (3.20) with the same η and with any small value of θ which is given in advance. Lemma 3.21 follows.

The next lemma provides us with another equivalent reformulation of the notion of a corona decomposition.

LEMMA 3.31. *Suppose that E has a corona decomposition. Then E satisfies the a priori stronger condition that you get by replacing "$x \in 2Q$" in (3.20) by "$x \in kQ$", where $k \geq 2$ is any constant given in advance.*

This is proved the same way as Lemma 3.21 using Lemma 3.26. The next variation of Definition 3.19 is somewhat trickier.

PROPOSITION 3.32. *Let $L > 0$ and $L' > 0$ be given. Suppose that there exists a coronization $(\mathscr{B}, \mathscr{G}, \mathscr{F})$ of E such that (3.20) holds with $\eta = L$ and $\theta = L'$. Then E admits a corona decomposition.*

We already know how to deal with the issue of θ, so we shall ignore that. The proof of Proposition 3.32 will rely on the following fact:

(3.33) If Γ is a Lipschitz graph with constant $\leq L$, then Γ has a corona decomposition, with constants for the Carleson packing conditions that depend on L, but not on the particular choice of Γ.

Before indicating why (3.33) is true, let us explain how it can be used to prove Proposition 3.32.

Let $(\mathscr{B}, \mathscr{G}, \mathscr{F})$ be a coronization of E that satisfies (3.20) with $\eta = L$ and $\theta = L'$. Thus, for each $S \in \mathscr{F}$, there is an L-Lipschitz graph $\Gamma = \Gamma(S)$ such that

(3.34) $\quad \operatorname{dist}(x, \Gamma(S)) \leq L' \operatorname{diam} Q \quad$ whenever $x \in 2Q$ for some $Q \in S$.

Let $\eta > 0$ be given. Using (3.33), you can get a corona decomposition of each $\Gamma(S)$. This means that for each S there is a set of dyadic cubes $\Delta(S)$ on $\Gamma(S)$ and then a coronization $(\mathscr{B}(S), \mathscr{G}(S), \mathscr{F}(S))$ of $\Gamma(S)$ such that, for every stopping-time region $T \in \mathscr{F}(S)$, there is a Lipschitz graph $\Gamma(S, T)$, with constant $\leq \eta$, such that

(3.35) $\quad \operatorname{dist}(y, \Gamma(S, T)) \leq \theta \operatorname{diam} R \quad$ whenever $x \in 2R$ for some $R \in T$.

[We should emphasize that $R \in T$ is a cube in $\Gamma(S)$, not E.]

Let A be a large constant (to be chosen later). For each $S \in \mathscr{F}$, we introduce a new bad set $\mathscr{B}_1(S)$ composed of all the cubes in $\Delta(S)$ which are A-close to some cube of $\mathscr{B}(S)$ or to some top cube of a stopping-time region in $\mathscr{F}(S)$ or which are in $\mathscr{G}(S)$ and are A-close to some other cube of $\mathscr{G}(S)$ that belongs to a different stopping-time region $T \in \mathscr{F}(S)$. We know from (the proof of) Lemma 3.26 that the $\mathscr{B}_1(S)$'s satisfy Carleson packing conditions, with constants that do not depend on S. Let $\mathscr{B}_2(S)$ be the subset of $\mathscr{B}_1(S)$ of cubes which are A-close to an element of S. [Although we originally defined A-closeness only for cubes which lie on the same regular set, the definition still makes sense when they do not.]

We claim that

(3.36) $$\sum_{S \in \mathscr{F}} \sum_{\substack{Q \in \mathscr{B}_2(S) \\ Q \subseteq B(x, \rho)}} |Q| \leq C\rho^d$$

for all $x \in \mathbf{R}^n$ and $\rho > 0$. To see this, we consider two cases. If $\operatorname{diam} Q(S) \leq \rho$, then we must have that $Q(S) \subseteq B(x, C(A)\rho)$ if S is to have a nontrivial contribution to the left side of (3.36). In this case we use the fact that

$$\sum_{Q \in \mathscr{B}_2(S)} |Q| \leq C|Q(S)|$$

(which follows from the Carleson packing condition on $\mathscr{B}_1(S)$) and then the Carleson packing condition on the $Q(S)$'s to control the total contribution to (3.36) by $C\rho^d$. If $\operatorname{diam} Q(S) \geq \rho$ and if S gives a nontrivial contribution to the left side of (3.36), then it is easy to see that S must lie in a subset of $\mathscr{F}$ with at most $C(A)$ elements. [Indeed, such an S must contain a cube R which satisfies $\operatorname{dist}(x, R) \leq C\rho$ and $\rho \leq \operatorname{diam} R \leq C\rho$, $C = C(A)$, and there is only a bounded number of such cubes.] For each such S we can control the contribution to (3.36) using again the fact that $\mathscr{B}_1(S)$ satisfies a Carleson packing condition. This proves the claim.

Now let $\mathscr{B}_1$ denote the set of cubes $Q \in \Delta$ (i.e., cubes on E) that are A-close to an element of $\mathscr{B}$, one of the $Q(S)$'s, $S \in \mathscr{F}$, or to a cube on $\Gamma(S)$ which lies in $\mathscr{B}_2(S)$, for any $S \in \mathscr{F}$. It is not hard to check that $\mathscr{B}_1 \subseteq \Delta$ satisfies a Carleson packing condition, using (3.36) and the same kind of argument as for the proof of Lemma 3.27. We now use Lemma 3.22 to find a coronization $(\mathscr{B}_2, \mathscr{G}_2, \mathscr{F}_2)$ of E such that $\mathscr{B}_1 \subset \mathscr{B}_2$.

Let $S_2 \in \mathscr{F}_2$ be given. We want to find a Lipschitz graph Γ_2 with constant $\leq \eta$ that satisfies

$$(3.37) \qquad \operatorname{dist}(x, \Gamma_2) \leq C\theta \operatorname{diam} Q \quad \text{whenever } x \in 2Q \text{ for some } Q \in S_2.$$

First notice that, by construction of $\mathscr{B}_1$, S_2 is entirely contained in some stopping-time region $S \in \mathscr{F}$. For each cube $Q \in S_2$, we can apply (3.34) and find a point y in $\Gamma(S)$ at a distance $\leq \operatorname{diam} Q$ from Q and then a cube $Q' \in \Delta(S)$ containing y and of roughly the same size as Q. If A is large enough, then Q' cannot be in $\mathscr{B}_1(S)$, and Q' and all its neighbors must even lie in some stopping-time region $T \in \mathscr{F}(S)$. Moreover, if $\widetilde{Q}$ is the parent or one of the children of Q and if $\widetilde{Q} \in S_2$, then $\widetilde{Q}$ must be associated to the same $T \in \mathscr{F}(S)$. By repeating this argument it follows that the same $T \in \mathscr{F}(S)$ is associated to all elements of S_2.

We choose $\Gamma_2 = \Gamma(S, T)$. The estimate (3.37) follows immediately from (3.34) and (3.35), and this concludes our proof of Proposition 3.32, modulo the result (3.33).

Let us now say why (3.33) is true. Of course, we could just say that (3.33) holds because Lipschitz graphs are uniformly rectifiable, and we know from [**DS2**] that uniformly rectifiable sets have a corona decomposition. The resulting proof of (3.33), though, would be much more complicated than needed. In fact, (3.33) is nothing more than a variation of Carleson's corona construction, but unfortunately we do not know a good reference for this version. We could refer to §3 of [**S4**], where an argument is given that is much less painful (or general) than the one in [**DS2**] but still more complicated than necessary for the present purpose. The method of [**S4**] is also only directly applicable to the codimension 1 case, although it is not hard to reduce to that, and it is a little unattractive aesthetically in this context, because it was designed to deal with a more geometric setting.

In Chapter 2 of Part IV we shall give a reasonably complete outline of a fairly direct proof of (3.33).

3.3. Generalized corona decompositions.

It turns out to be a good idea to consider versions of the corona decomposition in which you approximate the given set E by something besides Lipschitz graphs.

DEFINITION 3.38. Let $\mathscr{E}$ be a collection of d-dimensional regular sets in $\mathbf{R}^n$ that satisfy the regularity estimate (1.14) uniformly (i.e., with the same constant C_0). We define Corona($\mathscr{E}$) to be the collection of all

d-dimensional regular sets E in $\mathbf{R}^n$ which have a coronization $(\mathscr{B}, \mathscr{G}, \mathscr{F})$ with the property that

(3.39)   for each $S \in \mathscr{F}$ there is an $E_0 \in \mathscr{E}$ such that $\operatorname{dist}(x, E_0) \leq$ 10 diam Q whenever $x \in 2Q$ and $Q \in S$.

Just as in the preceding section, we have the following.

LEMMA 3.40. *Let $E \in \operatorname{Corona}(\mathscr{E})$, $\theta > 0$, and $\lambda \geq 2$ be given. Then there is a coronization $(\mathscr{B}, \mathscr{G}, \mathscr{F})$ of E such that for each $S \in \mathscr{F}$ there is a regular set $E_0 \in \mathscr{E}$ with the property that $\operatorname{dist}(x, E_0) \leq \theta \operatorname{diam} Q$ whenever $x \in \lambda Q$ for some $Q \in S$.*

We would also have an equivalent definition if we weakened it by replacing "$\operatorname{dist}(x, E_0) \leq 10 \operatorname{diam} Q$" in (3.39) by "$\operatorname{dist}(x, E_0) \leq L \operatorname{diam} Q$", where $L \geq 1$ is any given constant.

Let us relate Definition 3.38 to what we did in the last section. Let $\operatorname{LG}(\eta)$ be the set of d-dimensional Lipschitz graphs in $\mathbf{R}^n$ with constant $\leq \eta$. By definition, E admits a corona decomposition if and only if $E \in \bigcap_{\eta > 0} \operatorname{Corona}(\operatorname{LG}(\eta))$. Lemma 3.21 says that E has a corona decomposition as soon as it is in $\bigcup_{\eta > 0} \operatorname{Corona}(\operatorname{LG}(\eta))$.

The proof of Lemma 3.21 did not use any property of $\operatorname{LG}(\eta)$ other than the fact that its elements have corona decompositions with uniform bounds. The same proof actually gives the following result.

LEMMA 3.41. *Let $\mathscr{E}$ be a collection of d-dimensional regular sets in $\mathbf{R}^n$. Suppose that the elements of $\mathscr{E}$ satisfy the regularity condition with uniform estimates and have corona decompositions with uniform bounds. Then E has a corona decomposition as soon as $E \in \operatorname{Corona}(\mathscr{E})$.*

In Part II we shall use a special instance of this, which we state as a theorem.

THEOREM 3.42. *Let $\mathscr{E}$ be a class of d-dimensional regular sets which satisfy BPLG, with uniform bounds on the constants θ and C_1 of Definition 1.26 as well as on the regularity constant. If E is a d-dimensional regular set such that $E \in \operatorname{Corona}(\mathscr{E})$, then E is uniformly rectifiable.*

Theorem 3.42 follows from Lemma 3.41 because we know from [**DS2**] that a regular set is uniformly rectifiable if and only if it has a corona decomposition, and that regular sets which have BPLG are uniformly rectifiable.

Let us see more precisely what is involved here. It is usually surprisingly easy to go from the existence of a corona decomposition to the various conditions that characterize uniform rectifiability. See for instance [**DS2**, pp. 101–110] for a proof of the fact that E has BPBI (see Definition 1.33) when E has a corona decomposition and [**DS2**, pp. 111–123] for a proof of the fact that E is contained in the image of some ω-regular mapping (condition (1.62)) if it has a corona decomposition. It is also possible to go directly and without much pain from the existence of a corona decomposition to the geometric lemma (condition (1.59)) (see [**DS2**, pp. 93–99]), or to L^2-estimates on singular integral operators (condition (1.58)) (see [**S4**, pp. 1024–1031]).

If we rely on [**DS2**] to prove that all regular sets that have BPLG have a corona decomposition, we have to go through the boundedness of singular integral operators, and the proof is rather indirect and painful. We shall give a different, more direct, proof in Part IV, based on the fact that the existence of a corona decomposition is stable under the action of the big pieces functor. (See also the description of Part IV given in §2.4.)

It is interesting to compare the big pieces functor with the corona functor. The former is simpler and more intuitively accessible, while the latter seems to be more flexible and frequently easier to use. Neither is universally preferable to the other.

PART II

New Geometrical Conditions Related to Uniform Rectifiability

CHAPTER 1

One-Dimensional Sets

1.1. The weak connectedness condition.
For one-dimensional regular sets the notion of connectedness plays a very special role. Contrary to what happens in higher dimensions, the connected 1-dimensional regular sets are all uniformly rectifiable. Because of this, it will be fairly easy to state mild-looking conditions (such as the weak connectedness condition below) that imply uniform rectifiability.

Let us remind the reader of the definition of the WCC given in Part I, Definition 2.12.

DEFINITION 1.1. Let E be a one-dimensional regular set in $\mathbf{R}^n$. We say that E is weakly connected (or satisfies the WCC) if there exist constants $0 < r < 1$ and $M \geq 1$ such that $E \times \mathbf{R}^+ \setminus \mathscr{G}(r, M)$ is a Carleson set, where

(1.2)
$$\mathscr{G}(r, M) = \{(x, t) \in E \times \mathbf{R}^+ : \text{ for all } u, v \in E \cap B(x, t) \text{ such that}$$
$$|u - v| \geq \tfrac{t}{10}, \text{ there exists a chain of points}$$
$$y_0 = u, \; y_1, y_2, \ldots, y_N = v,$$
$$\text{with } y_i \in E \cap B(x, tM) \text{ and } |y_{i+1} - y_i| \leq r|u-v| \text{ for all } i\}.$$

(See Definition I.1.69 for Carleson sets.)

We owe the present formulation of this definition to P. Jones, who helped us to get rid of some unnecessary complications.

There is no explicit restriction on the number $N+1$ of points in the chain from u to v. On the other hand, since E is regular and all the points y_i belong to $E \cap B(x, tM)$, we can always assume that $N \leq N_0$, where N_0 depends on r, M, and the regularity constant for E, by getting rid of the useless points of the chain. In other words, if u and v can be connected by such a chain, then they can be connected by a minimal chain, and minimal chains have a bounded number of elements.

The choice of the constant $\tfrac{1}{10}$ in the definition above does not have any special meaning; we could have chosen any small enough constant instead, and the resulting definition would have been equivalent.

Also note that the weak connectedness condition (WCC) is not far from saying that E is bilaterally approximated by connected sets (more precisely, that E belongs to the class Approx($\mathscr{E}$) of Definition I.2.21, with $\mathscr{E} =$ the class of closed connected subsets of $B(0, 1)$). Definition 1.1 above is slightly less restrictive (a priori) because we also allow the points u and v to be connected by a path that goes outside of $B(x, t)$.

The main goal of this section is to prove that the weak connectedness condition implies uniform rectifiability. We shall actually prove the following result directly.

THEOREM 1.3. *Let E be a one-dimensional regular set in $\mathbf{R}^n$, and suppose that E is weakly connected. Then there is a connected regular set Γ that contains E. The regularity constant for Γ can be taken to depend only on the regularity constant for E, the parameters r and M from Definition 1.1, and the Carleson constant for $E \times \mathbf{R}_+ \setminus \mathscr{G}(r, M)$.*

The converse of this theorem is true (but not very exciting): if E is a regular set contained in a connected regular set, then E is weakly connected because it satisfies the bilateral weak geometric lemma (see Definition 2.2 of Part I). Notice, by the way, that it is not completely trivial that a regular curve is weakly connected (one still has to control the number of places where the curve comes back on itself after a long loop).

The proof of Theorem 1.3 will be surprisingly simple. Here is the main idea. If the complement of the set $\mathscr{G}(r, M)$ were actually empty, then it would be easy to check that E is connected. In the general case, we shall have to add, for each $(x, t) \notin \mathscr{G}(r, M)$, a finite bunch of line segments to make E essentially connected (i.e., within errors much smaller than t) near $B(x, t)$. The set Γ obtained as the union of E and all these line segments will then be connected, and we shall have a control on its total length because the total length of the added line segments will be controlled by the Carleson constant for $E \times \mathbf{R}_+ \setminus \mathscr{G}(r, M)$. There will only be a minor complication: we shall have to make sure that the line segments added at different scales do not get too close to each other (otherwise, Γ would not be regular).

To simplify our construction, we shall build a connected regular set Γ that is contained in $\mathbf{R}^{n+1}$ rather than $\mathbf{R}^n$. This does not matter if we are only interested in proving that E is uniformly rectifiable. Moreover, one can then find a regular curve $\Gamma' \supset \Gamma$ and use the argument in §§4 and 5 of [D2] to show that one can also find a regular curve in $\mathbf{R}^n$ that contains E. [The argument in [D2] can be simplified substantially for this application, because $d = 1$.]

Let us now begin the proof of Theorem 1.3. Let E be as in the statement of the theorem. For our construction, it will be easier to work with the dyadic cubes Q, $Q \in \Delta$, that were discussed in §3.1 of Part I. Let $\epsilon > 0$ be a rather small constant to be chosen later. (It will depend only on the regularity constant for E.) We let $\mathscr{G}$ be the good set of cubes $Q \in \Delta$ for which, whenever u, v are two points in $2Q = \{w \in E : \text{dist}(w, Q) \leq \text{diam } Q\}$, there exists a chain of points $y_0 = u, y_1, \ldots, y_N = v$ such that all the y_j lie in

$$C_0 M Q = \{x \in E : \text{dist}(x, Q) \leq (C_0 M - 1) \operatorname{diam} Q\}$$

and such that $|y_j - y_{j-1}| \leq \epsilon \operatorname{diam} Q$ for $j = 1, \ldots, N$. Here M is as in the statement of Theorem 1.3, and the constant C_0 will be chosen in the next lemma.

LEMMA 1.4. *If C_0 is large enough (depending on the constant r which comes from Definition 1.1 and also on the regularity constant for E), then the set $\mathscr{B} = \Delta \setminus \mathscr{G}$ satisfies the Carleson packing estimate*

$$(1.5) \qquad \sum_{\substack{Q \in \mathscr{B} \\ Q \subset R}} |Q| \leq C'|R| \quad \text{for all cubes } R \in \Delta.$$

Here the constant C' may depend on the various constants around (including r, M, ϵ, ...).

Indeed, it is easily checked that $Q \in \mathscr{G}$ as soon as all the points (x, t) with $x \in C_0 M Q$ and $\epsilon \operatorname{diam} Q \leq t \leq C \operatorname{diam} Q$ lie in $\mathscr{G}(r, 10M)$ (assuming that C_0 is large enough). Hence, to every cube $Q \in \mathscr{B}$, we can associate an $(x, t) \notin \mathscr{G}(r, 10M)$ which is not too far "hyperbolically" from Q. Then we can find $u, v \in E \cap B(x, t)$ such that $|u - v| \geq \frac{t}{10}$ and which cannot be connected by a chain of points as in (1.2). This implies that all the points $(x', t') \in E \times \mathbf{R}_+$ such that $|x' - u| \leq \frac{t}{100}$ and $2|u - v| \leq t' \leq 4|u - v|$, say, lie in $E \times \mathbf{R}_+ \setminus \mathscr{G}(r, M)$. Deducing (1.5) from the Carleson estimate on $E \times \mathbf{R}^+ \setminus \mathscr{G}(r, M)$ is now an easy exercise using the bounded overlap of the sets of (x', t') associated to the various cubes $Q \in \mathscr{B}$. We omit the details.

With Lemma 1.4 out of the way we are ready to begin the geometric construction. To each cube $Q \in \mathscr{B}$ we want to associate some line segments. Let k be the "generation of Q". (This means that $Q \in \Delta_k$, and implies in particular that $\operatorname{diam} Q \sim 2^k$.) Let c_Q be a center for Q, i.e., a point of Q such that $\operatorname{dist}(c_Q, E \setminus Q) \geq C^{-1} \operatorname{diam} Q$ for some geometric constant C. The existence of such a point is discussed in Part I, Lemma 3.5. Our line segments will start from the point $A_Q = (c_Q, \operatorname{diam} Q) \in \mathbf{R}^{n+1}$. (Of course we are identifying $\mathbf{R}^n$ with $\mathbf{R}^n \times \{0\} \subseteq \mathbf{R}^{n+1}$.)

Let l be the integer such that $2^{-l} \leq \epsilon < 2^{-l+1}$; we denote by $\mathscr{S}(Q)$ the set of all cubes S of generation $k - l$ such that $S \subset 3Q$. We would like to add to E the union of all the line segments that join A_Q to the centers c_S, $S \in \mathscr{S}(Q)$. We need to be a little careful, though, because we do not want these lines to interfere too much with what will be done at smaller scales. To prevent that from happening, we shall often cut off a little bit of these line segments near the bottom.

For each $S \in \mathscr{S}(Q)$, let $\mathscr{W}(S)$ be the set of cubes $W \subset S$ such that $\operatorname{diam} W \leq \delta \operatorname{diam} S$ (where δ is a small geometric constant that will be

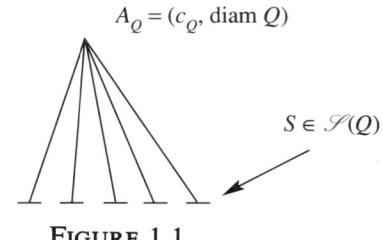

FIGURE 1.1

chosen soon), $W \in \mathscr{B}$, and

(1.6) $\qquad\qquad\qquad \text{dist}(W, c_S) \leq 10 \,\text{diam}\, W.$

If $\mathscr{W}(S)$ is empty, then we let $\Gamma_0(S)$ be the straight line segment from A_Q to c_S. (Thus $\Gamma_0(S)$ depends also on Q, but we shall not make that explicit in the notation.)

If not, let $H(S)$ be a cube of $\mathscr{W}(S)$ with maximal diameter, and let $\Gamma_0(S)$ be the straight line segment from A_Q to $A_{H(S)}$. Notice that it is reasonable to stop at the point $A_{H(S)}$ because $H(S)$ is another cube of $\mathscr{B}$; so $A_{H(S)}$ will be connected to E by future sets $\Gamma_0(S')$.

We now let $\Gamma(Q)$ be the union of all $\Gamma_0(S)$, $S \in \mathscr{S}(Q)$, and we take for Γ the union of E and all the $\Gamma(Q)$, $Q \in \mathscr{B}$.

LEMMA 1.7. *Γ is regular.*

Let $x \in \Gamma$ and $R > 0$ be given. The inequality

(1.8) $\qquad\qquad\qquad |\Gamma \cap B(x, R)| \geq C^{-1} R$

is obvious when $x \in E$ (because E is regular) or even when $\text{dist}(x, E) \leq \frac{R}{2}$. Otherwise, x is in some $\Gamma(Q)$ and, since $\Gamma(Q)$ always stays within $C \,\text{diam}\, Q$ of Q, we have $\text{diam}\, Q \geq \frac{R}{C}$. The inequality (1.8) now follows because $\Gamma(Q)$ contains a line segment of length $\geq \frac{R}{C}$ passing through x.

We now prove the inequality

(1.9) $\qquad\qquad\qquad |\Gamma \cap B(x, R)| \leq CR.$

Let us start with the most interesting case where $x \in E$. Of course, $|E \cap B(x, R)| \leq CR$, so we only need to consider the $\Gamma(Q)$'s.

Let $\mathscr{B}_0$ be the set of cubes $Q \in \mathscr{B}$ such that $Q \subset B(x, C_1 R)$, where the large constant C_1 will be chosen later. Then $\sum_{Q \in \mathscr{B}_0} |\Gamma(Q)| \leq C \sum_{Q \in \mathscr{B}_0} |Q| \leq C C_1 R$ because of Lemma 1.4. Hence we only need to consider the set $\mathscr{B}_1$ of cubes $Q \in \mathscr{B}$ such that $\Gamma(Q)$ meets $B(x, R)$ but Q is not contained in $B(x, C_1 R)$. We want to check that $\mathscr{B}_1$ has less than C elements, and for this it will be enough (as we shall see) to show that one cannot find two cubes Q_1 and $Q_2 \in \mathscr{B}_1$ such that

(1.10) $\qquad\qquad\qquad \text{diam}\, Q_2 \geq C_2 \delta^{-1} \,\text{diam}\, Q_1.$

1.1. THE WEAK CONNECTEDNESS CONDITION

We choose the constant C_2 so large that (1.10) implies that $\operatorname{diam} Q_1 < \delta \operatorname{diam} S$ for each cube $S \in \mathscr{S}(Q_2)$. Thus C_2 depends on ϵ, but we do not care.

So let Q_1 and Q_2 be two cubes of $\mathscr{B}_1$, and let us prove that (1.10) is impossible. Let $S \in \mathscr{S}(Q_2)$ be one of the cubes such that $\Gamma_0(S)$ meets $B(x, R)$, and let $y \in \Gamma_0(S) \cap B(x, R)$ be given. Since $x \in E \subset \mathbf{R}^n$, the last coordinate of y is less than R. If $\mathscr{W}(S)$ is not empty, $\Gamma_0(S)$ stops at the point $A_{H(S)}$, whose last coordinate is $\operatorname{diam} H(S)$, and therefore $\operatorname{diam} H(S) \leq R$. If we choose C_1 large enough, the fact that $Q_1 \in \mathscr{B}_1$ implies that $\operatorname{diam} Q_1 > R$, so Q_1 does not belong to $\mathscr{W}(S)$ (because the largest possible diameter for a cube of $\mathscr{W}(S)$ is $\operatorname{diam} H(S)$).

Let us assume that (1.10) holds and get a contradiction. We have that $\operatorname{diam} Q_1 \leq \delta \operatorname{diam} S$, by our choice of C_2. If Q_1 also satisfied (1.6), then we would have that $Q_1 \subset S$ by definition of the center c_S (and provided that we take δ small enough). This is impossible because $Q_1 \notin \mathscr{W}(S)$, so $\operatorname{dist}(Q_1, c_S) \geq 10 \operatorname{diam} Q_1$. We then get

$$(1.11) \qquad \operatorname{dist}(\Gamma(Q_1), c_S) \geq 5 \operatorname{diam} Q_1$$

by construction of $\Gamma(Q_1)$.

Let us check that $c_S \in B(x, CR)$. If $\mathscr{W}(S)$ is empty, then c_S lies on $\Gamma_0(S)$, so $|c_S - y| \leq CR$ since $\Gamma_0(S)$ was constructed so that it has bounded slope. Otherwise, we still have $|A_{H(S)} - y| \leq CR$ because $\Gamma_0(S)$ has bounded slope, and $|A_{H(S)} - c_S| \leq 20 \operatorname{diam} H(S)$ because $H(S)$ satisfies (1.6). In all cases, $c_S \in B(x, CR)$, so $\operatorname{dist}(\Gamma(Q_1), c_S) \leq C'R$ since $\Gamma(Q_1)$ meets $B(x, R)$. This contradicts (1.11) if we have chosen C_1 so large that $Q_1 \in \mathscr{B}_1$ implies $\operatorname{diam} Q_1 \geq C'R$.

Thus it is impossible to find cubes Q_1 and $Q_2 \in \mathscr{B}_1$ such that (1.10) holds. This means that all the cubes of $\mathscr{B}_1$ have comparable diameters. Since $x \in 10Q$ for each cube $Q \in \mathscr{B}_1$, we conclude that $\mathscr{B}_1$ has less than C elements. The estimate (1.9) follows immediately, because each $\Gamma(Q)$ satisfies (1.9) with bounded constants.

We now have to deal with the case when x is not on E. If we still have that $\operatorname{dist}(x, E) \leq CR$ for some constant C, we can deduce (1.9) from the previous case by using a ball of radius $2CR$ centered on E. So let us assume that $\operatorname{dist}(x, E) \geq CR$. Let $d(x)$ be the distance from x to $\mathbf{R}^n$. Because x belongs to Γ, it is in some $\Gamma_0(S)$, and since $\Gamma_0(S)$ has bounded slope and ends on some $c_{S'}$ or some $A_{S'}$, we have that $\operatorname{dist}(x, E) \leq C'd(x)$. Taking the constant C above equal to $3C'$ we get $d(x) \geq 3R$. Therefore, if Q is a cube such that $\Gamma(Q)$ meets $B(x, R)$, then $\operatorname{diam} Q \geq d(x) - R \geq \frac{d(x)}{2}$, because the highest point of $\Gamma(Q)$ is A_Q.

For each choice of C_1, there are less than $C(C_1)$ cubes Q such that $\frac{d(x)}{2} \leq \operatorname{diam} Q \leq C_1 d(x)$ and $\Gamma(Q)$ meets $B(x, R)$. (This is because the projection of $\Gamma(Q)$ stays at distance $\leq \operatorname{diam} Q$ from Q.) On the other hand, the same argument as in the case when $x \in E$ tells us that, if we choose C_1 large enough, there is only a bounded number of cubes Q such that $\operatorname{diam} Q > C_1 d(x)$ and $\Gamma(Q)$ meets $B(x, R)$ (or even $B(x, d(x))$). Altogether, $B(x, R)$ does not intersect more than a bounded number of the $\Gamma(Q)$'s, and (1.9) follows. This concludes our proof of Lemma 1.7.

LEMMA 1.12. *Γ is connected.*

Let us first show that if $Q \in \mathscr{B}$ and $S \in \mathscr{S}(Q)$, then there is a path $\gamma \subset \Gamma$ that connects A_Q to a point $y_0 \in S$ and has length less than $C \operatorname{diam} Q$. Indeed, if $\mathscr{W}(S) = \varnothing$, then we can take $\gamma = \Gamma_0(S)$. Otherwise, $\Gamma_0(S)$ only connects A_Q to the point $A_{H(S)}$. We let $\gamma_0 = \Gamma_0(S)$ and start iterating the construction. We consider the cube $Q_1 = H(S)$ and pick any cube $S_1 \in \mathscr{S}(Q_1)$. If $\mathscr{W}(S_1) = \varnothing$, we let $\gamma_1 = \Gamma_0(S_1)$ and take $\gamma = \gamma_0 \cup \gamma_1$. Otherwise, we still let $\gamma_1 = \Gamma_0(S_1)$, but we continue the construction with the cube $Q_2 = H(S_1)$ and any cube $S_2 \in \mathscr{S}(Q_2)$. We obtain a sequence of cubes Q_k, a sequence of cubes $S_k \in \mathscr{S}(Q_k)$, and a sequence of paths $\gamma_k = \Gamma_0(S_k)$, which are finite if we meet an S_k with $\mathscr{W}(S_k) = \varnothing$ and infinite otherwise. Take $\gamma = (\bigcup_k \gamma_k)^-$. (The closure is needed if the sequence is infinite.)

Notice that $\operatorname{diam} Q_1 = \operatorname{diam} H(S) \leq \delta \operatorname{diam} S \leq \delta \operatorname{diam} Q < \frac{1}{10} \operatorname{diam} Q$. (We shall take $\delta < \frac{1}{10}$.) Similarly, $\operatorname{diam} Q_k < \frac{1}{10} \operatorname{diam} Q_{k-1}$ for $k = 2, 3, \ldots$. Hence the length of γ is less than $C \operatorname{diam} Q$, and $\operatorname{diam} \gamma_k \leq C \operatorname{diam} Q_k \leq C 10^{-(k-1)} \operatorname{diam} Q_1 \leq C 10^{-k+1} \delta \operatorname{diam} S$.

The fact that γ connects A_Q to some point y_0 of E is obvious because A_{Q_k} is at distance less than $C \operatorname{diam} Q_k$ from E. The fact that the point y_0 lies in S will follow from

(1.13) $\qquad \operatorname{dist}(y_0, A_{H(S)}) \leq C\delta \operatorname{diam} S$

and

(1.14) $\qquad \operatorname{dist}(A_{H(S)}, c_S) \leq 11\delta \operatorname{diam} S,$

provided that we choose δ so small that $(C+11)\delta \operatorname{diam} S \leq \frac{1}{2} \operatorname{dist}(c_S, E \setminus S)$. (We shall not impose any further constraint on δ, and so we can choose δ now.) The estimate (1.13) follows directly from our estimates on $\operatorname{diam}(\gamma_k)$, whereas (1.14) is a direct consequence of (1.6). Hence our path γ connects A_Q to a point of S.

Since we now know that every point of $\Gamma \setminus E$ can be connected to E by a path of finite length, Lemma 1.12 will follow if we prove that for all points $x, y \in E$ there is a path $\gamma \subset \Gamma$, of finite length, that joins x to y.

Since we shall use discrete paths first, it will be convenient to use the following convention. If $\tau > 0$ is given, a τ-link from x to y is a sequence $y_0 = x, y_1, \ldots, y_N = y$ of points such that $|y_{j+1} - y_j| \leq \tau$ for $j = 1, \ldots, N-1$. We say that the τ-link lies in a set A if all the points y_j lie in A. The main step of the construction of γ is the following lemma.

1.1. THE WEAK CONNECTEDNESS CONDITION

LEMMA 1.15. *There is a constant $C_2 > 0$ such that, for each $\tau > 0$ and every $x, y \in E$, there is a τ-link in $\Gamma \cap B(x, C_2 M|x - y|)$ that joins x to y.*

It will be fairly easy to deduce Lemma 1.12 from this lemma, but for the moment let us construct the τ-links. Let $\tau > 0$ and $x, y \in E$ be given. Let Q be a cube of diameter $\leq C|x-y|$ such that x and y both belong to $2Q$. First suppose that Q is in the good set $\mathscr{G}$. By the definition of $\mathscr{G}$ (a little before Lemma 1.4) there is an $\epsilon \operatorname{diam} Q$-link from x to y that lies in $C_0 M Q$. If, on the other hand, $Q \in \mathscr{B}$, then we have to use $\Gamma(Q)$. Let S be the cube of $\mathscr{S}(Q)$ that contains x, and similarly let $T \in \mathscr{S}(Q)$ be the cube that contains y. There is a path $\gamma \subset \Gamma$ of length $\leq C \operatorname{diam} Q$ that connects some point $x' \in S$ to some point $y' \in T$ (via the point A_Q). We can thus join x to y by first a (short) link from x to x', then the path γ, and then a short link from y' to y. The two short links are $C\epsilon \operatorname{diam} Q$-links, since $\operatorname{diam} S \leq C\epsilon \operatorname{diam} Q$ (and similarly for T).

In both cases, we were able to join x to y by a combination of $C\epsilon \operatorname{diam} Q$-links and paths of length $\leq C \operatorname{diam} Q$. We now choose ϵ so small that $C\epsilon \operatorname{diam} Q < \frac{1}{2}|x - y|$. We have proved that any two points $x, y \in E$ can be joined by a combination of $\frac{1}{2}|x - y|$-links and paths, with the following properties:

- the links are within distance $\leq CC_0 M|x-y|$ from x and are composed of points of E;
- the paths are contained in Γ, have finite length, and stay at a distance $\leq CM|x-y|$ from x.

We can now easily iterate the construction, replacing each elementary $\frac{1}{2}|x-y|$-link by a combination of $\frac{1}{4}|x-y|$-links and paths. We do this enough times (i.e., about $\log \frac{1}{\tau}$ times), and we get Lemma 1.15 by replacing the obtained paths with τ-links in Γ. The diameter of the final τ-link is controlled by a geometric series. Note that we do not need any estimate on the number of points in the link; this will come for free because Γ is regular.

Let us now deduce Lemma 1.12 from Lemma 1.15. Given two points $x, y \in \Gamma$, we want to find a path γ contained in Γ that connects them. Because we already know that each point of $\Gamma \backslash E$ can be connected to E by a path in Γ, we can assume that x and y belong to E. The path γ will be obtained as a limit of approximating paths γ_τ. For each small $\tau > 0$, we start with the τ-link from x to y given by Lemma 1.15. Let $y_0, \ldots, y_N$ be the points of this τ-link. We may as well suppose that $|y_i - y_j| > \tau$ whenever $j > i+1$; if this were not true, we could simply throw away the intermediate points y_l, $i+1 \leq l \leq j-1$, and still get a τ-link, and we could repeat this process as many times as necessary. Once we have that $|y_i - y_j| > \tau$ whenever $j > i+1$, the fact that all the y_j's lie in $\Gamma \cap B(x, C_2 M|x-y|)$ and the regularity of Γ imply that our τ-link has less than $C\tau^{-1}|x-y|$ points.

By connecting the successive points of our τ-link by line segments, we get a path γ_τ which joins x to y, stays within a distance $\leq \tau$ from Γ, and has length $\leq C|x-y|$. We can find a parameterization $f_\tau : [0, |x-y|] \mapsto \gamma_\tau$ of this path which is C-Lipschitz and satisfies $f_\tau(0) = x$ and $f_\tau(|x-y|) = y$. The Arzela-Ascoli theorem implies that we can find a sequence τ_k that tends to 0 such that the f_{τ_k} converge uniformly to a C-Lipschitz function f on $[0, |x-y|]$. The path $\gamma = f([0, |x-y|])$ is contained in Γ, has a length $\leq C|x-y|$, and connects x to y.

This completes the proof of Lemma 1.12, and hence Theorem 1.3 is established.

In some applications, the weak connectedness condition will be obtained using the following slightly simpler criterion.

COROLLARY 1.16. *Let E be a one-dimensional regular set in $\mathbf{R}^n$. Suppose that there is a constant $0 < r < 1$ such that $E \times \mathbf{R}_+ \setminus \mathscr{G}(r)$ is a Carleson set, where*

(1.17)
$$\mathscr{G}(r) = \{(x, t) \in E \times \mathbf{R}_+ : \text{for all } u, v \in E \cap B(x, t) \text{ with } |u - v| \geq \tfrac{t}{10}$$
there is a point $w \in E$
such that $|w - u| \leq r|u - v|$ and $|w - v| \leq r|u - v|\}$.

Then there is a connected regular set Γ that contains E.

As in Theorem 1.3, the regularity constant for Γ can be taken to depend only on r, the regularity constant for E, and the Carleson constant for $E \times \mathbf{R}^+ \setminus \mathscr{G}(r)$. Also notice that we could ask that Γ be a regular curve, since we know that every connected regular set of dimension 1 is contained in a regular curve. (See the remark after (I.1.64).)

The corollary is an immediate consequence of Theorem 1.3, since $\mathscr{G}(r) \subset \mathscr{G}(r, 2)$.

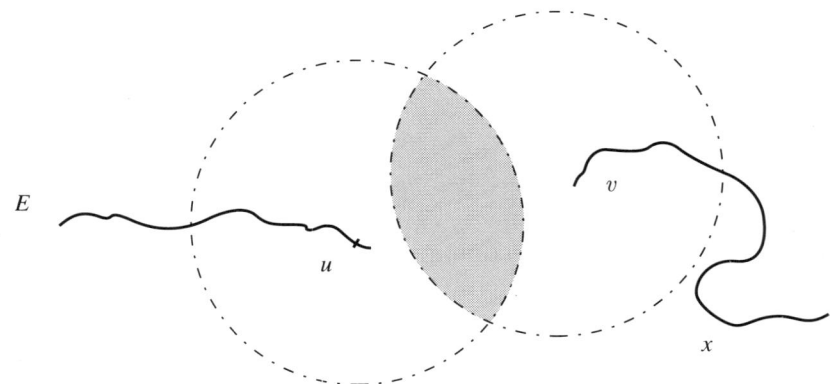

FIGURE 1.2. If $(x, t) \in \mathscr{G}(r)$, then there should be a point of E in the shaded region.

1.2. The weaker local symmetry condition $(d = 1)$.

[This section is not essential to the understanding of the monograph and may be skipped.]

Remember that the local symmetry condition LS (see Definition 1.79 in Part I) implies uniform rectifiability, because it implies the bilateral weak geometric lemma BWGL (see Definition 2.2 of Part I). The fact that the BWGL implies uniform rectifiability will be proved in full generality in the next chapter, but we already have a proof when $d = 1$, since the BWGL clearly implies the WCC in that case. The fact that LS implies the BWGL is not hard. A fairly self-contained proof is given in [**DS2**] (see Proposition 5.5 there), but we can give a very rapid sketch of the proof here.

Let $(x, t) \in E \times \mathbf{R}_+$ be such that all (x', t') with $|x' - x| \leq C_0 t$ and $C_0^{-1} t \leq t' \leq C_0 t$ are "good for the local symmetry condition". It is enough to prove that $(x, C_0^{1/2} t)$ is "good for the BWGL", provided that we take C_0 large enough. We pick $(d+1)$ points $x_0, \ldots, x_d$ in $E \cap B(x, t)$ that are affinely independent (with an estimate). Such points exist because otherwise $E \cap B(x, t)$ would stay too close to a $(d-1)$-plane to be regular of dimension d. (A slightly more detailed explanation for this point will be given in the proof of Lemma 2.75.) An easy induction then shows that the points $x_0 + k(x_1 - x_0)$, where k is a nonnegative integer such that $k \leq C_0/2$, are very close to E. [The point $x_0 + k(x_1 - x_0)$ is obtained from $x_0 + (k-2)(x_1 - x_0)$ by a symmetry with respect to $x_0 + (k-1)(x_1 - x_0)$.] The same thing holds with x_1 replaced by $x_2, x_3, \ldots, x_d$. Applying the local symmetry condition with various choices of points $x_0 + k(x_j - x_o)$ then gives the existence of a substantial part of a d-dimensional lattice which stays very close to E. The corresponding d-plane then has all its points (at least, those in $B(x, C_0^{1/2} t)$) fairly close to E. The nearby points of E must also be close to the d-plane, because otherwise we could find a $(d+2)$nd affinely independent point x_{d+1} and then (by the argument above) a substantial piece of $(d+1)$-dimensional lattice that would stay very close to E. This would contradict the fact that E is regular of dimension d.

When $d = 1$, we shall be able to use the notion of weak connectedness to prove a result that is a little more general. The local symmetry condition will be replaced by the slightly weaker condition stated below.

For the definition of the weaker local symmetry condition (WLS) we let a constant τ, $0 < \tau \leq 1$, be given. (The case $\tau = 1$ will correspond to the local symmetry condition.) For $y, z \in \mathbf{R}^n$, let

(1.18) $\qquad L(y, z) = $ the closed line segment with endpoints
$\qquad\qquad\qquad z + \tau(z - y)$ and $z + (z - y)$.

(See Figure 1.3, next page.)

For $\epsilon > 0$ small enough ($\epsilon < \frac{\tau}{100}$ will be enough), we define a good set $\mathscr{G}(\tau, \epsilon)$ by

FIGURE 1.3. 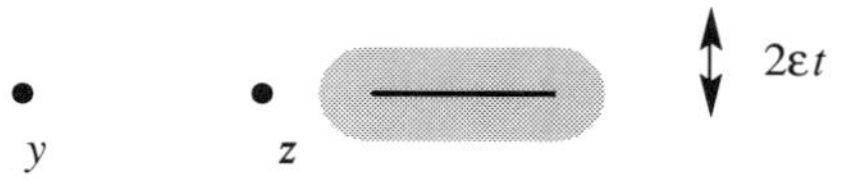   FIGURE 1.4.

(1.19)
$$\mathscr{G}(\tau, \epsilon) = \{(x, t) \in E \times \mathbf{R}_+ : \text{ for all points } y, z \in E \cap B(x, t)$$
$$\text{such that } |z - y| \geq \frac{t}{10} \text{ we have } \operatorname{dist}(L(y, z), E) \leq \epsilon t\}.$$

Thus, for $(x, t) \in \mathscr{G}(\tau, \epsilon)$ and $y, z \in E \cap B(x, t)$ not too close to each other, there is a point of E in the shaded region in Figure 1.4.

DEFINITION 1.20. We say that the one-dimensional regular set E satisfies the weaker local symmetry condition (WLS) if there exists a $\tau \in [0, 1]$ such that, for all sufficiently small $\epsilon > 0$, the set $\mathscr{B}(\tau, \epsilon) = E \times \mathbf{R}_+ \setminus \mathscr{G}(\tau, \epsilon)$ is a Carleson set.

Notice that $\tau = 1$ gives back the condition LS of Definition I.1.79. Here we introduced the condition $|y - z| \geq \frac{t}{10}$ in the definition of $\mathscr{G}(\tau, \epsilon)$ to make things clearer, but we could have removed it (while still getting an equivalent definition) because $L(y, z)$ is always at distance $\leq \epsilon t$ from z (and hence E) when $|y - z| \leq \epsilon t$, while the other scales of $|y - z|$ can be obtained by an easy game with Carleson sets.

THEOREM 1.21. *If the one-dimensional regular set E satisfies the WLS, then it is weakly connected (and is therefore contained in a regular curve).*

REMARK 1.22. We asked that $\mathscr{B}(\tau, \epsilon)$ be a Carleson set for all small enough values of ϵ, but the proof will only need one value of ϵ (which depends on τ and the regularity constant for E). On the other hand, it will in practice be as easy to check the property for all ϵ rather than just one (very small) value of ϵ. Similar remarks will apply to essentially all our weak conditions (in particular, the BWGL, LS, etc.).

Let E be given, as in Theorem 1.21, and let τ be as in Definition 1.20. The proof of the theorem will be based on the same sort of ideas as for the proof that LS implies BWGL. Notice that WLS becomes weaker as τ decreases, and so we may as well assume that τ is very small.

We intend to use the weak connectedness condition through Corollary 1.16. More precisely, we shall prove the following lemma.

LEMMA 1.23. *If $\epsilon > 0$ is small enough, then there exists a constant r, $0 < r < 1$, and a constant $C_0 \geq 1$ (both of which may depend on τ and the regularity constants for E) such that (x, t) belongs to the set $\mathscr{G}(r)$ of (1.17) as soon as all $(x', t') \in E \times \mathbf{R}_+$ such that*

(1.24) $$|x' - x| \leq C_0 t \quad \text{and} \quad C_0^{-1} t \leq t' \leq C_0 t$$

lie in $\mathscr{G}(\tau, 4\epsilon)$.

1.2. THE WEAKER LOCAL SYMMETRY CONDITION

The proof of this lemma will occupy us for most of this section. Let us first see why Theorem 1.21 follows from Lemma 1.23. The argument is the same as for Lemma 1.4 in the previous section. If (x, t) is in the bad set $E \times \mathbf{R}_+ \setminus \mathscr{G}(r)$, then there is an $(x', t') \in E \times \mathbf{R}_+$ that satisfies (1.24) and does not belong to $\mathscr{G}(\tau, 4\epsilon)$. This implies that there exist points y, $z \in E \cap B(x', t')$ and such that $|y - z| \geq t'/10$ and $\operatorname{dist}(L(y, z), E) > 4\epsilon t'$. Then all the points $(x'', t'') \in E \times \mathbf{R}_+$ such that $|x'' - y| \leq t'/100$ and $3|z - y|/2 \leq t'' \leq 2|z - y|$ lie in the complement of $\mathscr{G}(\tau, \epsilon)$, because the points y and z satisfy $y, z \in E \cap B(x'', t'')$ and $\operatorname{dist}(L(y, z), E) > \epsilon t''$ for such (x'', t'').

Thus, to each point $(x, t) \in \mathscr{B}(r) = E \times \mathbf{R}_+ \setminus \mathscr{G}(r)$, we have associated a set $\mathscr{A}_{x,t} \subset \mathscr{B}(\tau, \epsilon) = E \times \mathbf{R}_+ \setminus \mathscr{G}(\tau, \epsilon)$ in such a way that $\mathscr{A}_{x,t}$ has $(dx''dt''/t'')$-measure $\geq C^{-1}t$ and is contained in a set of the form

$$\{(x'', t'') \in E \times \mathbf{R}_+ : |x'' - x| \leq 10C_0 t \text{ and } (10C_0)^{-1}t \leq t'' \leq 10C_0 t\}.$$

It is now easy to deduce the desired Carleson measure estimates on $\mathscr{B}(r)$ from the corresponding estimates on $\mathscr{B}(\tau, \epsilon)$. Although the argument is surpriseless, we shall sketch it for the convenience of the reader. Let a ball $B(X, R)$ centered on E be given. We want to estimate

$$I = \iint_{\{B(X,R)\times(0,R]\}\cap\mathscr{B}(r)} \frac{dx\,dt}{t}.$$

Let A be a maximal set of points of $\{B(X, R) \times (0, R]\} \cap \mathscr{B}(r)$ with large enough mutual hyperbolic distance, i.e., such that two points (x, t) and (x', t') of A never satisfy $|x'-x| \leq C_1 t$ and $C_1^{-1} t \leq t' \leq C_1 t$, where C_1 is a large constant that we can choose. The set A gives a covering of $\{B(X, R) \times (0, R]\} \cap \mathscr{B}(r)$ by large hyperbolic balls, so $I \leq C \sum_{(x,t)\in A} t$. On the other hand, we can easily choose C_1 so large that all the $\mathscr{A}_{x,t}$ corresponding to $(x, t) \in A$ are disjoint, and then we get

$$\sum_{(x,t)\in A} t \leq C \sum_{(x,t)\in A} \iint_{\mathscr{A}_{x,t}} \frac{dx''\,dt''}{t} \leq C \iint_F \frac{dx''\,dt''}{t''},$$

where $F = \{B(X, 10C_0 R) \times (0, 10C_0 R]\} \cap \mathscr{B}(\tau, \epsilon)$. Hence

$$I \leq C \iint_F \frac{dx''\,dt''}{t''} \leq C'R,$$

because $\mathscr{B}(\tau, \epsilon)$ is a Carleson set.

Thus it is enough to prove Lemma 1.23. We shall proceed by contradiction, so we assume that we have a point (x, t) which is not in $\mathscr{G}(r)$ but which has the property that all the (x', t') satisfying (1.24) lie in $\mathscr{G}(\tau, 4\epsilon)$, and we want to reach a contradiction. We may as well assume also that $t = 1$.

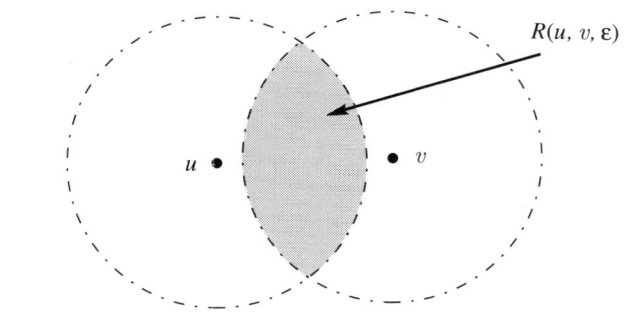

FIGURE 1.5

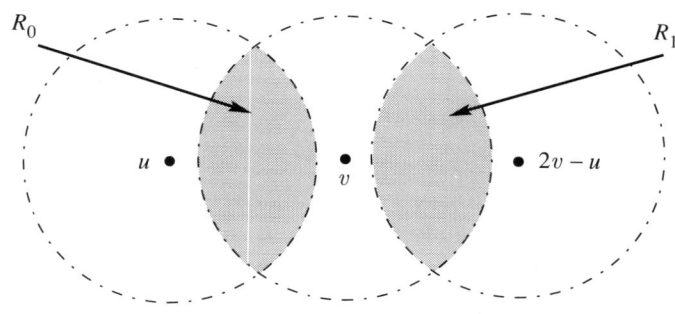

FIGURE 1.6

Let $\epsilon > 0$ be small, how small to be specified later. We take $r = 1 - \epsilon$.
Since $(x, 1) \notin \mathscr{G}(r)$, there is a pair of points $u, v \in E \cap B(x, 1)$ such that $|u - v| \geq \frac{1}{10}$ and $E \cap R(u, v, \epsilon) = \varnothing$, where we set

(1.25)
$$R(u, v, \epsilon) = \{w \in \mathbf{R}^n : |w - u| \leq (1-\epsilon)|u-v| \text{ and } |w-v| \leq (1-\epsilon)|u-v|\}.$$

(See Figure 1.5.)

Set $R_0 = R(u, v, \epsilon)$ and $R_1 = R(v, 2v - u, \frac{10\epsilon}{\tau})$ so that R_1 is a diminished version of R_0 on the other side of v. (See Figure 1.6.)

We want to check that R_1 does not intersect E either. Notice first that if $w \in R_1$, then the line segment $L(w, v)$ is entirely contained in R_0, and it even stays at distance $> 8\epsilon|u-v|$ from its boundary (and hence from E). If we could find a point $w \in E \cap R_1$, we would get a contradiction by applying the WLS to the ball $B(x', t')$, with $x' = v$ and $t' = \frac{11}{10}|w - v|$. Indeed, (x', t') would satisfy (1.24) if C_0 is chosen large enough, and so we would have that $(x', t') \in \mathscr{G}(\tau, 4\epsilon)$ by hypothesis. Using the definition of $\mathscr{G}(\tau, 4\epsilon)$ (see (1.19)), we would get $\text{dist}(L(v, w), E) \leq 4\epsilon t' \leq \frac{44}{10}\epsilon|u - v|$. We know that this is impossible because $L(w, v)$ stays at distance $> 8\epsilon|u - v|$ from E for all $w \in R_1$. Thus R_1 does not intersect E, and we have proved that

(1.26)
$$E \cap (R_0 \cup R_1) = \varnothing.$$

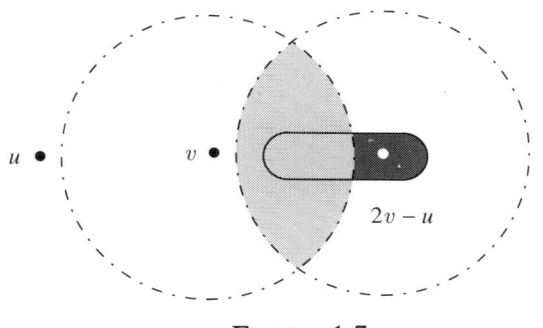

FIGURE 1.7

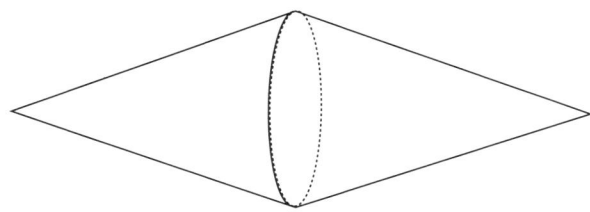

FIGURE 1.8. $S(y, z)$ is the inside of the carrot.

Notice that the fact that R_1 does not intersect E implies that $\operatorname{dist}(2v - u, E) \leq \frac{20\epsilon}{\tau}|u - v|$. Indeed, there is a point of E at distance $\leq 10\epsilon|u-v|$ from $L(u, v)$ (because $(u, \frac{11}{10}|u - v|)$ satisfies (1.24) and thus belongs to $\mathscr{G}(\tau, 4\epsilon)$); if ϵ is small enough, this point can only be very close to $2u - v$, because it cannot be in R_1. (See Figure 1.7.)

We want to iterate the argument above to show that the points $u + j(v - u)$, $j = 1, 2, \ldots$, are very close to E. Since the same construction will also be used later in the argument, we start with a little bit of extra notation.

For $y, z \in \mathbf{R}^n$, we let $S(y, z)$ be the intersection of the two open half-cones pictured in Figure 1.8:

$$(1.27) \quad S(y, z) = \{w \in \mathbf{R}^n : \operatorname{dist}(w, [y, z]) < \tfrac{1}{10}\min(|w - y|, |w - z|)\},$$

where $[y, z]$ denotes the line segment from y to z.

Next we define, for each small enough $\eta > 0$, a slightly smaller region $S(y, z, \eta)$ given by

$$(1.28) \quad S(y, z, \eta) = \{w \in S(y, z) : \operatorname{dist}(w, \partial S(y, z)) \geq \eta\}.$$

LEMMA 1.29. *For each $j \in \mathbf{N}$ there exist constants $C_1(j)$ and $C_2(j)$ such that the following is true. Let y_0 and y_1 be two points of $E \cap B(x, 2)$ such that $|y_1 - y_0| \geq \frac{1}{100}$ and $E \cap S(y_0, y_1, 2\epsilon) = \varnothing$. Set $y_j = y_0 + j(y_1 - y_0)$. Then we have*

$$(1.30) \quad \operatorname{dist}(y_j, E) \leq C_1(j)\epsilon$$

and

$$(1.31) \quad E \cap S(y_{j-1}, y_j, C_2(j)\epsilon) = \varnothing,$$

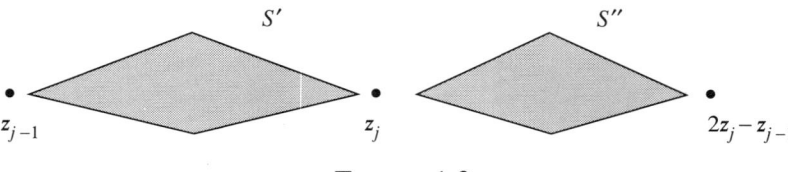

FIGURE 1.9

provided that $C_0 \geq 4j + 100$ *and* ϵ *is small enough (depending only on* j).

We are mostly interested in the estimate (1.30), but proving (1.31) at the same time will help us prove Lemma 1.29 by induction. The estimate (1.30) is obviously true for $j = 0$ and $j = 1$, with $C_1(0) = C_1(1) = 0$, and (1.31) holds for $j = 1$ with $C_2(1) = 2$. Let us now assume that (1.30) holds for $j' = 0, 1, \ldots, j$ and that (1.31) holds for $j' = 1, \ldots, j$, and let us establish (1.30) and (1.31) for $j + 1$.

Let z_{j-1} (respectively z_j) be a point of E whose distance to y_{j-1} (resp. y_j) is at most $C_1(j-1)\epsilon$ (resp. $C_1(j)\epsilon$). Note that, by taking ϵ small enough, we can assume that z_{j-1} is as close as we want to y_{j-1}, and similarly for z_j.

Choose C' so large that $S' = S(z_{j-1}, z_j, C'\epsilon)$ is contained in

$$S(y_{j-1}, y_j, C_2(j)\epsilon).$$

For instance, $C' = 100(C_2(j) + C_1(j) + C_1(j-1))$ will do. Since we have assumed that (1.31) holds for j, we get $E \cap S' = \emptyset$. Next let $S'' = S(z_j, 2z_j - z_{j-1}, C''\epsilon)$ be a diminished version of S' on the other side of z_j. (See Figure 1.9.)

We choose C'' so large that, if $w \in S''$, then the line segment $L(w, z_j)$ is contained in S', and even stays at distance $\geq 20\epsilon$ from the boundary of S'. For instance, we can choose $C'' = \frac{1}{\tau}\{C' + 20\}$. Let us check that S'' does not meet E. If we could find a point $w \in E \cap S''$, then $(x', t') = \left(z_j, \frac{11}{10}|w - z_j|\right)$ would satisfy (1.24) and therefore be in $\mathscr{G}(\tau, 4\epsilon)$, and hence we would have that

$$\operatorname{dist}(L(w, z_j), E) \leq 4\epsilon t' = \tfrac{44}{10}\epsilon |w - z_j| \leq \tfrac{44}{10}\epsilon |z_j - z_{j-1}| \leq 10\epsilon$$

because z_j and z_{j-1} are as close as we want to y_j and y_{j-1}. This is impossible because we know that $L(w, z_j)$ is contained in S' and even stays away from its boundary. Thus S'' does not intersect E.

Because $|z_j - y_j| \leq C_1(j)\epsilon$ and $|z_{j-1} - y_{j-1}| \leq C_1(j-1)\epsilon$, and hence $|(2z_j - z_{j-1}) - (2y_j - y_{j-1})| \leq [2C_1(j) + C_1(j-1)]\epsilon$, we can choose $C_2(j+1)$

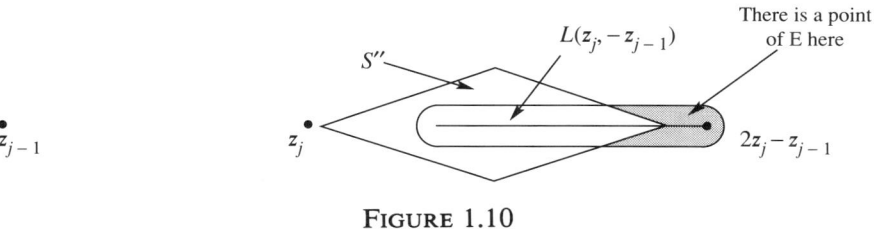

FIGURE 1.10

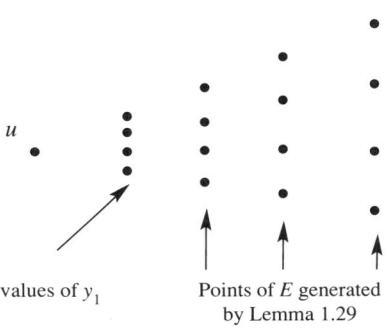

FIGURE 1.11

so large that $S(y_j, y_{j+1}, C_2(j+1)\epsilon) \subset S''$. (For instance, $C_2(j+1) = 100C'' + 100C_1(j) + 100C_1(j-1)$ will work.) This proves (1.31) for $j+1$.

Let us now prove (1.30) for $(j+1)$. Consider the pair (x', t') given by $\left(z_j, \frac{11}{10}|z_j - z_{j-1}|\right)$. If $C_0 \geq 4j + 100$, (x', t') satisfies (1.24), so $(x', t') \in \mathscr{G}(\tau, 4\epsilon)$. Using the analogue of (1.19) with the points z_{j-1} and z_j, we find a point on the line segment $L = L(z_{j-1}, z_j)$ which is at distance $\leq 4\epsilon t' = \frac{44}{10}\epsilon|z_j - z_{j-1}| \leq 10\epsilon$ from E. Now, if ϵ is small enough, most of L is well inside the domain S'' defined above, so the only points of L that could be within 10ϵ from E are at distance $\leq 100C''\epsilon$ from $2z_j - z_{j-1}$. (See Figure 1.10.) Hence we get $\text{dist}(2z_j - z_{j-1}, E) \leq 200C''\epsilon$, and (1.30) for $j+1$ follows because $2z_j - z_{j-1}$ is close to $2y_j - y_{j-1}$. This proves Lemma 1.29.

We intend to apply Lemma 1.29 to various choices of points y_0, y_1. Actually, we shall always take $y_0 = u$, and we shall take various values of y_1 in a small ball centered on v. This will give, for each choice of y_1, a lot of points y_j that are very close to E, and we hope that taking enough choices of y_1 will give us enough points that are close to E to get a contradiction. See Figure 1.11.

For this it will be helpful to know that we can take many choices of y_1 which are very close to a "vertical" hyperplane (rather than in a messy cloud).

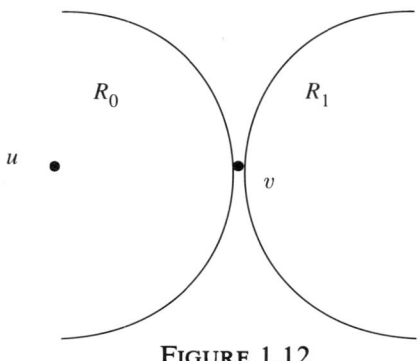

FIGURE 1.12

To prove this, we intend to use the fact that E does not intersect $R_0 \cup R_1$ (see (1.26)), which implies (if ϵ is small enough) that, near v, E stays very close to a "vertical" hyperplane. (See Figure 1.12.)

Let K be a large constant. The value of K will be decided later, and it will not depend on our choices of ϵ or C_0 (which will be made afterward). We shall take our points y_1 from $E \cap B_1$, where $B_1 = B(v, K^{-2/3})$. First select a maximal set of points v_m, $m \in M$, all belonging to $E \cap \frac{1}{2}B_1$ and at mutual distances $\geq \frac{10}{K}$. Because E is regular, we can find at least $C^{-1}K^{1/3}$ such points.

For each v_m, we select a point $y^m \in E$ as follows. First let

$$\overline{S}_m = \{w \in \mathbf{R}^n : \operatorname{dist}(w, [u, v_m]) \leq \tfrac{1}{5}\min(|w-u|, |w-v_m|)\}.$$

($\overline{S}_m$ is a little wider than the region $S(u, v_m)$ defined above.)

We take for y^m a point of $E \cap \{\overline{S}_m \setminus B(u, 10\epsilon)\}$ whose distance to u is minimal. First note that, because of (1.26), all the points of $E \cap \{\overline{S}_m \setminus B(u, 10\epsilon)\}$ are very close to v_m. More precisely, we can choose K sufficiently large and then ϵ sufficiently small so that the facts

(1.32) $\qquad v_m \in B(v, K^{-2/3}),$

(1.33) $\qquad v_m \notin R_0 \cup R_1,$

and

(1.34) $\qquad y^m \in \overline{S}_m \setminus (R_0 \cup B(u, 10\epsilon))$

imply

(1.35) $\qquad |y^m - v_m| \leq C\epsilon + \left(CK^{-2/3}\right)^2 \leq K^{-1}.$

(See Figure 1.13.)

Our next observation is that, by taking $\overline{S}_m$ to be a little wider than $S(u, v_m)$, we have ensured that $S(u, y^m, 2\epsilon)$ does not intersect E. To

1.2. THE WEAKER LOCAL SYMMETRY CONDITION 85

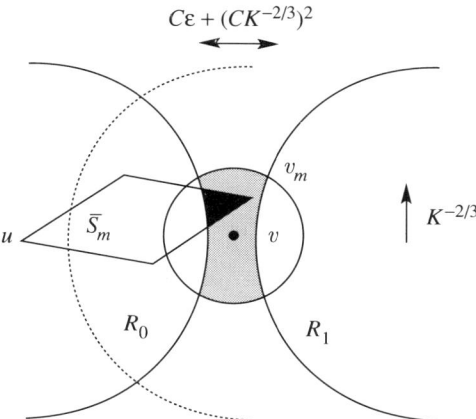

FIGURE 1.13. The shaded region is the possible location of the point v_m. The black region is the possible location of the point y^m; if ϵ is small enough, it has a diameter $\leq CK^{-4/3}$.

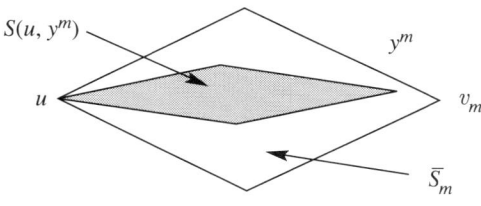

FIGURE 1.14

see this, we consider separately the part near u and the part near y^m. The part near u lies inside R_0 — and hence outside E — because

$$S(u, y^m, 2\epsilon) \subset B(v, (1-\epsilon)|u-v|).$$

(To verify this last inclusion, it is helpful to reduce to showing that $S(u, y^m)$ is contained in $B(v, |u-v|)$. Remember that $|u-v| \leq 2$.) On the other hand, suppose that z lies in the part of $S(u, y^m)$ near y^m. Then $z \in \overline{S}_m$, because $S(u, y^m) \subset \overline{S}_m$, and z is closer to u than y^m is. (See Figure 1.14.) This implies that z cannot lie in E, because of the way that we chose y^m.

We can now apply Lemma 1.29 with $y_0 = u$ and $y_1 = y^m$, and we get that the points $y_j^m = u + j(y^m - u)$, $1 \leq j \leq \frac{K}{100}$, are at distance $\leq C(K)\epsilon$ from E. (For this to work, we have to take C_0 large enough and ϵ small enough, both depending on K.)

Let us consider only the points y_j^m such that $\frac{K}{1000} \leq j \leq \frac{K}{100}$. We want to prove that these points are sufficiently far from each other to contradict the regularity of E. Notice that, when $m \neq m'$, we have that $|v_m - v_{m'}| \geq \frac{10}{K}$ by construction. Let π_m (resp. $\pi_{m'}$) be the orthogonal projection of v_m (resp. $v_{m'}$) onto the hyperplane H that contains v and is perpendicular to the line through u and v. Because v_m satisfies (1.32) and (1.33), we have that $|\pi_m - v_m| \leq C\left(K^{-2/3}\right)^2 + C\epsilon$. We shall take K so large, and (then) ϵ so small, that this implies that $|\pi_m - v_m| \leq \frac{1}{K}$. Similarly, $|\pi_{m'} - v_{m'}| < \frac{1}{K}$, so $|\pi_m - \pi_{m'}| \geq \frac{8}{K}$. Because of (1.35), we have that $|y^m - \pi_m| \leq \frac{2}{K}$ and $|y^{m'} - \pi_{m'}| \leq \frac{2}{K}$, so the angle between the line L_m through u and y^m and the line $L_{m'}$ through u and $y^{m'}$ is $\geq \frac{1}{K}$. Consequently, the distance from any y_j^m to any $y_{j'}^{m'}$ is always $\geq 10^{-6}$ because j and j' are $\geq \frac{K}{1000}$ and $|u - y^m| \geq \frac{1}{2}|u - v| \geq \frac{1}{20}$. Since we also have that $|y_j^m - y_{j'}^m| \geq 10^{-2}$ for $j \neq j'$, we see that y_j^m is at distance $\geq 10^{-6}$ from all the other points $y_{j'}^{m'}$. Let $B_{m,j} = B(y_j^m, 10^{-7})$. These balls are disjoint, and they are contained in $B(u, K)$. Consequently, $\sum_m \sum_j |E \cap B_{m,j}| \leq CK$.

On the other hand, if we take ϵ small enough (depending on K), each $\frac{1}{2}B_{m,j}$ intersects E. (Remember that we used Lemma 1.29 to prove that dist$(y_j^m, E) \leq C(K)\epsilon$.) Hence $|E \cap B_{m,j}| \geq C^{-1}$. There are more than $C^{-1}K^{1/3}$ choices of m and, for each of those, more than $\frac{K}{200}$ choices of j, so that $\sum_m \sum_j |E \cap B_{m,j}| \geq C^{-1}K^{4/3}$, where C depends only on the regularity constant for E (and not K!). This estimate contradicts the previous one, so we have shown that the points u and v whose existence we postulated (i.e., points of $E \cap B(x, 1)$ such that $|u - v| \geq \frac{1}{10}$ and the set $R(u, v, \epsilon)$ defined by (1.25) does not meet E) cannot exist if all the (x', t') that satisfy (1.24) lie in $\mathscr{G}(\tau, 4\epsilon)$.

This completes our proof of Lemma 1.23. As we explained after the statement of Lemma 1.23, Theorem 1.21 follows from this.

We should mention that there are a number of other variants of the local symmetry condition (and the local convexity condition) that appear to be reasonable candidates for characterizations of uniform rectifiability, but for which we do not know whether they actually imply uniform rectifiability. We shall give a description of some of these in §2.7.

1.3. Weak constant density for one-dimensional sets.

Let us recall the definition of the weak constant density condition WCD. Let E be, as usual, a d-dimensional regular set in $\mathbf{R}^n$. For each $C_0 > 0$ and $\epsilon > 0$ let $\mathscr{G}_d(C_0, \epsilon)$ be the set of $(x, t) \in E \times \mathbf{R}_+$ for which there exists a C_0-regular measure μ (i.e., a measure that satisfies the condition (I.2.51)) with supp $\mu = E$ such that

(1.36) $$|\mu(E \cap B(y, r)) - r^d| \leq \epsilon t^d$$

for all $y \in E \cap B(x, t)$ and all $0 < r \leq t$.

1.3. WEAK CONSTANT DENSITY ($d = 1$)

As in Definition I.2.55, we say that E satisfies the WCD if there exists a $C_0 > 0$ such that the complement in $E \times \mathbf{R}_+$ of $\mathscr{G}_d(C_0, \epsilon)$ is a Carleson set for each $\epsilon > 0$.

If E is uniformly rectifiable, then it satisfies the WCD and even the more precise condition given by Theorem I.2.52. This is not hard to check once we know the BWGL and the weak Littlewood-Paley condition (I.1.85). See §6 of [**DS2**] for more details.

The converse is also true when $d = 1, 2$, and $n - 1$. A proof of this will be given in Chapter 5 of Part III. This proof will use a compactness argument together with a description of the sets which support a measure with "constant density".

In this section we give a more direct proof of the converse which does not use compactness but which only works when $d = 1$. More precisely, we shall prove that the WCD implies the BWGL (bilateral weak geometric lemma). The best way to deduce uniform rectifiability from this is probably to notice that the BWGL implies the WCC and apply Theorem 1.3.

PROPOSITION 1.37. *For a one-dimensional regular set, the weak constant density condition* WCD *implies the bilateral weak geometric lemma* BWGL.

Obviously, Proposition 1.37 will follow as soon as we prove the next lemma.

LEMMA 1.38. *For each $0 < \tau_0 < 1$ there is an $\epsilon > 0$ such that if $(x_0, 2t_0) \in \mathscr{G}_d(C_0, \frac{\epsilon}{2})$ for some C_0, then there is a line L passing through x_0 such that*

(1.39) *every point of $E \cap B(x_0, t_0)$ is at distance $\leq \tau_0 t_0$ from L, and*

every point of $L \cap B(x_0, t_0)$ is at distance $\leq \tau_0 t_0$ from E.

Notice that the value of ϵ does not depend on C_0 or on the regularity constant on E. This is not surprising, because the estimate (1.36) will give us all the information that we need on the $\mu(B(y, t))$. [We shall not need to know precisely the value of $\mu(B(y, t))$ when y is far from x_0 or t is large.] The proof of Lemma 1.38 will occupy the rest of this section. Let E and $(x_0, 2t_0) \in \mathscr{G}_d(C_0, \frac{\epsilon}{2})$ be given. We want to prove first that we have an estimate similar to (1.39) but with the line L replaced by a union of two half-lines emanating from x_0.

To simplify our notation, we shall assume (without loss of generality) that $x_0 = 0$ and $t_0 = 1$. Our assumption that $(x_0, 2t_0) \in \mathscr{G}_d(C_0, \frac{\epsilon}{2})$ for some C_0 then means that there is a regular measure μ such that $\operatorname{supp} \mu = E$ and

(1.40) $$|\mu(E \cap B(y, r)) - r| \leq \epsilon$$

for all $y \in E \cap B(0, 2)$ and all $0 < r \leq 2$.

It will be convenient for us to give a name to the annular shell around the sphere $C(r) = \{z \in \mathbf{R}^n : |z| = r\}$ that has width $2e$, so we set

(1.41) $\qquad A(r, e) = \{z \in \mathbf{R}^n : r - e \leq |z| < r + e\}.$

Since $A(r, e)$ is the difference of two balls centered on $0 \in E$, (1.40) immediately yields

(1.42) $\qquad\qquad |\mu(A(r, e)) - 2e| \leq 2\epsilon$

for all $0 < r < 1$ and all $0 < e < r$.

LEMMA 1.43. *For every $0 < r \leq 1$, we have* $\operatorname{dist}(C(r), E) \leq \epsilon$.

This is trivial when $r \leq \epsilon$, because $0 \in E$. Otherwise, note that for every $e \in (\epsilon, r)$ the annulus $A(r, e)$ meets E because of (1.42). The lemma follows.

Lemma 1.43 says that, for each r, $C(r)$ gets very close to E at least once. To continue with the proof, we shall distinguish between two cases, depending on whether it is possible or not to find a radius r_1 such that, when one runs along $C(r_1)$, one gets very close to E at least twice. (We shall be more precise in a moment.) To do so, we introduce two new parameters, τ and η, with $\tau_0 > \tau > \eta > \epsilon$. The constant τ, which will be considered as given to us for most of the argument, will be chosen near the end of the proof. A choice of η will then follow from that choice, and ϵ will be chosen even smaller than η.

Our first case is the one where there exists a radius r_1 and two points x_1 and x_2 on the sphere $C(r_1)$ with the properties

(1.44) $\tau \leq r_1 \leq 1$,
(1.45) $\operatorname{dist}(x_1, x_2) \geq \tau r_1$, and
(1.46) $\operatorname{dist}(x_i, E) \leq \eta$ for $i = 1, 2$.

We wish to show that, in this case, $E \cap B(0, 1)$ is very close to $L_1 \cup L_2$, where L_i is the half-line emanating from 0 and passing through x_i. The following lemma is a first step in this direction. [It gives us control on the part of E which lies on a small annulus around $C(r_1)$.]

LEMMA 1.47. *Let r_1, x_1, and x_2 be as above, and let δ be any number such that*

(1.48) $\qquad\qquad 100\eta \leq \delta \leq \dfrac{\tau^2}{3}.$

Then

(1.49) $\qquad E \cap A(r_1, \delta) \subset B(x_1, \delta + 10\eta) \cup B(x_2, \delta + 10\eta)$

(*assuming that ϵ is small enough*). [*See Figure* 1.15.]

We prove the lemma by contradiction. Suppose there exists a point $x_3 \in E \cap A(r_1, \delta)$ which is at distance $\geq \delta + 10\eta$ from both x_1 and x_2. Set $B_1 = B(x_1, \delta + 5\eta)$, $B_2 = B(x_2, \delta + 5\eta)$, and $B_3 = B(x_3, 5\eta)$. Note that

1.3. WEAK CONSTANT DENSITY ($d = 1$)

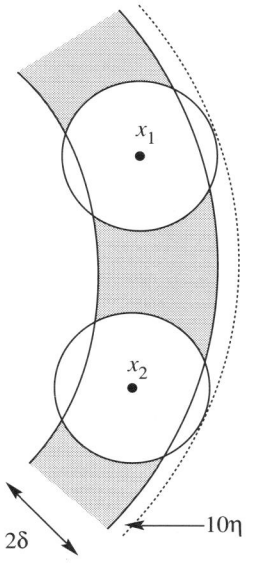

There is no point of E in the shaded areas

FIGURE 1.15

the B_i are all contained in $A(r_1, \delta + 5\eta)$, so $\mu(B_1 \cup B_2 \cup B_3) \leq 2\delta + 10\eta + 2\epsilon$ by (1.42). Next, the ball B_3 is disjoint from B_1 and B_2 by definition of x_3, and B_1 is disjoint from B_2 because $|x_1 - x_2| \geq \tau r_1 > \tau^2 \geq 3\delta > 2\delta + 10\eta$ [by (1.45), (1.44), and (1.48)]. Hence $\mu(B_1) + \mu(B_2) + \mu(B_3) \leq 2\delta + 10\eta + 2\epsilon$.

On the other hand, it follows from (1.46) that B_1 contains a ball centered on $E \cap B(0, 2)$ with radius $\delta + 4\eta$. Consequently, $\mu(B_1) \geq \delta + 4\eta - \epsilon$. Similarly, $\mu(B_2) \geq \delta + 4\eta - \epsilon$. Finally, since B_3 is already centered on $E \cap B(0, 2)$, $\mu(B_3) \geq 5\eta - \epsilon$, so altogether we have $\mu(B_1) + \mu(B_2) + \mu(B_3) \geq 2\delta + 13\eta - 3\epsilon$. We get the desired contradiction as soon as $\epsilon < \eta/2$.

LEMMA 1.50. *Let r_1, x_1, and x_2 be as above, and let $A = A(r_1, \tau^2/10)$. Then*

(1.51) *every point of $E \cap A$ is at distance $\leq 10\tau\sqrt{\eta}$ from $L_1 \cup L_2$, and*
(1.52) *every point of $(L_1 \cup L_2) \cap A$ is at distance $\leq 100\tau\sqrt{\eta}$ from E,*

provided that we take η small enough (depending on τ), and that ϵ is sufficiently small.

We first prove (1.51). Let $y \in E \cap A$ be given, and set $r = |y|$. We can apply Lemma 1.47 with $\delta = \max(|r - r_1|, 100\eta)$ (so that (1.48) is satisfied even if $r - r_1$ is very small). We get that $y \in B(x_i, |r - r_1| + 110\eta)$ for some $i \in \{1, 2\}$. Let us check that $\text{dist}(y, L_i) \leq 10\tau\sqrt{\eta}$ for this same i. If $|r - r_1| \leq 10^5 \eta$, say, we have $|y - x_i| \leq 2 \cdot 10^5 \eta$, which is much better than (1.51) (if η is small enough compared to τ^2). Otherwise, we can use the fact that $x_i \in C(r_1)$, $y \in C(r)$, and $|y - x_i| \leq |r - r_1| + 110\eta$ to get the estimate

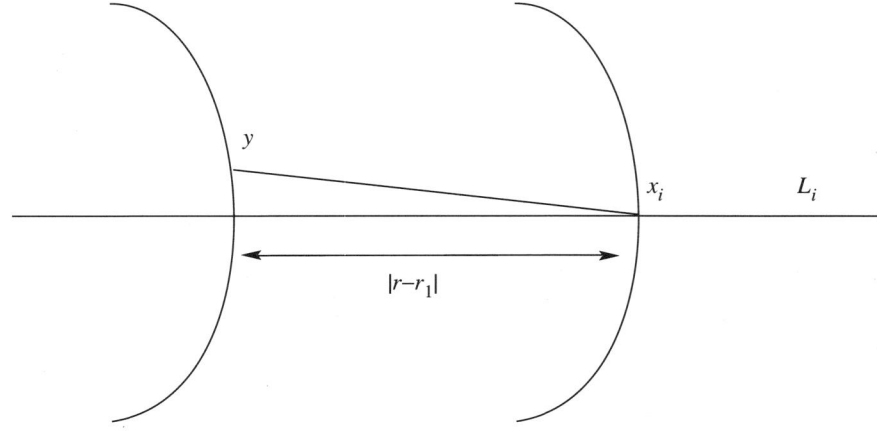

FIGURE 1.16

$$\operatorname{dist}(y, L_i) \leq 2\{(|r - r_1| + 110\eta)^2 - (r - r_1)^2\}^{1/2}$$
$$\leq 2\{220|r - r_1|\eta + (110\eta)^2\}^{1/2}$$
$$\leq 2\{22\tau^2\eta + (110\eta)^2\}^{1/2} \leq 10\tau\sqrt{\eta}$$

if η is small enough compared to τ^2. (See Figure 1.16 for the first inequality.)

This proves (1.51). To prove (1.52), we let r be given, $|r - r_1| \leq \tau^2/10$, and denote by $y_i(r)$ the intersection of L_i with $C(r)$. We have to check that both points $y_i = y_i(r)$ are within $100\tau\sqrt{\eta}$ of E. Let us proceed by contradiction and assume, for instance, that $\operatorname{dist}(y_2, E) > 100\tau\sqrt{\eta}$.

Set $A_1 = A(r, 90\tau\sqrt{\eta})$. Because of (1.51), $E \cap A_1$ stays at distance $\leq 10\tau\sqrt{\eta}$ from $L_1 \cup L_2$. [To be fair, we have to admit that A_1 is not necessarily contained in $A(r, \tau^2/10)$, but this does not cause any trouble, because $A_1 \subset A(r_1, \tau^2/9)$ if η is small enough, and our proof of (1.51) also works with $A = A(r_1, \tau^2/9)$.] Because $\operatorname{dist}(y_2, E) > 100\tau\sqrt{\eta}$, no point of $E \cap A_1$ can be at distance $\leq 10\tau\sqrt{\eta}$ from L_2, so all the points of $E \cap A_1$ are at distance $\leq 10\tau\sqrt{\eta}$ from L_1 alone.

Note that, by Lemma 1.43, there is a point $z \in E$ at distance $\leq \epsilon$ from $C(r)$. If ϵ is small enough, then z lies in $A_1 \cap E$, so $\operatorname{dist}(z, L_1) \leq 10\tau\sqrt{\eta}$. Consequently, $|y_1 - z| \leq 2\epsilon + 20\tau\sqrt{\eta}$, and $E \cap A_1 \subset B(y_1, 100\tau\sqrt{\eta}) \subset B(z, 121\tau\sqrt{\eta})$ if ϵ is small enough. This last ball is centered on $E \cap B(0, 2)$, so (1.40) yields $\mu(E \cap A_1) \leq 121\tau\sqrt{\eta} + \epsilon \leq 122\tau\sqrt{\eta}$. On the other hand, we can also apply (1.42) directly and get $\mu(E \cap A_1) \geq 180\tau\sqrt{\eta} - 2\epsilon$. This gives the desired contradiction, so $\operatorname{dist}(y_2, E) \leq 100\tau\sqrt{\eta}$. The same argument works with $y_1 = y_1(r)$, and Lemma 1.50 follows.

LEMMA 1.53. *Let r_1, x_1, x_2, L_1, and L_2 be as above, and take B' to be $B(0, 1) \setminus B(0, \tau)$. Then*

(1.54) *every point of $E \cap B'$ is at distance $\leq \theta(\tau, \eta)$ from $L_1 \cup L_2$,*
and every point of $(L_1 \cup L_2) \cap B'$ is at distance $\leq \theta(\tau, \eta)$
from E, where $\theta(\tau, \eta)$ is some constant that depends on
τ and η, and which tends to 0 as η tends to 0 for each (fixed)
value of τ.

The proof is a rather easy induction argument. Let

$$I_1 = \left[r_1 - \frac{\tau^2}{10}, r_1 + \frac{\tau^2}{10}\right] \cap [\tau, 1].$$

By (1.52), we know that the points $y_i(r) = C(r) \cap L_i$ are at distance $\leq 100\tau\sqrt{\eta}$ from E for every $r \in I_1$. Thus these points satisfy (1.46) with η replaced by $f(\eta) = 100\tau\sqrt{\eta}$. Note that $\operatorname{dist}(y_1(r), y_2(r)) \geq \tau r$ (the analogue of (1.45)), because (1.45) is invariant by dilation. Thus we can apply Lemma 1.50 with r_1 replaced by r and x_i replaced by $y_i(r)$. We get the estimates (1.51) and (1.52) with A replaced by the larger annulus $A_2 = A(r_1, 2\tau^2/10) \cap B'$ but with the constant $100\tau\sqrt{\eta} = f(\eta)$ replaced by $f \circ f(\eta)$.

An iteration of this argument gives the inequalities (1.51) and (1.52) for the annulus $A_m = A(r_1, m\tau^2/10) \cap B'$, but with $100\tau\sqrt{\eta}$ replaced by $f^{(m)}(\eta) = f \circ \cdots \circ f(\eta)$. All this works as long as we take η so small that the $f^j(\eta)$, $1 \leq j \leq m$, are all small enough. Taking a suitable value of m (such as $m = 1 + [10/\tau^2]$), we obtain Lemma 1.53, where $\theta(\tau, \eta)$ is the appropriate iterate of the function f.

We are now ready to prove the conclusion of Lemma 1.38 in our first case. [Notice that, if we wished merely to prove that the WCD implies the weak connectedness condition, Lemma 1.53 would be enough in this case.] Since we already know that E is well approximated by $L_1 \cup L_2$, it will be enough to prove that the angle between L_1 and $-L_2$ is as small as we want.

To do so, pick a point $x_3 \in E$ at a distance $\leq \theta(\tau, \eta)$ from the point on L_1 which is at a distance $\frac{1}{4}$ from 0. (See Figure 1.17, next page.) Our intention is to apply the argument that led to Lemma 1.53 but with the origin replaced by the point x_3. We first choose points x_1' and x_2' that will play the role of x_1 and x_2: we take the two intersections of L_1 with the sphere centered at x_3 and with radius $\frac{1}{5}$.

We can apply the same argument as before, with r_1 replaced by $\frac{1}{5}$, 0 replaced by x_3, x_1 and x_2 replaced by x_1' and x_2', and η replaced by $5\theta(\tau, \eta)$. We get that, if η is small enough, every point of $E \cap B(x_3, \frac{1}{2})$ is at distance $\leq \frac{\tau}{100}$ from $L_1' \cup L_2'$, where L_i' is the half-line emanating from x_3 and passing through x_i'. Comparing this with (1.54), we obtain Angle $(L_1, -L_2) \leq \frac{\tau}{2}$ (see Figure 1.18), so the conclusion of Lemma 1.38 is true in our first case. [We even have (1.39) with τ instead of τ_0.]

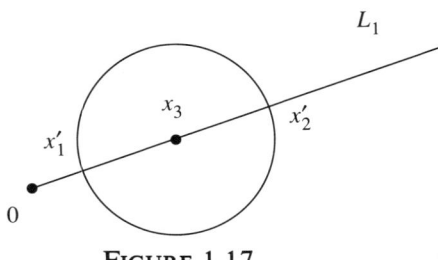

FIGURE 1.17

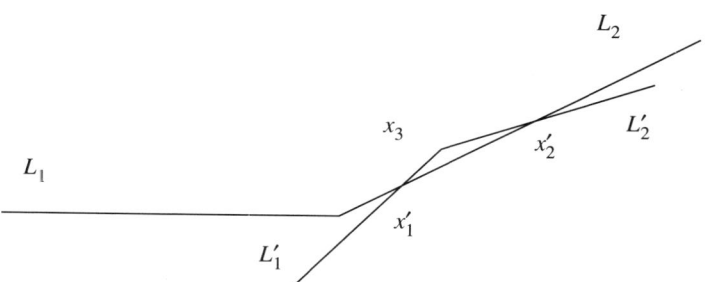

FIGURE 1.18. This case is impossible because $L_1 \cup L_2$ must stay close to $L'_1 \cup L'_2$ inside $B(0, \frac{1}{2})$.

The situation is now as follows. We have (implicitly) chosen, for each small τ, a value of η such that the conclusion of Lemma 1.38 holds whenever there exist a radius r_1 and a pair of points x_1 and x_2 which satisfy (1.44)–(1.46) (and ϵ is sufficiently small). We are now going to show that, if τ is chosen to be small enough, we can also prove the conclusion of Lemma 1.38 in the remaining case.

So let us suppose that r_1, x_1, and x_2 do not exist. We first construct a chain of points of E as follows. For $m = 0, 1, \ldots, \left[\frac{10}{\eta}\right]$, we let y_m be a point of E such that $\left| |y_m| - \frac{m\eta}{10} \right| \leq \epsilon$. Such a point exists because of Lemma 1.43. We claim that, if m is so large that $|y_m| \geq \tau$, then $|y_m - y_{m-1}| \leq 2\tau$. Indeed, if this were not the case, we could consider the radius $r_1 = |y_m|$ and take $x_1 = y_m$ and $x_2 = y_{m-1}|y_m|/|y_{m-1}|$, and a trivial computation shows that they would satisfy (1.44)–(1.46) if η is small enough (depending on τ).

At this stage, if we wanted only to prove that E is weakly connected, we would be able to conclude by remarking that every point of $E \cap B(0, 1)$ is at a distance $\leq 2\tau$ from one of the y_m's. [If not, there would be a $x_1 \in E \cap B(0, 1)$ such that $|x_1| \geq \tau$ and $\text{dist}(x_1, y_m) \geq 2\tau$ for all m, including the integer for which $\left| |y_m| - |x_1| \right|$ is minimal. This would be impossible for the same reason as before.] Since we want to obtain the conclusion of Lemma 1.38, we continue a little longer.

1.4. WEAK TWO POINTS ON SPHERES

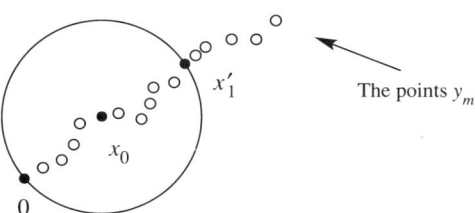

FIGURE 1.19

Let x_0 be one of the points y_m such that $||x_0| - \frac{1}{4}| \leq \frac{1}{100}$, say. We can use the estimates $|y_m - y_{m-1}| \leq 2\tau$ to find a later point y_m (i.e., a point y_m with $m \geq m_0$, where $x_0 = y_{m_0}$) such that $||y_m - x_0| - |x_0|| \leq \tau$. Let S denote the sphere centered at x_0 and with radius $|x_0|$, and let x_1' be the point of S obtained by projecting y_m radially on S. Then $\operatorname{dist}(x_1', E) \leq \tau$, and $\operatorname{dist}(x_1', 0) \geq \frac{1}{5}$ because $|y_m| \geq |x_0|$ (since y_m comes after x_0 in the sequence of y_j's). See Figure 1.19.

We are now in essentially the same position as in our first case. We have points x_0 (instead of the origin), x_1', and $x_2' = 0$ (instead of x_1 and x_2), and a radius $|x_0|$ (replacing r_1), with properties analogous to (1.44)–(1.46). This time, η has been replaced by τ, but this hardly matters because we can choose τ as small as we want now. Thus the conclusion of Lemma 1.38 can be obtained exactly as in our first case. Only one additional detail should be checked here: since we start the argument with a new origin, we have to make sure that we never need to use (1.40) for a ball $B(y, r)$ that is centered out of $B(0, 2)$ or with radius > 2. We claim that this is the case, but we shall not bother to check it here, because even if it were false, only minor modifications to the argument would be required. This concludes our proof of Lemma 1.38 and Proposition 1.37.

1.4. The weak "two points on spheres" condition.

We conclude this chapter with an amusing consequence of Theorem 1.3. Let E be a one-dimensional regular set in $\mathbf{R}^n$. We shall say that E satisfies the weak two points on spheres condition (in short, the WTPS) if there exist positive constants τ and η, with $0 < 6\eta < \tau < \frac{1}{20}$, such that the following set $\mathscr{B}(\tau, \eta) \subset E \times \mathbf{R}_+$ is a Carleson set.

A point $(x, t) \in E \times \mathbf{R}_+$ is *not* in (the bad set) $\mathscr{B}(\tau, \eta)$ if there exists a point x_0 in $E \cap B(x, t)$ such that, for all radii $\frac{t}{20} \leq r \leq 2t$, one can find two points $x_1(r)$ and $x_2(r)$ such that

(1.55) $|x_1(r) - x_0| = |x_2(r) - x_0| = r$,
(1.56) $|x_1(r) - x_2(r)| \geq \tau t$,
(1.57) $\operatorname{dist}(x_i(r), E) \leq \eta t$ for $i = 1, 2$, and
(1.58) every point of the sphere $\{y \in \mathbf{R}^n : |y - x_0| = r\}$ which is at distance $\leq 2\eta t$ from E is at distance $\leq \frac{\tau t}{3}$ from $x_1(r)$ or $x_2(r)$.

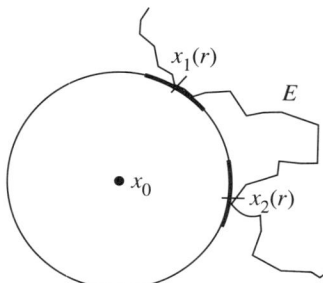

FIGURE 1.20. Both $x_i(r)$ are ηt-close to E, and all points of the sphere that are $2\eta t$-close to E are in one of the two marked disks around the $x_i(r)$.

(See Figure 1.20.) In fairly imprecise terms, we are demanding that the spheres centered at x_0 and with radius $r \in [\frac{t}{20}, 2t]$ essentially meet E at two separate places but no more.

Note that if E is uniformly rectifiable (and hence satisfies the bilateral weak geometric lemma) then it satisfies the WTPS. It is not clear that our formulation of the WTPS is the most natural one; we could probably dispense with the additional generality provided by using two constants τ and η (instead of just taking $\eta = \frac{\tau}{6}$, say) and by allowing the origin x_0 to be any point of $E \cap B(x, t)$ (rather than x itself). Of course, the constant $\frac{1}{20}$ is not sharp: it was chosen simply to permit a smooth application of Theorem 1.3.

One of the reasons for considering the WTPS condition is that it may possibly be useful for showing that the L^2-boundedness of the Cauchy integral operator on a 1-dimensional regular set implies that the set is uniformly rectifiable. It is also not completely unrelated to some conditions that arise in connection with harmonic measure estimates. (See [**BCGJ**], [**B**], and [**BJ**].)

In this section we shall verify that the WTPS implies that E is weakly connected and, hence, uniformly rectifiable.

To do this, we consider a point $(x, t) \in E \times \mathbf{R}_+$ such that $(x, t) \notin \mathscr{B}(\tau, \eta)$, and we try to show that $E \cap B(x, t)$ is well-enough connected. To simplify our notation we shall assume that $t = 1$ and that the point x_0 provided by the WTPS is simply the origin.

LEMMA 1.59. *If* $\frac{1}{20} < r < r' < r + \eta \leq 2$, *then the point* $x_1(r)$ *is at distance* $< \frac{\tau}{2}$ *from one of the points* $x_1(r')$, $x_2(r')$, *and the point* $x_2(r)$ *is at distance* $< \frac{\tau}{2}$ *from the other one.*

Indeed, let y_1 be the projection of $x_1(r)$ on the sphere $C(r') = \{z \in \mathbf{R}^n : |z| = r'\}$. By (1.57), y_1 is at distance $< 2\eta$ from E, and hence y_1 is within $\frac{\tau}{3}$ of one of the $x_i(r')$ because of (1.58). Suppose, for instance, that $|y_1 - x_1(r')| < \frac{\tau}{3}$. Since we have $\eta \leq \frac{\tau}{6}$, we get $|x_1(r) - x_1(r')| < \frac{\tau}{2}$. Using the same argument with $x_2(r)$, we obtain $|x_2(r) - x_i(r')| < \frac{\tau}{2}$ for some i.

1.4. WEAK TWO POINTS ON SPHERES

This i cannot be equal to 1 because of (1.56) applied to r, so $i = 2$. The lemma follows.

Let us now consider the radii $r_m = \frac{1}{20} + m\frac{\eta}{2}$, where m runs through the set of nonnegative integers such that $r_m \leq 2$. We may as well assume that $|x_i(r_m) - x_i(r_{m+1})| < \frac{\tau}{2}$ for $i = 1, 2$ and for $m = 0, 1, \ldots$, by exchanging the names of the two points $x_i(r_m)$ whenever necessary. For each of the $x_i(r_m)$'s, select a point $y_i(r_m) \in E$ such that $|y_i(r_m) - x_i(r_m)| \leq \eta$. Let z_k, $k \in K$, be the sequence obtained by taking all the points $y_1(r_m)$ in reverse order, then 0, and then all the points $y_2(r_m)$ in the initial order. The z_k's generate a chain of points of $E \cap B(0, 2 + \eta)$ that satisfy $|z_{k+1} - z_k| \leq \frac{1}{15}$ for all k. Moreover, we claim that every point of $E \cap B(0, 2)$ is at distance $< \frac{1}{20}$ from one of the z_k's. This is obvious for any point $y \in E \cap B(0, \frac{1}{20})$; otherwise, let m be such that $||y| - r_m| < \eta$, and let $\tilde{y}$ be the projection of y on the sphere $\{|z| = r_m\}$. Since $|y - \tilde{y}| < \eta$, we have that $\operatorname{dist}(\tilde{y}, x_i(r_m)) < \frac{\tau}{3}$ for $i = 1$ or 2 (by (1.58)), and hence y is at distance $\leq \eta + \frac{\tau}{3} + \eta$ from $y_i(r_m)$. So every point of $E \cap B(0, 2)$ is at distance $\leq \frac{1}{20}$ from one of the z_k's.

Consequently, every pair of points of $E \cap B(0, 2)$ (and hence of $E \cap B(x, 1)$) can be joined by a link composed of points of $E \cap B(0, 2 + \eta)$ that are at consecutive distances $\leq \frac{1}{15}$. In particular, our original point belongs to the good set $\mathscr{G}(\frac{3}{4}, 3)$ of Definition 1.1. This means that the complement of the bad set $\mathscr{B}(\tau, \eta)$ for the WTPS is contained in $\mathscr{G}(\frac{3}{4}, 3)$.

Hence, the WTPS implies the weak connectedness condition (and, by Theorem 1.3, uniform rectifiability).

CHAPTER 2

The Bilateral Weak Geometric Lemma and its Variants

2.1. Introduction; the corona method.
Recall from Definition 2.2 of Part I that a regular set E satisfies the bilateral weak geometric lemma (BWGL) if, for each $\epsilon > 0$, the complement in $E \times \mathbf{R}_+$ of the set

(2.1)
$$\mathscr{G}(\epsilon) = \{(x, t) \in E \times \mathbf{R}^+ : \text{ there exists a } d\text{-plane } P \text{ such that}$$
$$\operatorname{dist}(y, P) \leq \epsilon t \text{ for all } y \in E \cap B(x, t),$$
$$\text{and } \operatorname{dist}(p, E) \leq \epsilon t \text{ for all } p \in P \cap B(x, t)\}$$

is a Carleson set. (See Definition I.1.69 for the definition of a Carleson set.) The main goal of this chapter is to prove the following proposition, which is the "hard" part of Theorem 2.4 in Part I.

PROPOSITION 2.2. *If a regular set E satisfies the bilateral weak geometric lemma, then it is uniformly rectifiable.*

Before we discuss the proof of this result, let us mention some conditions which imply the BWGL (and hence uniform rectifiability). As we pointed out in Part I, the local symmetry condition LS (see Definition I.1.79) implies the BWGL. (See Proposition I.2.5.) A similar condition that implies the BWGL is the local convexity condition (LCV) of Definition I.2.7. The fact that the LCV implies the BWGL will be proved in §4. We shall even give in §5 an apparently weaker condition (the WLCV) that still implies the BWGL. Note that the BWGL easily implies all these conditions, so they are all equivalent to uniform rectifiability.

Proposition 2.2 will be proved using Theorem 3.42 of Part I. More precisely, given a regular set E which satisfies the BWGL, we shall prove that E has a "generalized corona decomposition" in terms of sets which contain big pieces of Lipschitz graphs (uniformly). This means that we shall find a coronization $(\mathscr{B}, \mathscr{G}, \mathscr{F})$ of E (see Definition I.3.13) and, for each "stopping time region" $S \in \mathscr{F}$, a set $E(S)$ that satisfies the BPLG with uniform estimates, and also

(2.3) $\operatorname{dist}(x, E(S)) \leq 4 \operatorname{diam} Q$ whenever $x \in 2Q$ for some cube $Q \in S$.

(Compare with Definition I.3.38.) The uniform rectifiability of E will then follow from Theorem I.3.42.

Let us say a few words now about the organization of the proof. Let E be a regular set that satisfies the BWGL. Let $\epsilon > 0$ be small and $k > 1$ be large, to be chosen later, and let $\mathscr{G}(\epsilon)$ be as in (2.1). We shall first use Lemma I.3.22 to find a coronization $(\mathscr{B}, \mathscr{G}, \mathscr{F})$ of E such that, for each cube $Q \in \mathscr{G}$,

(2.4) there is a d-plane $P(Q)$ such that every point of kQ is at distance
$\leq \epsilon \operatorname{diam} Q$ from $P(Q)$, and every point of $P(Q)$ at distance
$\leq k \operatorname{diam} Q$ from Q is at distance $\leq \epsilon \operatorname{diam} Q$ from E.

(See (I.3.7) for the definition of λQ.) We shall then construct, for each stopping time region $S \in \mathscr{F}$, a regular set $E(S)$ that satisfies (2.3). This part of the argument does not use our hypothesis that E satisfies the BWGL, and the same construction will be used in Chapter 3 in somewhat different circumstances.

To prove that each $E(S)$ has BPLG, our strategy will be to check that it satisfies the WGL and to prove that it has big projections. (See Definitions I.1.71 and I.1.74.) The fact that the $E(S)$'s have BPLG will then follow from Theorem 1.14 of [DS3], i.e., Theorem I.1.76. The construction of the $E(S)$'s and the verification that they inherit the WGL from E will be performed in this section. The fact that the $E(S)$'s have big projections will be proved in §2 for the codimension 1 case and in §3 for higher codimensions.

Let us conclude these preliminary remarks with some comments on the proof. The general technique of proving that a regular set E is uniformly rectifiable by showing that it has a generalized corona decomposition in terms of some class of uniformly rectifiable sets (here, sets that have BPLG) seems to be quite powerful and will be used again in Chapters 3 and 4. Let us try to explain why this technique works so well. The point is that the sets $E(S)$ are much easier to study that our initial set E. In the proof of Proposition 2.2, for instance, it will be easy to show that each $E(S)$ inherits the BWGL from E, but it is actually much better than that. Indeed, all the $(x, t) \in E(S) \times \mathbf{R}_+$, except for a few places near the top cube and the minimal cubes of S, lie in the analogue of $\mathscr{G}(\epsilon)$ for $E(S)$. In practical terms, Theorem I.3.42 comes very close to permitting us to reduce to the case where the bad set $E \times \mathbf{R}_+ \setminus \mathscr{G}(\epsilon)$ is empty.

REMARK 2.5. The proof of Proposition 2.2 will give us a little more than what is announced in its statement. We shall prove that there is a small constant $\epsilon > 0$, that depends only on the dimensions n and d and on the regularity constant for E, such that if the complement in $E \times \mathbf{R}_+$ of $\mathscr{G}(\epsilon)$ is a Carleson set, then E is uniformly rectifiable. Also, precise uniform rectifiability estimates (such as the Carleson norm in the geometric lemma (I.1.59) or the constants θ and C_1 in the BPBI condition of Definition I.1.33) can be computed in terms of this ϵ, the regularity constant for E, the Carleson

2.1. THE CORONA METHOD

constant for the complement of $\mathscr{G}(\epsilon)$, and the dimensions n and d.

Let us now begin the proof of Proposition 2.2 in earnest. Let E be a d-dimensional regular set in $\mathbf{R}^n$ that satisfies the BWGL. Let us choose, once and for all, a set Δ of cubes as in the beginning of §I.3.1. Our first task is to choose a coronization of E.

Let $\epsilon > 0$ be small and $k > 1$ be large, to be chosen later. Let $\mathscr{A}$ be the set of cubes Q that do not satisfy the condition (2.4). The set $\mathscr{A}$ satisfies a Carleson packing condition (as in Definition I.3.9) because E satisfies the BWGL. (We omit the very simple proof, but the reader can go back to Lemma I.3.10 for a similar argument.) By Lemma I.3.22, there is a coronization $(\mathscr{B}, \mathscr{G}, \mathscr{F})$ of E such that $\mathscr{A} \subset \mathscr{B}$, i.e., such that each cube $Q \in \mathscr{G}$ satisfies (2.4).

The next step in our proof of Proposition 2.2 is the construction, for each stopping-time region $S \in \mathscr{F}$, of a regular set $E(S)$. This part of the argument does not use our hypothesis that E satisfies the BWGL, or our special choice of a coronization, and it will be used again, with different hypotheses on the regular set E, in the next chapter. Because of this we are going to suspend temporarily the assumptions that we have made so far and do this part of the argument in greater generality.

To each regular set E (equipped with a set Δ of cubes as in §I.3.1) and to each coherent region $S \subset \Delta$ (i.e., to each subset of Δ which satisfies (I.3.17)), we shall now associate a regular set $E(S)$.

Fix E and a coherent region $S \subset \Delta$, and, for each $x \in \mathbf{R}^n$, set

$$(2.6) \qquad d(x) = \inf_{Q \in S}\{\operatorname{dist}(x, Q) + \operatorname{diam} Q\}.$$

This function will be used to measure how well $E(S)$ approximates E. Notice that d is Lipschitz with norm ≤ 1. Next let

$$(2.7) \qquad Z = \{x \in E : d(x) = 0\}.$$

Our set $E(S)$ will be (essentially) the union of Z and certain star-shaped surfaces $\Sigma(Q)$ associated to the minimal cubes of S.

We first define a "unit star" Σ_0; it is the intersection of the closed unit ball of $\mathbf{R}^n$ with a finite union of d-planes through the origin. We choose the finite collection of d-planes so that, for every d-plane $P \subset \mathbf{R}^n$,

$$(2.8) \qquad \Pi(\Sigma_0) \supset P \cap B(\Pi(0), \tfrac{1}{2}),$$

where Π denotes the orthogonal projection onto P.

Next, for each minimal cube Q of S, we choose a center $b_Q \in Q$, i.e., a point of Q such that

$$(2.9) \qquad \operatorname{dist}(b_Q, E \setminus Q) \geq C_1^{-1} \operatorname{diam} Q,$$

where the constant C_1 depends only on n, d, and the regularity constant for E. Such a point exists because our cubes have "small boundaries". (See Lemma 3.5 in Part I.) We now let $\Sigma(Q)$ be the image of the unit star Σ_0

by the composition of a translation and a dilation, where the translation and dilation are chosen so that $\Sigma(Q)$ is centered at b_Q and has a "radius" equal to $\frac{1}{3} \operatorname{Min}\{\operatorname{diam} Q, \operatorname{dist}(b_Q, E \setminus Q)\}$.

Notice that, by our choice of $\Sigma(Q)$, we have that

$$(2.10) \qquad \operatorname{dist}(\Sigma(Q), \Sigma(Q')) \geq C^{-1} \operatorname{Max}(\operatorname{diam} Q, \operatorname{diam} Q')$$

for all choices of distinct minimal cubes $Q, Q' \in S$, and

$$(2.11) \qquad \operatorname{dist}(\Sigma(Q), Z) \geq C^{-1} \operatorname{diam} Q$$

for all minimal cubes Q of S. We set

$$(2.12) \qquad E(S) = Z \cup \left\{ \bigcup_{Q \in m(S)} \Sigma(Q) \right\} \cup P_0,$$

where $m(S)$ denotes the set of minimal cubes of S and P_0 is any d-plane such that, say,

$$(2.13) \qquad \operatorname{dist}(P_0, Q(S)) = \operatorname{diam} Q(S).$$

(Recall that $Q(S)$ is the maximal cube of S.) We added the d-plane P_0 to ensure that $E(S)$ will satisfy the lower estimate for the regularity condition even for large radii. Except for that, P_0 will not play any significant role.

LEMMA 2.14. *Let E be a regular set, and let S be a coherent region of Δ (i.e., a subset of Δ that satisfies the condition (I.3.17)). Let $E(S)$ be as above. Then $E(S)$ is regular, and we have*

$$(2.15) \qquad \operatorname{dist}(x, E(S)) \leq 2d(x)$$

for all $x \in \mathbf{R}^n$ (where d is as in (2.6)). Moreover, the regularity constant for $E(S)$ depends only on n, d, and the regularity constant for E.

We start the proof of Lemma 2.14 by checking that $E(S)$ is closed. Let $\{x_i\}$ be a sequence of points of $E(S)$, and suppose that $x = \lim x_i$ exists. We want to prove that $x \in E(S)$. If infinitely many x_i lie in Z, then $x \in Z$ because Z is closed (remember that d is continuous). The case when infinitely many x_i lie in P_0, or a single $\Sigma(Q)$, is treated the same way. Thus we can reduce to the case where each x_i belongs to some $\Sigma(Q_i)$ and $\lim \operatorname{diam} Q_i = 0$. Because $d(x_i) \leq \operatorname{dist}(x_i, Q_i) + \operatorname{diam} Q_i \leq 2 \operatorname{diam} Q_i$, which tends to 0, we have that $d(x) = 0$, and hence $x \in Z \subset E(S)$. This proves that $E(S)$ is closed.

Next we prove that

$$(2.16) \qquad |E(S) \cap B(x, R)| \leq CR^d \quad \text{for all } x \in E(S) \text{ and } R > 0.$$

Let $x \in E(S)$ and $R > 0$ be given. The left-hand side of (2.16) is less than

$$|P_0 \cap B(x, R)| + |Z \cap B(x, R)| + \sum_{\substack{Q \in m(S) \\ \operatorname{diam} Q \leq R}} |\Sigma(Q) \cap B(x, R)|$$

$$+ \sum_{\substack{Q \in m(S) \\ \operatorname{diam} Q > R}} |\Sigma(Q) \cap B(x, R)|.$$

The first term is trivially less than CR^d. The second term is no greater than $|E \cap B(x, R)| \leq CR^d$, because E is regular. For the third term, notice that all the cubes Q that give a nonzero contribution to the sum are contained in $B(x, 3R)$. Consequently, the third term is less than

$$C \sum_{\substack{Q \in m(S) \\ Q \subset B(x, 3R)}} |Q|,$$

and this is less than CR^d, because the minimal cubes of S are disjoint and E is regular. For the last term, observe that we cannot have two different cubes $Q \in m(S)$ with diameters $\geq 10CR$ (with C as in (2.10)) such that $\Sigma(Q)$ intersects $B(x, R)$, because of (2.10). This implies that the fourth term is a sum of less than C' terms, each of which is less than $C''R^d$, because each $\Sigma(Q)$ satisfies the upper estimate for regularity, with estimates that do not depend on Q. This proves (2.16).

Let us now prove that

(2.17) $|E(S) \cap B(x, R)| \geq C^{-1}R^d$ for all $x \in E(S)$ and $R > 0$.

If $x \in P_0$, then (2.17) holds because $|P_0 \cap B(x, R)|$ is already large enough. This is also the case when $R > 3 \operatorname{diam} Q(S)$, so we may assume that $x \notin P_0$ and $R \leq 3 \operatorname{diam} Q(S)$.

We start with the case where $x \in Z$. Since $d(x) = 0$ and $R \leq 3 \operatorname{diam} Q(S)$, we can find a cube $Q_1 \in S$ such that $\operatorname{dist}(x, Q_1) \leq \frac{R}{100}$ and $C^{-1}R \leq \operatorname{diam} Q_1 \leq \frac{R}{100}$. Then

$$|E(S) \cap B(x, R)| \geq |Z \cap B(x, R)| + \sum_{\substack{Q \in m(S) \\ Q \subset Q_1}} |\Sigma(Q) \cap B(x, R)|$$

(the $\Sigma(Q)$ are disjoint from each other and from Z because of (2.10) and (2.11)), and this is

$$\geq |Z \cap B(x, R)| + C^{-1} \sum_{\substack{Q \in m(S) \\ Q \subset Q_1}} |Q| \geq C^{-1}|Q_1| \geq C^{-1}R^d,$$

as desired.

Now suppose that $x \in \Sigma(Q)$ for some $Q \in m(S)$. If $\operatorname{diam} Q \leq \frac{R}{100}$, then we can find an ancestor Q_1 of Q that lies in S and satisfies $C^{-1}R \leq \operatorname{diam} Q_1 \leq \frac{R}{100}$, and we can apply the same argument as above. Otherwise,

$R < 100 \operatorname{diam} Q$ and we have $|E(S) \cap B(x, R)| \geq |\Sigma(Q) \cap B(x, R)| \geq C^{-1} R$ because $\Sigma(Q)$ satisfies the lower estimate in the regularity condition for all scales smaller than its diameter. This completes the proof of (2.17).

Altogether, we have proved that $E(S)$ is a regular set, with estimates that depend only on n, d, and the regularity constants for E. Let us now prove (2.15).

Let $x \in \mathbf{R}^n$ be given. When $d(x) = 0$, (2.15) is trivial, because $x \in Z \subset E(S)$. Otherwise, let $Q \in S$ be such that $\operatorname{dist}(x, Q) + \operatorname{diam} Q \leq \frac{3}{2} d(x)$, and then choose a point $y \in Q$ such that $\operatorname{dist}(x, y) + \operatorname{diam} Q \leq 2 d(x)$. If $y \in Z$, then $\operatorname{dist}(x, E(S)) \leq \operatorname{dist}(x, y) \leq 2 d(x)$, and (2.15) holds. Otherwise, y belongs to some minimal cube Q_1 (see (∗) after Definition I.3.13) and, by construction of $\Sigma(Q_1)$, we have that $\operatorname{dist}(y, \Sigma(Q_1)) \leq \operatorname{diam} Q_1 \leq \operatorname{diam} Q$. Therefore, $\operatorname{dist}(x, E(S)) \leq \operatorname{dist}(x, y) + \operatorname{dist}(y, \Sigma(Q_1)) \leq \operatorname{dist}(x, y) + \operatorname{diam} Q \leq 2 d(x)$. This completes our proof of Lemma 2.14.

Let us now come back to the proof of Proposition 2.2. Let E be, as before, our regular set which satisfies the BWGL. We apply Lemma 2.14 to each of the stopping time regions S, $S \in \mathcal{F}$, where $(\mathcal{B}, \mathcal{G}, \mathcal{F})$ is the coronization of E that was chosen earlier. This gives us a collection of regular sets $E(S)$, $S \in \mathcal{F}$. Notice that (2.15) implies in particular that $E(S)$ satisfies the estimate (2.3) above. (This is because $d(x) \leq 2 \operatorname{diam} Q$ whenever $x \in 2Q$ for some $Q \in S$.) Consequently, in order to deduce Proposition 2.2 from Theorem 3.42 in Part I, it is enough to show that, if $\epsilon > 0$ and $k > 1$ are chosen correctly (depending on n, d, and the regularity constant for E), we have that the $E(S)$'s satisfy the BPLG condition of Definition I.1.26, with constants C_1 and θ that do not depend on S.

As we mentioned earlier, our plan for proving that the $E(S)$'s have BPLG is to show that they satisfy the weak geometric lemma and then that they have big projections. This last will be proved in §§2 and 3, but, for the moment, we shall derive the WGL for $E(S)$ from the corresponding property for E.

LEMMA 2.18. *Each $E(S)$ satisfies the weak geometric lemma (with constants that do not depend on S).*

To prove the lemma, we fix a stopping time region S, and set

$$\tilde{\beta}(x, t) = \inf_P \left\{ \sup_{y \in E(S) \cap B(x, t)} t^{-1} \operatorname{dist}(y, P) \right\}$$

for all $x \in E(S)$ and $t > 0$, where the infimum is taken, as usual, over all d-planes. We have to show that, for each $\tau > 0$, there is a constant $C(\tau) > 0$ such that the set

(2.19) $\qquad \{(x, t) \in E(S) \times \mathbf{R}_+ : \tilde{\beta}(x, t) > \tau\}$

is a Carleson set with constant $\leq C(\tau)$. (See Definition I.1.71 and (I.1.46) for the definition of the WGL.)

So let $\tau > 0$ be given. Notice that if $t \geq C\tau^{-1} \operatorname{diam} Q(S)$ (for a sufficiently large C), then $\tilde{\beta}(x, t) \leq \tau$ for all $x \in E(S)$; indeed, for these large values of t, $E(S)$ looks like a minor perturbation of P_0. The pairs (x, t) for which $C^{-1}\tau \operatorname{diam} Q(S) \leq t < C\tau^{-1} \operatorname{diam} Q(S)$ form a Carleson set, so we do not need to look at them.

Next let $A_0 = A_0(\tau)$ be a large constant, to be chosen soon, and let $\mathscr{B}_0$ be the set of cubes Q such that

$$\beta_0(Q) = \inf_P \{\operatorname{diam} Q^{-1} \sup_{y \in A_0 Q} \operatorname{dist}(y, P)\}$$

is larger than A_0^{-1}. Then $\mathscr{B}_0$ satisfies a Carleson packing condition because E satisfies the WGL. (See Lemma 3.10 in Part I.) Also, $m(S)$ (the set of minimal cubes of S) satisfies a Carleson packing condition, because the minimal cubes are disjoint. Consequently, the set

$$\mathscr{B} = \{(x, t) \in E(S) \times \mathbf{R}_+ : \text{ there exists a cube } Q \in \mathscr{B}_0 \cup m(S) \text{ with }$$
$$A_0^{-1} t \leq \operatorname{diam} Q \leq A_0 t \text{ and } \operatorname{dist}(x, Q) \leq A_0 t\}$$

is a Carleson set.

Altogether, we find that we need only consider those $(x, t) \in E(S) \times \mathbf{R}_+$ such that $0 < t < C^{-1}\tau \operatorname{diam} Q(S)$ and $(x, t) \notin \mathscr{B}$. We may also assume that $B(x, t)$ meets $E(S) \setminus P_0$, because otherwise $\tilde{\beta}(x, t) = 0$. Since $t < C^{-1}\tau \operatorname{diam} Q(S)$, this implies that $B(x, t)$ does not meet P_0 (if C is large enough; see (2.13)).

First suppose that $x \in Z$. Let Q be a minimal cube of S such that $\Sigma(Q)$ intersects $B(x, t)$ (if such a cube exists). Because $(x, t) \notin \mathscr{B}$, we must have $\operatorname{diam} Q > A_0 t$ or $\operatorname{diam} Q < A_0^{-1} t$. The first case is impossible because of (2.11) (if we take A_0 large enough), so we have $\operatorname{diam} Q < A_0^{-1} t$, which implies that $\Sigma(Q)$ stays within $CA_0^{-1} t$ of $E \cap B(x, 2t)$. This is true for every minimal cube Q such that $\Sigma(Q)$ meets $B(x, t)$, and since

$$E(S) \cap B(x, t) = \{Z \cap B(x, t)\} \cup \left\{\bigcup (\Sigma(Q) \cap B(x, t))\right\},$$

we have that every point of $E(S) \cap B(x, t)$ is at a distance $\leq CA_0^{-1} t$ from a point of $E \cap B(x, 2t)$. Using again the fact that $(x, t) \notin \mathscr{B}$, we see that every point of $E \cap B(x, 2t)$ is at distance $\leq CA_0^{-1} t$ from some d-plane P. This implies that $\tilde{\beta}(x, t) < \tau$ (if we take A_0 large enough).

We are thus left with the case where $x \in \Sigma(Q)$ for some minimal cube Q. The case when $t \geq A_0 \operatorname{diam} Q$ can be treated exactly like the case when $x \in Z$, so we may assume that $t \leq A_0^{-1} \operatorname{diam} Q$ (because $(x, t) \notin \mathscr{B}$). Notice that, in this case, $B(x, t) \cap E(S) = B(x, t) \cap \Sigma(Q)$ because of (2.10) and (2.11). Thus $\tilde{\beta}(x, t) > \tau$ if and only if $\beta_{\Sigma(Q)}(x, t) > \tau$, where $\beta_{\Sigma(Q)}$ is the analogue of $\tilde{\beta}$ but defined with respect to the set $\Sigma(Q)$. The desired Carleson measure estimate on our set (2.19) now follows from the corresponding (uniform) estimate on the $\beta_{\Sigma(Q)}$'s (which comes from the fact that the $\Sigma(Q)$'s satisfy the weak geometric lemma uniformly) and the fact that the $\Sigma(Q)$'s are disjoint. This concludes the proof of Lemma 2.18.

We announced earlier (see Remark 2.5) that our proof of Proposition 2.2 would only require that the complement of the set $\mathscr{G}(\epsilon)$ defined by (2.1) be a Carleson set for a single value of ϵ (which can be computed in terms of n, d, and the regularity constant for E). Our proof of Lemma 2.18 uses the fact that all the sets $\{(x, t) \in E \times \mathbf{R}_+ : \beta(x, t) > \eta\}$, $\eta > 0$, are Carleson sets. However, in order to prove that the $E(S)$'s have BPLG with uniform estimates, we shall not need the full WGL but only the information that the sets (2.19) are Carleson sets (uniformly in S) for some precise value of τ which can be computed in terms of n, d, and the regularity constant for E. Let us be more precise. The following lemma will be proved in the next section in the codimension 1 case and in §2.3 in the general case.

LEMMA 2.20. *Let E be a regular set and $S \subset \Delta$ be a coherent region (i.e., a subset of Δ that satisfies (I.3.17)). Suppose that every cube Q of S satisfies the condition (2.4). Then, if ϵ is small enough and k large enough (depending on n, d, and the regularity constant for E), the approximating regular set $E(S)$ defined by (2.12) has big projections. More precisely, $E(S)$ satisfies the condition of Definition I.1.74 with a value of θ that depends only on n, d, and the regularity constant for E.*

Lemma 2.20 is the only missing piece in our proof of Proposition 2.2. Given a regular set E that satisfies the BWGL, we first choose ϵ and k according to Lemma 2.20, then let $(\mathscr{B}, \mathscr{G}, \mathscr{F})$ be the coronization of E defined above, and then construct the regular set $E(S)$ for each region $S \in \mathscr{F}$. The set E has a generalized corona decomposition in terms of the $E(S)$'s; the $E(S)$'s are regular with uniform estimates by Lemma 2.14, have big projections (uniformly) by Lemma 2.20, and satisfy the WGL (uniformly) by Lemma 2.18. Proposition 2.2 then follows from Theorems I.1.76 and I.3.42. Because the regularity constant for $E(S)$ and the size of its projections are estimated in terms of n, d, and the regularity constant for E only, the proof of Theorem I.1.76 (see [**DS3**]) only requires a Carleson measure estimate on the set (2.19) for a single value of τ rather than the full WGL on $E(S)$. Remark 2.5 follows from this.

2.2. Big projections in codimension 1.

We now turn to the proof of Lemma 2.20. Altogether, we shall give three different proofs of this lemma in this monograph. The first one, which is the object of this section, will only work in codimension 1, but it is the simplest. The second one works in all cases and will be given in the next section. We shall also give a third argument in Chapter 3, which is a little less natural but has the advantage of working under even more general hypotheses (namely, the WHIP and the WTP). Indeed, we shall see in the beginning of §3.2 that the BWGL easily implies the WHIP and the WTP, and Lemma 2.20 is essentially the same as Lemma 3.58 in that context. (See also Remark 3.79.) We still give special proofs of Lemma 2.20 in the case of the BWGL because we think they have some geometrical interest of their own. Also, there are other natural ways to generalize the BWGL—for instance, the conditions studied in Chapter 4—for which the arguments we shall give in this chapter can be more suitable.

The reason that the codimension 1 case is special is that it will be possible to define, near $Q(S)$, two regions in the complement of $E(S)$ that (morally) lie on different sides of $E(S)$. The fact that E satisfies the BWGL (or, more precisely, that each cube of S satisfies (2.4)) will be used to show that $E(S)$ cannot have holes at scales larger than the size of its minimal cubes, and this fact will be used to prove that any "vertical" line segment that joins one of the sides of the complement of $E(S)$ to the other must come very close to $E(S)$.

Let E, Δ, and S be given, as in Lemma 2.20. Our first step will be to define compatible orientations on all the d-planes $P(Q)$, $Q \in S$, given by (2.4). For this step, we do not need our codimension 1 hypothesis yet.

Let A be a large positive constant, to be chosen later. We say that two cubes Q_1, $Q_2 \in \Delta$ are neighbors if

(2.21) $\operatorname{dist}(Q_1, Q_2) \leq A(\operatorname{diam} Q_1 + \operatorname{diam} Q_2)$ and $|j_1 - j_2| \leq A$,

where j_i is the integer such that $Q_i \in \Delta_{j_i}$, $i = 1, 2$.

[We choose this slightly awkward definition of neighbors to make sure that if Q_1 and Q_2 are neighbors, then their respective parents are neighbors too.]

Note that if Q_1 and Q_2 are two cubes of S which are neighbors, and if k is large enough, then $\operatorname{Angle}(P(Q_1), P(Q_2)) \leq C\epsilon$, where C is some constant that may depend on A (but not on ϵ). This is proved in [**DS2**] (see Lemma 5.13 there); the idea is that the d-planes $P(Q_1)$ and $P(Q_2)$ are essentially determined by the fact that they are both very close to $d+1$ points in $kQ_1 \cap kQ_2$ which can be chosen to be affinely independent (with precise estimates). Thus, if ϵ is small enough, it makes sense to talk about having compatible orientations on $P(Q_1)$ and $P(Q_2)$. When $d = n - 1$, giving an orientation on $P(Q_i)$ is the same as choosing a unit normal $N(Q_i)$ to $P(Q_i)$, and then the orientations are compatible if and only if $|N(Q_1) - N(Q_2)| \leq \frac{1}{10}$, say.

LEMMA 2.22. *We can choose orientations on all the $P(Q)$'s, $Q \in S$, in such a way that the orientations on $P(Q_1)$ and $P(Q_2)$ are compatible whenever Q_1 and Q_2 are two cubes of S which are neighbors.*

To choose the orientations, we start with the top cube $Q(S)$, choose any orientation on $P(Q(S))$, and then determine the orientation on all the other d-planes $P(Q)$, $Q \in S$, by demanding that the orientation on $P(Q)$ be compatible with the orientation on $P(\widetilde{Q})$, where $\widetilde{Q}$ is the parent of Q. To prove that these orientations satisfy the requirements of the lemma, let a neighborly pair of cubes Q_1, $Q_2 \in S$ be given. If either Q_1 or Q_2 is the

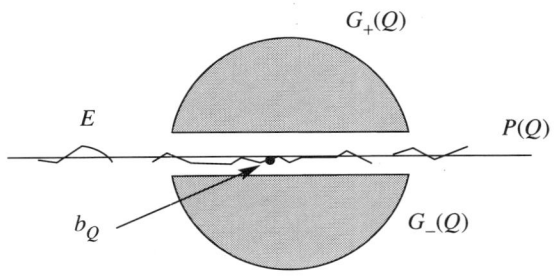

FIGURE 2.1

top cube $Q(S)$, then the orientations on $P(Q_1)$ and $P(Q_2)$ are compatible (if ϵ^{-1} and k are large enough) because the d-plane $P(Q)$ remains very close to $P(Q_1)$ when Q changes "continuously" from Q_1 to Q_2, generation by generation. If neither of the cubes Q_1 and Q_2 is the top cube $Q(S)$, then we observe that the parents $\tilde{Q}_1$ and $\tilde{Q}_2$ are still neighbors and that the orientations on $P(Q_1)$ and $P(Q_2)$ are compatible if and only if the orientations on $P(\tilde{Q}_1)$ and $P(\tilde{Q}_2)$ are compatible. This allows us to reduce to the previous case, which proves the lemma.

So far everything we have said works in any codimension, but now we assume that $d = n - 1$. For $Q \in S$, let $N(Q)$ denote the unit normal to $P(Q)$ that determines the orientation on $P(Q)$ chosen as in Lemma 2.22. (We assume that an orientation on $\mathbf{R}^n$ has been chosen once and for all.) Note that

$$(2.23) \qquad |N(Q_1) - N(Q_2)| \leq C\epsilon$$

whenever Q_1, $Q_2 \in S$ are neighbors, because $\text{Angle}(P(Q_1), P(Q_2)) \leq C\epsilon$ and the orientations on $P(Q_1)$ and $P(Q_2)$ are compatible. The fact that C may depend on A will not disturb us, because ϵ will be chosen after A.

To each cube $Q \in S$ we associate regions $G_+(Q)$ and $G_-(Q)$ as follows. We choose a point $b_Q \in Q$—for instance, we can take b_Q to be a "center for Q" for the sake of always using the same notation, but now the distance from b_Q to the other cubes will not play any role—and we set

$$(2.24) \quad G_+(Q) = \{x \in B(b_Q, A_0 \operatorname{diam} Q) : A_0^{-1} \operatorname{diam} Q < \langle x - b_Q, N(Q) \rangle\},$$

where A_0 is some large constant, to be chosen later, and $\langle\, ,\, \rangle$ denotes the usual Euclidian scalar product on $\mathbf{R}^n$. (See Figure 2.1.) Similarly, set
$$(2.25)$$
$$G_-(Q) = \{x \in B(b_Q, A_0 \operatorname{diam} Q) : \langle x - b_Q, N(Q) \rangle < -A_0^{-1} \operatorname{diam} Q\}.$$

Also let $G(Q) = G_+(Q) \cup G_-(Q)$, $G_\pm = \bigcup_{Q \in S} G_\pm(Q)$, and $G = G_+ \cup G_-$.

LEMMA 2.26. *If ϵ and k^{-1} are small enough, we have that $\frac{1}{2} A_0^{-1} \operatorname{diam} Q \leq \operatorname{dist}(x, E) \leq A_0 \operatorname{diam} Q$ for all $x \in G(Q)$.*

This follows from the definition (2.4) of $P(Q)$.

2.2. BIG PROJECTIONS IN CODIMENSION 1

LEMMA 2.27. $G_+ \cap G_- = \varnothing$.

Suppose that $G_+(Q_1) \cap G_-(Q_2) \neq \varnothing$ for some choices of cubes Q_1, $Q_2 \in S$, and let x be a point of the intersection. We now choose the constant A so large (with respect to A_0) that Lemma 2.26 implies that Q_1 and Q_2 are neighbors. [That is, we shall not impose any further constraint on A, but it will be a little while before we choose A_0.] Thus $|N(Q_1) - N(Q_2)| \leq C\epsilon$, by (2.23). Also, $|\langle b_{Q_2} - b_{Q_1}, N(Q_1)\rangle| \leq 2\epsilon \operatorname{diam} Q_1$ because b_{Q_1} and b_{Q_2} both belong to kQ_1. (See (2.4).) Hence

$$\langle x - b_{Q_1}, N(Q_1)\rangle = \langle x - b_{Q_2}, N(Q_2)\rangle + \langle x - b_{Q_2}, N(Q_1) - N(Q_2)\rangle$$
$$+ \langle b_{Q_2} - b_{Q_1}, N(Q_1)\rangle$$
$$\leq -A_0^{-1} \operatorname{diam} Q_2 + C\epsilon A_0 \operatorname{diam} Q_2 + 2\epsilon \operatorname{diam} Q_1 \leq 0$$

if ϵ is small enough. This contradicts the fact that $x \in G_+(Q_1)$, and the lemma follows.

LEMMA 2.28. *We have that*

$$(2.29) \qquad \operatorname{dist}(x, E) \leq 4A_0^{-1} d(x)$$

for all $x \in \Omega = \{x \in \mathbf{R}^n : x \notin G \text{ and } \operatorname{dist}(x, Q(S)) \leq (A_0 - 2) \operatorname{diam} Q(S)\}$.

(Remember that $d(x)$ was defined in (2.6).)

Let $x \in \Omega$ be given, and choose a cube $Q \in S$ such that

$$(2.30) \qquad \operatorname{dist}(x, Q) + \operatorname{diam} Q \leq \tfrac{4}{3} d(x).$$

Let Q' be any ancestor of Q (including possibly Q itself) such that $Q' \in S$. Since $x \notin G(Q')$, we have

$$(2.31) \qquad |\langle x - b_{Q'}, N(Q')\rangle| \leq A_0^{-1} \operatorname{diam} Q'$$

if

$$(2.32) \qquad \operatorname{dist}(x, b_{Q'}) < A_0 \operatorname{diam} Q'.$$

Because $\operatorname{dist}(x, Q(S)) \leq (A_0 - 2) \operatorname{diam} Q(S)$, we know that (2.32) holds for $Q' = Q(S)$. Thus we may consider the smallest cube Q_0 such that $Q \subset Q_0 \subset Q(S)$ and for which (2.32) holds. Notice that $Q_0 \in S$, so that (2.31) holds with $Q' = Q_0$. Consequently, $\operatorname{dist}(x, P(Q_0)) \leq 2A_0^{-1} \operatorname{diam} Q_0$ (if ϵ is small enough).

By definition of $P(Q_0)$, every point of $P(Q_0) \cap B(b_{Q_0}, 2A_0 \operatorname{diam} Q_0)$ is at a distance $\leq \epsilon \operatorname{diam} Q_0$ from E (if we take k large enough compared to A_0). (Notice that we are using this part of the BWGL for the first time.) Applying this to the point of $P(Q_0)$ that is closest to x (and taking into account the fact that Q_0 satisfies (2.32)), we get that $\operatorname{dist}(x, E) \leq 3A_0^{-1} \operatorname{diam} Q_0$.

Now, if $Q_0 = Q$, we have that $\operatorname{dist}(x, E) \leq 3A_0^{-1} \operatorname{diam} Q \leq 4A_0^{-1} d(x)$ (by (2.30)). Otherwise, let Q_1 be the child of Q_0 that contains Q. Then

$Q_1 \in S$, and, since (2.32) fails for Q_1, we have that $\text{dist}(x, Q_1) \geq (A_0 - 1) \text{diam } Q_1 \geq C^{-1}(A_0 - 1) \text{diam } Q_0 \geq \text{diam } Q_0$, if we choose A_0 large enough. Thus we get

$$\text{dist}(x, E) \leq 3A_0^{-1} \text{diam } Q_0 \leq 3A_0^{-1} \text{dist}(x, Q_1)$$
$$\leq 3A_0^{-1} \text{dist}(x, Q) \leq 4A_0^{-1} d(x)$$

(again by (2.30)). This proves Lemma 2.28.

We are now ready to start proving Lemma 2.20 (that $E(S)$ has big projections). The main ingredient in the proof will be the following lemma.

LEMMA 2.33. *For each cube $Q_0 \in S$, let $\mathscr{T} = \mathscr{T}(Q_0)$ be the collection of all minimal cubes Q of S such that $Q \subset Q_0$. Set*

$$(2.34) \qquad F = \{Z \cap Q_0\} \cup \left\{ \bigcup_{Q \in \mathscr{T}} B(b_Q, 2 \text{diam } Q) \right\}.$$

Then we have

$$(2.35) \qquad |\Pi(F)| \geq C_0^{-1} (\text{diam } Q_0)^d,$$

where Π denotes the orthogonal projection onto the d-plane $P(Q_0)$ and C_0 is some constant that may depend on A_0 but not on Q_0 or S.

We should emphasize that A_0 has to be chosen large enough for the lemma to be true.

To prove the lemma we first introduce three points x_0, x_+, and x_- as follows. We choose x_0 to be a "center for Q_0", i.e., a point of Q_0 whose distance to $E \setminus Q_0$ is $\geq C_1^{-1} \text{diam } Q_0$ for some constant C_1. (See Lemma I.3.5.) Next let $x_\pm = x_0 \pm (10A_0^{-1} \text{diam } Q_0) N(Q_0)$, where $N(Q_0)$ still denotes the same choice of unit normal to $P(Q_0)$ and A_0 is still the same constant as in (2.24) and (2.25). Also let $B_\pm = B(x_\pm, A_0^{-1} \text{diam } Q_0)$. Notice that $B_\pm \subset G_\pm(Q_0) \subset G_\pm$.

We claim that any line segment L parallel to $N(Q_0)$ which connects B_- to B_+ must intersect the set F.

To prove our claim, let such a line segment L be given. Since L meets both G_+ and G_- and since G_+ and G_- are two disjoint open sets (see Lemma 2.27), L must also meet the complement of $G = G_+ \cup G_-$. Let y be a point in $L \setminus G$; we want to prove that $y \in F$.

First notice that $\text{dist}(y, Q(S)) \leq \text{diam } Q(S)$, because $\text{dist}(y, Q_0) \leq \text{dist}(y, x_0) \leq \text{diam } Q_0$ (if $A_0 \geq 20$, say). Lemma 2.28 now yields

$$(2.36) \qquad \text{dist}(y, E) \leq 4A_0^{-1} d(y) \leq \frac{d(y)}{10}.$$

Next let z be a point of E which minimizes the distance to y. Notice that

$$\text{dist}(y, E \setminus Q_0) \geq \text{dist}(x_0, E \setminus Q_0) - 11A_0^{-1} \text{diam } Q_0 \geq 100A_0^{-1} \text{diam } Q_0$$

if we choose A_0 large enough (compared to the constant C_1 used to define x_0). Hence $\text{dist}(y, E \setminus Q_0) > \text{dist}(y, x_0) \geq \text{dist}(y, Q_0)$, which implies that $z \in Q_0$. Since we know from (2.36) that $|z - y| = \text{dist}(y, E) \leq \frac{d(y)}{10}$, we also have that $d(z) \geq d(y) - |z - y| \geq \frac{9d(y)}{10}$. Thus z cannot be contained in a cube $Q \in S$ such that $\text{diam } Q < \frac{9d(y)}{10}$.

Assume first that $d(y) > 0$. Then z belongs to some minimal cube Q of S, contained in Q_0, with $\text{diam } Q \geq \frac{9d(y)}{10}$. In this case, $\text{dist}(y, Q) \leq |z - y| \leq \frac{d(y)}{10} \leq \frac{\text{diam } Q}{9}$, so $y \in B(b_Q, 2\,\text{diam } Q)$, which is contained in F. On the other hand, if $d(y) = 0$, then $z = y$ and, since we know that $z \in Q_0$, we also have that $z \in Q_0 \cap Z \subset F$. This proves our claim that every line segment L which is parallel to $N(Q_0)$ and meets both B_+ and B_- must also intersect F.

The estimate (2.35) now follows immediately, because the claim implies that $|\Pi(F)| \geq |\Pi(B_+)| = cA_0^{-d}\,\text{diam } Q_0^{-d}$. This completes the proof of Lemma 2.33.

We shall not impose any further conditions on A_0, so it can now be chosen, once and for all. This gives rise to a choice of A, as we discussed earlier (in the proof of Lemma 2.27). We can now choose ϵ and k also, subject to the various requirements that we have already imposed.

Let us now derive Lemma 2.20 (the fact that $E(S)$ has big projections) from Lemma 2.33. Note that this part of the proof also works in the higher-codimensional case.

We have to prove that there is a constant $C_2 > 0$ so that, for all $x \in E(S)$ and all $R > 0$, we can find a d-plane P such that

(2.37) $$|\Pi(E(S) \cap B(x, R))| \geq C_2^{-1} R^d,$$

where Π is (as usual) the orthogonal projection onto P.

If $x \in P_0$ or if $R > 10\,\text{diam } Q(S)$, we can take $P = P_0$ and (2.37) follows because $P_0 \cap E(S) \cap B(x, R)$ is already large enough. [For the case when $R > 10\,\text{diam } Q(S)$, we are using (2.13).] Next, if x lies in $\Sigma(Q)$ for some minimal cube Q of S and if $\text{diam } Q \geq \frac{R}{100}$, we have (2.37) because $\Sigma(Q)$ contains a large enough piece of a d-plane going through x.

We can therefore assume that either $x \in Z$ or $x \in \Sigma(Q)$ for some minimal cube Q such that $\text{diam } Q < \frac{R}{100}$. In both cases there is a cube $Q_0 \in S$ such that $\text{diam } Q_0 < \frac{R}{100}$ and $\text{dist}(x, Q_0) \leq \frac{R}{100}$. (Remember that $\Sigma(Q)$ is centered at a point of Q.) Taking an ancestor of Q_0 if necessary, we can even assume that

(2.38) $$\frac{R}{C} \leq \text{diam } Q_0 \leq \frac{R}{100}.$$

We choose $P = P(Q_0)$. Lemma 2.33 then tells us that

$$|\Pi(F)| \geq C_0^{-1}(\text{diam } Q_0)^{-d} \geq C^{-1} C_0^{-1} R^d.$$

If $|\Pi(Z \cap Q_0)| \geq \frac{1}{2}C^{-1}C_0^{-1}R^d$, then (2.37) holds (because $Z \subset E(S)$ and $Q_0 \subset B(x, R)$). Otherwise, we have

$$\left|\Pi\left\{\bigcup_{Q \in \mathscr{T}} B(b_Q, 2\operatorname{diam} Q)\right\}\right| \geq (2CC_0)^{-1}R^d,$$

where $\mathscr{T} = \mathscr{T}(Q_0)$. This can be rewritten as

(2.39) $$\left|\bigcup_{Q \in \mathscr{T}} B(Q)\right| \geq C_0'^{-1}R^d,$$

where we set $B(Q) = P \cap B(\Pi(b_Q), 2\operatorname{diam} Q) = \Pi(B(b_Q, 2\operatorname{diam} Q))$ and $|\cdot|$ stands for d-dimensional Lebesgue measure on P.

We must now replace the balls $B(Q)$ by the smaller $\Pi(\Sigma(Q))$ and check that we do not lose too much of the projection. By the Vitali covering lemma (see p. 9 of [St], for instance), there is a subset $\mathscr{T}'$ of $\mathscr{T}$ such that the $B(Q)$'s for $Q \in \mathscr{T}'$ are pairwise disjoint and $\sum_{Q \in \mathscr{T}'} |B(Q)| \geq 5^{-d} |\bigcup_{Q \in \mathscr{T}} B(Q)|$, which is $\geq C_0''^{-1}R^d$ by (2.39).

Let us come back to the definition of the stars $\Sigma(Q)$. Recall that the unit star Σ_0 was chosen so that, for any d-plane P, $\Pi(\Sigma_0)$ contains the ball $P \cap B(\Pi(0), \frac{1}{2})$. (See (2.8).) Thus $\Pi(\Sigma(Q))$ contains the ball $D(Q) = P \cap B(\Pi(b_Q), r_Q)$, where

$$r_Q = \frac{1}{6}\operatorname{Min}(\operatorname{diam} Q, \operatorname{dist}(b_Q, E \setminus Q)) \geq \frac{\operatorname{diam} Q}{C}.$$

Notice that $D(Q) \subset B(Q)$, so the $D(Q)$, $Q \in \mathscr{T}'$, are pairwise disjoint. Therefore,

$$\left|\Pi\left(\bigcup_{Q \in \mathscr{T}'} \Sigma(Q)\right)\right| \geq \left|\bigcup_{Q \in \mathscr{T}'} D(Q)\right|$$
$$= \sum_{Q \in \mathscr{T}'} |D(Q)| \geq C^{-1} \sum_{Q \in \mathscr{T}'} |B(Q)| \geq C_2^{-1}R^d.$$

This implies (2.37), because each of the $\Sigma(Q)$, $Q \in \mathscr{T}'$, is contained in $E(S)$ and also in $B(x, R)$. [Indeed, each such Q is contained in Q_0 by definition of $\mathscr{T}$, and Q_0 has a diameter $\leq \frac{R}{100}$ and is at distance $\leq \frac{R}{100}$ from x.] We have now established (2.37) in all cases and thus proved Lemma 2.20. [Notice that, as promised, the constant C_2 in (2.37) does not depend on ϵ or k.] Our proof of Proposition 2.2 in the codimension 1 case is now complete.

2.3. Big projections in the higher codimension case.

As we said earlier, much of the proof of Proposition 2.2 is the same in the general case as when the codimension equals 1. The argument which reduces Proposition 2.2 to Lemma 2.20, as well as the argument for deriving

Lemma 2.20 from Lemma 2.33, can be used in general, so we only have to prove the analogue of Lemma 2.33.

Our method for proving Lemma 2.33 in the general case will be slightly different. The general idea will be to approximate $E(S)$, near Q_0, by a smooth d-dimensional surface with boundary equal to a $(d-1)$-dimensional sphere in $P = P(Q_0)$ and then to use degree theory to prove that this surface has a big projection on P.

To make this more precise, we need more notation. Let E, S, etc., be as before. Let $Q_0 \in S$ be given as in the statement of Lemma 2.33, let $P = P(Q_0)$ be the d-plane given by (2.4), and let Π be the orthogonal projection onto P. Let x_0 be a center for Q_0, i.e., a point of Q_0 such that $\text{dist}(x_0, E \setminus Q_0) \geq C_1^{-1} \operatorname{diam} Q_0$, as in Lemma I.3.5. Set

$$(2.40) \qquad r = \tfrac{1}{3} \operatorname{Min}(\operatorname{diam} Q_0, \operatorname{dist}(x_0, E \setminus Q_0))$$

so that in particular $(3C_1)^{-1} \operatorname{diam} Q_0 \leq r \leq \tfrac{1}{3} \operatorname{diam} Q_0$. The main ingredient in our proof of Lemma 2.33 will be the following lemma.

LEMMA 2.41. *For each $\delta > 0$ there is an orientable d-dimensional properly embedded submanifold of $\mathbf{R}^n$, of class C^2, which we shall denote by $H = H(\delta)$, with boundary $\partial H = P \cap \partial B(\Pi(x_0), 5 \operatorname{diam} Q_0)$, and which satisfies:*

(2.42) $\text{dist}(z, P) \leq C_2 \epsilon r$ *and* $|\Pi(z) - \Pi(x_0)| \leq 5 \operatorname{diam} Q_0$ *for all $z \in H$;*

(2.43) $z \in P$ *for all $z \in H$ such that $|\Pi(z) - \Pi(x_0)| \geq 4 \operatorname{diam} Q_0$;*

(2.44) $\text{dist}(z, E) \leq C_2 \epsilon \{d(z) + \delta r\}$ *for all $z \in H$ such that $|\Pi(z) - \Pi(x_0)| \leq r$.*

Here C_2 is a constant that may depend on n, d, and the regularity constant for E but not on ϵ, δ, S, or Q_0.

We shall explain how to deduce the analogue of Lemma 2.33 from this lemma shortly, but first let us comment on its statement. We shall not need any control on the way the smoothness constants for $H(\delta)$ depend on δ; the only reason for insisting that $H(\delta)$ be smooth is that we want to apply degree theory to each $H(\delta)$. Of course, the additional δr in (2.44) will help us in constructing a smooth H. [Otherwise, (2.44) with $\delta = 0$ would almost force us to construct a surface H that contains $Z \cap B(x_0, r)$, but there is no reason for Z to be smooth.] The price we pay for δ will be low; we shall only have to let δ tend to 0 at the right moment.

Let us now assume Lemma 2.41 for the moment and show how it implies the analogue of Lemma 2.33 in this case. We say "the analogue" because Lemma 2.33 does not quite make sense in this case, since there is no longer an A_0. This is of course not a serious issue; we need only forget about A_0 and check that the conclusions of Lemma 2.33 hold if ϵ and k^{-1} are sufficiently small.

Consider the restriction of the projection Π to $H(\delta)$. Call this mapping ρ. Thus ρ maps $H(\delta)$ into the d-dimensional ball

$$B = P \cap \overline{B}(\Pi(x_0), 5\operatorname{diam} Q_0),$$

$\rho^{-1}(\partial B) = \partial H(\delta)$, and ρ equals the identity on a neighborhood of $\partial H(\delta)$. It follows that ρ has degree 1 (if the orientation of $H(\delta)$ is chosen correctly), so ρ must be surjective. (See p. 5-9 of [N] for the relevant facts from degree theory.)

Now let p be any point of $P \cap B(\Pi(x_0), r))$. We know that for each $\delta > 0$ there is a point $z \in H(\delta)$ such that $\Pi(z) = p$. Taking a sequence of δ's which tend to 0 and for which the points z have a limit, we get a new point z such that $\Pi(z) = p$, $\operatorname{dist}(z, P) \leq C_2 \epsilon r$, and $\operatorname{dist}(z, E) \leq C_2 \epsilon d(z)$. We want to show that $z \in F$ so that we can then conclude that $\Pi(F)$ contains a ball of radius r and, hence, satisfies (2.35).

Notice that $|z - x_0| \leq |\Pi(z) - \Pi(x_0)| + \operatorname{dist}(z, P) + \operatorname{dist}(x_0, P) \leq r + C_2 \epsilon r + \epsilon \operatorname{diam} Q_0 \leq \frac{11r}{10}$ if ϵ is small enough. In particular, any point of E which minimizes the distance to z must lie in Q_0. (See (2.40).)

If $d(z) = 0$, then $z \in Z \cap Q_0$, so z lies in F. Otherwise, let w be a point of E which minimizes the distance to z. Thus $w \in Q_0$, and we have that $|z - w| = \operatorname{dist}(z, E) \leq C_2 \epsilon d(z) \leq \frac{d(z)}{10}$ (if ϵ is small enough), and hence $d(w) \geq d(z) - |z - w| \geq \frac{9d(z)}{10}$. Therefore, w cannot be contained in any cube of S with a diameter $< \frac{9d(z)}{10}$. Since w lies in Q_0, there is a minimal cube $Q \subset Q_0$ of S that contains w, and $\operatorname{diam} Q \geq \frac{9d(z)}{10}$. On the other hand, $\operatorname{dist}(z, Q) \leq |z - w| \leq \frac{d(z)}{10} \leq \frac{\operatorname{diam} Q}{9}$, so z lies in $B(b_Q, 2\operatorname{diam} Q)$, which is contained in F. (We used a very similar argument in the last section.)

We have shown that $z \in F$ in all cases, and this completes our proof of the fact that Lemma 2.41 implies the analogue of Lemma 2.33 in this case. Thus Proposition 2.2 will follow as soon as we have proven Lemma 2.41.

We should probably admit now that we are going to employ the notion of neighbors (2.21) again in this section, with a large constant A that we get to choose anew, but which will depend only on n, d, and the regularity constant for E and not on ϵ or k. As before, if ϵ and k^{-1} are small enough, then Angle $(P(Q_1), P(Q_2)) \leq C\epsilon$ whenever Q_1, $Q_2 \in S$ are neighbors, and Lemma 2.22 remains valid.

Let Q_0, P, x_0, r, and δ be as before, and let us construct the surface H. To measure how close our surface H is to E, it will be convenient to use the following variant of the function d. For $z \in \mathbf{R}^n$ let

(2.45) $\qquad \tilde{d}(z) = \operatorname{Max}\{d(z), \delta r, f(|\Pi(z) - \Pi(x_0)|)\},$

where $f(t)$ is defined for $t \geq 0$ by

(2.46) $\qquad f(t) = \begin{cases} 0 & \text{for } 0 \leq t \leq r, \\ t - r & \text{for } t > r. \end{cases}$

It is easy to see that $\tilde{d}$ is Lipschitz with norm 1. Our surface H will satisfy

(2.47) $\qquad \text{dist}(z, E) \leq C_2 \epsilon \tilde{d}(z) \quad \text{for all } z \in H,$

from which (2.44) will of course follow.

Denote by D_j, $1 \leq j \leq 5$, the cylindrical domains defined by

(2.48) $\quad D_j = \{z \in B(x_0, 10 \operatorname{diam} Q_0) : |\Pi(z) - \Pi(x_0)| < j \operatorname{diam} Q_0\}.$

Let Y be a maximal subset of $E \cap \overline{D}_3$ for which

(2.49) $\quad |y_1 - y_2| > 10^{-3} \operatorname{Max}\{\tilde{d}(y_1), \tilde{d}(y_2)\}$

$\qquad\qquad\qquad$ whenever y_1, y_2 are two distinct points of Y.

Notice that Y is finite, because $\tilde{d}(y) \geq \delta r$ for all y.

If z is any point in $E \cap \overline{D}_3$, then there is a $y \in Y$ such that $|y - z| \leq 10^{-3} \operatorname{Max}\{\tilde{d}(y), \tilde{d}(z)\}$. Since $|\tilde{d}(y) - \tilde{d}(z)| \leq |y - z|$, we also have that

(2.50) $\qquad\qquad |y - z| \leq \frac{1}{999} \operatorname{Min}\{\tilde{d}(y), \tilde{d}(z)\}.$

For each $y \in Y$, let $B(y)$ be the closed ball with center y and radius $\frac{\tilde{d}(y)}{999}$. Thus

(2.51) $\qquad$ the balls $B(y)$, $y \in Y$, cover $E \cap \overline{D}_3$,

and

(2.52) $\qquad$ the balls $\frac{1}{3} B(y)$, $y \in Y$, are pairwise disjoint.

If $y \in Y$ and z is a point of $499 B(y)$, then $|\tilde{d}(z) - \tilde{d}(y)| \leq \frac{499}{999} \tilde{d}(y) < \frac{1}{2} \tilde{d}(y)$, so

(2.53) $\quad \frac{1}{2} \tilde{d}(y) \leq \tilde{d}(z) \leq \frac{3}{2} \tilde{d}(y)$ whenever $y \in Y$ and $z \in 499 B(y)$.

From this it follows in particular that if two balls $499 B(y)$ meet, then the corresponding $d(y)$ are comparable, so the $B(y)$ have roughly the same size. Consequently, there is an integer $M = M(n)$ such that the number of $499 B(y)$, $y \in Y$, which intersect a given $499 B(y_0)$, $y_0 \in Y$, is always $\leq M$. This allows us to express Y as a disjoint union $Y = \bigcup_{i=1}^{M} Y_i$, where each Y_i satisfies

(2.54) $\quad 499 B(y_1) \cap 499 B(y_2) = \varnothing$

$\qquad\qquad\qquad$ whenever y_1 and y_2 are two distinct points of Y_i.

For each $y \in Y$ there is a reasonable choice of a surface (or even a d-plane) which passes through $B(y)$ and is sufficiently close to E near $B(y)$. The problem is of course that we need to make all the various bits of surface match up with each other. The point of the decomposition of Y into the Y_i's is that we do not want to consider each $y \in Y$, one after the other, and modify a provisional surface near y; the estimates for the surface would get worse and worse, and we do not have a good estimate on the number of points in Y. What we shall do instead is define a provisional piece of surface which is reasonable near all the points of Y_1, then modify it so that it becomes also reasonable near all the $y \in Y_2$, and so on. Because the different points y in a single Y_i are so far away from each other, we shall be able to do all the modifications concerning the points of each Y_i independently (and at the same time). Thus we shall construct a collection H_i, $i = 0, 1, \ldots, M$, of d-dimensional sets, with H_i obtained from H_{i-1} by adding to it a few pieces near the points y, $y \in Y_i$. Our final set H_M will be the desired surface H.

For each $y \in Y$, there is a cube $Q \in S$ such that $\operatorname{dist}(y, Q) + \operatorname{diam} Q \leq 2d(y)$. By replacing Q by an ancestor if necessary, we can select a cube $Q(y) \in S$ such that

(2.55) $\quad \operatorname{dist}(y, Q(y)) \leq 2d(y)$ and $\frac{1}{5}\tilde{d}(y) \leq \operatorname{diam} Q(y) \leq C\tilde{d}(y)$.

Note that $Q(y)$ and $Q(y')$ are neighbors, in the sense of (2.21), whenever $499B(y)$ and $499B(y')$ intersect, at least if we take A to be sufficiently large.

Set $P(y) = P(Q(y))$. Our first approximation to H is

(2.56) $\quad H_{-1} = [P \cap (\overline{D}_5 \setminus D_4)] \cup \left[\bigcup_{y \in Y} \left(P(y) \cap \frac{1}{10} B(y) \right) \right]$.

Notice that H_{-1} is composed of many pieces that are far away from each other; the $\frac{1}{10}B(y)$ are disjoint from each other by (2.52) and also from $\overline{D}_5 \setminus D_4$ because, for $y \in Y$,

$$\tilde{d}(y) \leq d(y) + \delta r + f(3 \operatorname{diam} Q_0) \leq 5 \operatorname{diam} Q_0 + \delta r + 3 \operatorname{diam} Q_0 \leq 9 \operatorname{diam} Q_0$$

(we may assume that δ is as small as we want). We must now fill in the holes between the various pieces to obtain the desired surface.

In the following, whenever P' is a d-plane, we shall mean by "a graph over P'" a set Γ of the form $\{z + \rho(z) : z \in P'\}$, where ρ is some function from P' to the $(n-d)$-plane through the origin that is orthogonal to P'. This function ρ is of course determined by Γ. A graph over P' is said to be Lipschitz (respectively Lipschitz with norm $\leq C$ or of class C^2) if the function ρ is Lipschitz (respectively Lipschitz with norm $\leq C$ or of class C^2). In order to save time, we shall even say that Γ is a graph over P' which is "C-good" to mean that it is both Lipschitz with norm $\leq C$ and of class C^2.

Let us now define the set H_0. We claim that there is a graph Γ_0 over P which is $C\epsilon$-good and which contains

(2.57) $\quad [P \cap (\overline{D}_5 \setminus D_4)] \cup \left[\bigcup_{y \in Y \cap (\overline{D}_3 \setminus D_{3/2})} P(y) \cap \frac{1}{10} B(y) \right]$.

2.3. BIG PROJECTIONS IN HIGHER CODIMENSIONS 115

To prove the claim, let $y \in Y \cap (\overline{D}_3 \setminus D_{3/2})$ be given. Then $\tilde{d}(y) \geq \inf_{\overline{D}_3 \setminus D_{3/2}} \tilde{d}(z) \geq f\left(\frac{3}{2} \operatorname{diam} Q_0\right) \geq \operatorname{diam} Q_0$. Hence $Q(y)$ is a neighbor of Q_0, provided that we take the constant A in the definition (2.21) of neighbors large enough. Therefore, $\operatorname{Angle}(P(y), P) \leq C\epsilon$, and, for every point $z \in \frac{1}{10}B(y) \cap P(y)$, we have $\operatorname{dist}(z, P) \leq C\epsilon \operatorname{diam} Q_0$, which is less than $C\epsilon$ times the distance from $\frac{1}{10}B(y)$ to $P \cap (\overline{D}_5 \setminus D_4)$ or to any of the $\frac{1}{10}B(y')$, $y' \neq y$. [See (2.52).] The existence of Γ_0 follows (as long as ϵ is small enough). We set

$$H_0 = H_{-1} \cup \left[\Gamma_0 \cap (\overline{D}_5 \setminus D_2)\right].$$

We are now ready to add pieces to H_0 to obtain the sets $H_1, H_2, \ldots, H_M$. We shall do this in such a way that, for each $1 \leq j \leq M$, we have that:

(2.58) $H_j \subseteq H_0 \cup \bigcup_{i=1}^{j} \bigcup_{y \in Y_i} 5B(y)$;

(2.59) for each $y \in Y_j$, $H_j \cap 5B(y)$ is the intersection of $5B(y)$ with some graph over $P(y)$ which is $C_j\epsilon$-good;

(2.60) for each $y \in Y$, there is a graph $\Gamma_j(y)$ over $P(y)$ which is $C_j\epsilon$-good and which contains $H_j \cap 20B(y)$.

[The constants C_j will become larger and larger as j increases, but this is not important because we shall stop after M steps. Note that the third condition concerns more points y and larger balls $20B(y)$ but is less restrictive than the second one in that it allows holes in $H_j \cap 20B(y)$.]

Let us check that H_0 satisfies the condition (2.60). If $y \in Y$ is such that $20B(y)$ is contained in D_2, then $H_0 \cap 20B(y) = H_{-1} \cap 20B(y)$ is contained in the union of the $P(y') \cap \frac{1}{10}B(y')$ where $y' \in Y$ is such that $\frac{1}{10}B(y')$ intersects $20B(y)$. For these points y', $Q(y')$ is a neighbor of $Q(y)$, and the existence of the graph $\Gamma_0(y)$ follows because the $\frac{1}{10}B(y')$'s are sufficiently far from each other, while all the $P(y')$'s are very close to $P(y)$. If $20B(y)$ is not contained in D_2, then $\operatorname{dist}(y, \mathbf{R}^n \setminus D_2) \leq \frac{20\tilde{d}(y)}{999} \leq \frac{\operatorname{diam} Q_0}{10}$, so $y' \in \overline{D}_3 \setminus D_{3/2}$ whenever $\frac{1}{10}B(y')$ touches $20B(y)$. Consequently, $H_0 \cap 20B(y)$ is contained in Γ_0, and (2.60) holds for $j = 0$.

Now assume that we are given j, $1 \leq j \leq M$, and that H_{j-1} was already constructed and satisfies (2.58), (2.59), and (2.60). (We adopt the convention that $Y_0 = \emptyset$ so that (2.59) also holds when $j = 0$.) To construct H_j, we shall add to H_{j-1} a certain number of pieces contained in the balls $5B(y)$, $y \in Y_j$. The fact that the balls $499B(y)$, $y \in Y_j$, are pairwise disjoint will make the modification of H_{j-1} in each $5B(y)$ pleasantly independent from the others.

For each $y \in Y_j$ our induction hypothesis (2.60) provides us with a graph $\Gamma_{j-1}(y)$ over $P(y)$ which is $C_{j-1}\epsilon$-good and which contains $H_{j-1} \cap 20B(y)$. Since we want $H_j \cap 5B(y)$ to be a good graph over $P(y)$, it is natural to take $H_j \cap 5B(y) = \Gamma_{j-1}(y) \cap 5B(y)$, because then H_j will automatically satisfy

(2.59) with the constant C_{j-1}. Thus we take

$$(2.61) \qquad H_j = H_{j-1} \cup \left\{ \bigcup_{y \in Y_j} [\Gamma_{j-1}(y) \cap 5B(y)] \right\}.$$

We now have to check that H_j also satisfies (2.60). We begin with a few preliminary observations.

LEMMA 2.62. *We have that*

$$(2.63) \qquad \operatorname{dist}(z, E) \leq C\epsilon \tilde{d}(z) \quad \text{for all } z \in H_j$$

and

$$(2.64) \qquad \operatorname{dist}(z, P(w)) \leq C\epsilon \tilde{d}(w)$$

whenever $z \in H_j \cap 5B(w)$ for some $w \in Y$. (Here and in what follows, we shall denote by C any constant that depends only on C_{j-1}, the regularity constant for E, n, and d.)

Let us first show how to derive (2.63) from (2.64). First, (2.63) holds when $z \in H_0$. Next, if $z \in H_j \setminus H_0$, then z lies in a ball $5B(w)$ for some $w \in Y$, and (2.64) implies that

$$\operatorname{dist}(z, E) \leq C\epsilon \tilde{d}(w) + \sup_{u \in P(w) \cap 10B(w)} \operatorname{dist}(u, E) \leq C\epsilon \tilde{d}(w)$$

(because $P(w) = P(Q(w))$ and $Q(w)$ satisfies (2.55)); (2.63) follows because $\tilde{d}(w) \leq 2\tilde{d}(z)$, by (2.53).

Let us now prove (2.64). First suppose that $z \in H_{j-1} \cap 5B(w)$ for some $w \in Y$. Let $\Gamma = \Gamma_{j-1}(w)$ be the $C_{j-1}\epsilon$-Lipschitz graph over $P(w)$ given by (2.60) for $j-1$. By definition, Γ contains $H_{-1} \cap 5B(w)$, which itself contains $P(w) \cap \frac{1}{10}B(w)$. Hence $\Gamma \cap 5B(w)$ stays at distance $\leq C_{j-1}\epsilon \operatorname{diam}(5B(w)) \leq C\epsilon \tilde{d}(w)$ from $P(w)$. [Remember that Γ is Lipschitz with norm $\leq C_{j-1}\epsilon$.] This proves (2.64) when $z \in H_{j-1}$.

If $z \in H_j \setminus H_{j-1}$, then there is a $y \in Y_j$ such that $z \in 5B(y)$. By construction of H_j, $H_j \cap 5B(y) = \Gamma_{j-1}(y) \cap 5B(y)$, where $\Gamma_{j-1}(y)$ is the graph over $P(y)$ provided by (2.60). Since $\Gamma_{j-1}(y) \supset H_j \cap 5B(y) \supset P(y) \cap \frac{1}{10}B(y)$ and $z \in \Gamma_{j-1}(y)$, we have

$$(2.65) \qquad \operatorname{dist}(z, P(y)) \leq C\epsilon \tilde{d}(y),$$

because $\Gamma_{j-1}(y)$ is a $C_{j-1}\epsilon$-Lipschitz graph over $P(y)$. Also, the balls $5B(y)$ and $5B(w)$ intersect, so $\frac{1}{3}\tilde{d}(w) \leq \tilde{d}(y) \leq 3\tilde{d}(w)$ by (2.53). Thus $Q(y)$ and $Q(w)$ are neighbors, and (2.64) then follows from (2.65) and the fact that $P(y)$ is very close to $P(w)$. This proves Lemma 2.62.

LEMMA 2.66. *For each $w \in Y$ there is at most one $y \in Y_j$ such that $20B(w)$ intersects $5B(y)$.*

Let $w \in Y$ and $y \in Y_j$ be such that $20B(w) \cap 5B(y) \neq \emptyset$. It follows from (2.53) that $\tilde{d}(w) \leq 3\tilde{d}(y)$, so the diameter of $20B(w)$ is no greater than 120 times the radius of $B(y)$. Thus $20B(w) \subset 125B(y)$. The lemma follows because the $499B(y)$, $y \in Y_j$, are disjoint (see (2.54)).

We are now ready to prove that H_j satisfies (2.60). Let $w \in Y$ be given. If $20B(w)$ is disjoint from $5B(y)$ for all $y \in Y_j$, we are immediately reduced to (2.60) for H_{j-1}, because $H_j \cap 20B(w) = H_{j-1} \cap 20B(w)$. Thus we can assume that there is a $y \in Y_j$ such that $20B(w)$ intersects $5B(y)$. By Lemma 2.66, there is only one such y, so the only place where $H_j \cap 20B(w)$ may be different from $H_{j-1} \cap 20B(w)$ is inside of $5B(y)$.

Let $\Gamma_1 = \Gamma_{j-1}(w)$ be the graph over $P(w)$ given by (2.60) and $\Gamma_2 = \Gamma_{j-1}(y)$ the graph over $P(y)$ that was used to define $H_j \cap 5B(y)$. Notice that, since $20B(w)$ meets $5B(y)$, the cubes $Q(w)$ and $Q(y)$ are neighbors, and therefore $\text{Angle}(P(w), P(y)) \leq C\epsilon$. Thus Γ_2 is also a graph over $P(w)$, which is $C\epsilon$-good (if ϵ is small enough).

Let $P^\perp(w)$ be the $(n-d)$-plane through the origin that is perpendicular to $P(w)$, and let Π_w and $\Pi_w^\perp$ be the orthogonal projections onto $P(w)$ and $P^\perp(w)$ respectively. Let ρ_i, $i = 1, 2$, be the Lipschitz functions from $P(w)$ to $P^\perp(w)$ whose graphs are Γ_1 and Γ_2. Thus both ρ_i are Lipschitz with norm $\leq C\epsilon$ and of class C^2. We want to find a Lipschitz function ρ such that $H_j \cap 20B(w)$ is contained in the graph of ρ. A reasonable choice is

$$(2.67) \qquad \rho(x) = \phi(x)\rho_1(x) + (1 - \phi(x))\rho_2(x),$$

where ϕ is a function of class C^2 such that $0 \leq \phi \leq 1$, $\phi \equiv 0$ on $\Pi_w(6B(y))$, $\phi \equiv 1$ on the complement of $\Pi_w(19B(y))$, and $|\nabla \phi| \leq C(\text{diam } B(y))^{-1}$.

It is clear that ρ is of class C^2 because ρ_1 and ρ_2 are. Let us now check that ρ is also Lipschitz with norm $\leq C\epsilon$ (where we continue with our convention that C may be a combination of C_{j-1} and geometric constants). Notice that the inequality

$$(2.68) \qquad |\rho(u) - \rho(v)| \leq C\epsilon|u - v|$$

is true when both u and v lie in the set where $\phi \equiv 1$. Thus we may content ourselves with proving (2.68) when, say, $v \in \text{supp}(1-\phi) \subset \Pi_w(19B(y))$. We write

$$\rho(u) = \rho_2(u) + \phi(u)\left(\rho_1(u) - \rho_2(u)\right) = \rho_2(u) + \phi(u)D(u),$$

where $D = \rho_1 - \rho_2$, so that

$$|\rho(u) - \rho(v)| = |[\rho_2(u) - \rho_2(v)]$$
$$+ [\phi(u)D(u) - \phi(u)D(v) + \phi(u)D(v) - \phi(v)D(v)]|$$
$$\leq |\rho_2(u) - \rho_2(v)| + \phi(u)|D(u) - D(v)| + |\phi(u) - \phi(v)||D(v)|$$
$$\leq C\epsilon|u - v| + C(\text{diam } B(y))^{-1}|u - v||D(v)|.$$

Thus (2.68) will follow once we know that

(2.69) $$|D(v)| \le C\epsilon \operatorname{diam} B(y).$$

This is basically trivial, because the graphs of ρ_1 and ρ_2 are trying to approximate almost the same set, but the details are slightly cumbersome. We leave (2.69) as an exercise to the reader.

Let Γ denote the graph of ρ. We have proved that Γ is a graph over $P(w)$ which is $C_j \epsilon$-good, so we need only check that $H_j \cap 20B(w)$ is contained in Γ. If z is a point of $H_j \cap [20B(w) \setminus 20B(y)]$, then $\Pi_w^\perp(z) = \rho_1(\Pi_w(z))$ by definition of Γ_1, so it is enough in this case to check that $\phi(\Pi_w(z)) = 1$. Notice that $|\Pi_w^\perp(z) - \Pi_w^\perp(y)| \le C\epsilon \operatorname{diam} B(y)$. [Indeed, both z and y are at a distance $\le C\epsilon \operatorname{diam} B(y)$ from $P(w)$.] Hence the fact that $z \notin 20B(y)$ implies that $\Pi_w(z) \notin \Pi_w(19B(y))$, and consequently $\phi(\Pi_w(z)) = 1$, as desired.

Our next case is when $z \in H_j \cap 5B(y)$. In this case, $\Pi_w(z) \in \Pi_w(5B(y))$, so $\phi(\Pi_w(z)) = 0$, and hence $\rho(\Pi_w(z)) = \rho_2(\Pi_w(z)) = \Pi_w^\perp(z)$, since $z \in \Gamma_2$. In the remaining case—where $z \in H_j \cap 20B(w) \cap [20B(y) \setminus 5B(y)]$—$z$ lies on Γ_1 and on Γ_2 (because $z \in H_{j-1} \cap 20B(w) \cap 20B(y)$), and so $z \in \Gamma$.

This completes our proof of the property (2.60) for H_j. Thus we can construct our sets H_j, $1 \le j \le M$, with the properties (2.58), (2.59), and (2.60). We now take $H = H_M$, and we must check that it has all the properties required in Lemma 2.41.

The fact that $\operatorname{dist}(z, P) \le C_2 \epsilon r$ for all $z \in H$ follows from (2.63) and the definition of P. The fact that $|\Pi(z) - \Pi(x_0)| \le 5 \operatorname{diam} Q_0$ (or, in other words, that $z \in \overline{D}_5$) for all $z \in H$, as well as the fact that $z \in P$ for all $z \in H \setminus D_4$, follows directly from the construction. [Notice that $H_M \setminus D_4 = H_{-1} \setminus D_4 = P \cap (\overline{D}_5 \setminus D_4)$.] This proves (2.42) and (2.43); (2.44) follows immediately from (2.63) and the definition (2.45) of $\tilde{d}$. Thus we only need to check that H is an orientable embedded submanifold of class C^2 with boundary $P \cap \partial B(\Pi(x_0), 5 \operatorname{diam} Q_0)$.

To prove that H is an embedded submanifold of class C^2, we just need to check that for each $z \in H \cap D_5 = H \setminus \partial H$ there is a small ball B centered at z such that $H \cap B$ is the intersection of B with the graph of a C^2-function over some d-plane. If z belongs to any $5B(y)$, $y \in Y$, we can apply (2.59) (applied to y and the integer j such that $y \in Y_j$) to get a graph of class C^2 over $P(y)$ that contains $H(j) \cap 5B(y)$. We then apply (2.60) with $j = M$ to conclude that $H \cap 5B(y)$ contains no additional points. This takes care of the case when z belongs to some $5B(y)$. Assume now that z does not belong to any $5B(y)$, so that it belongs to H_0, by (2.58). Thus either $z \in H_{-1}$, in which case $z \in P \cap (\overline{D}_5 \setminus D_4)$ (see (2.56)), or $z \in \Gamma_0 \cap (D_5 \setminus D_2)$.

If z lies outside $D_{5/2}$, then z is well inside of Γ_0, and $B \cap H \supset B \cap \Gamma_0$ for all small enough balls B centered at z. We still have to check that if B

has a radius $< 10^{-3}\tilde{d}(z)$, say, then $B \cap H$ contains no other points. If this were not the case, then $B \cap H$ would contain a point in $H \setminus H_0$, so $B \cap H$ would intersect some $5B(y_0)$, $y_0 \in Y$. Because B and $5B(y_0)$ meet, we would have that $\tilde{d}(z) \leq \frac{3}{2}\tilde{d}(y_0)$ (by (2.53)), and then B would be contained in $20B(y_0)$. Applying (2.60), we get that $H \cap 20B(y_0)$ would be contained in some good graph over $P(y_0)$, which implies that $H \cap B = \Gamma_0 \cap B$ because Angle$(P(y_0), P) \leq C\epsilon$. [Notice that $\tilde{d}(y_0) \geq \frac{2}{3}\tilde{d}(z) \geq \frac{2}{3}f(\frac{5}{2}\text{diam } Q_0) \geq$ diam Q_0.]

We are left with the case when $z \in D_{5/2}$. Let ζ be the point of E which is closest to z. By (2.63), $|z - \zeta| \leq C\epsilon\tilde{d}(z) < \frac{1}{10}\text{diam }Q_0$ (if ϵ is small enough), so $\zeta \in D_3$. By definition of Y, there is a $y \in Y$ such that $\zeta \in B(y)$ (see (2.51)). Then $\tilde{d}(\zeta) \leq \frac{3}{2}\tilde{d}(y)$ by (2.53), and since $|\tilde{d}(z)-\tilde{d}(\zeta)| \leq |z-\zeta| \leq C\epsilon\tilde{d}(z) < \frac{\tilde{d}(z)}{10}$, we also have that $\tilde{d}(z) \leq 2\tilde{d}(y)$. Hence $|z - \zeta| \leq C\epsilon\tilde{d}(y)$ and $z \in 5B(y)$. This contradicts our hypothesis on z and shows that our last case never occurs. Hence H is a C^2 submanifold with boundary as promised.

It remains to check that H is orientable. We first let the $P(Q)$'s, $Q \in S$, be given orientations in accordance with Lemma 2.22. Thus we have orientations on P and the $P(y)$'s, $y \in Y$, which in turn induce orientations on $\Gamma_0 \cap (\overline{D}_5 \setminus D_2)$ and on each $H \cap 5B(y)$. It is not hard to check that these orientations are compatible, using Lemma 2.22. We omit the details.

We have now proved Lemma 2.41, so, as we explained before, Proposition 2.2 follows.

We conclude this section with comments about alternative approaches to the proof of Proposition 2.2.

It is reasonable to think that, for each stopping time region S, there is a nice set $E'(S)$—for instance a chord-arc surface or a surface satisfying Condition B (see [S1], [S4], [DS3]) in codimension 1—which approximates E at the scale of the smallest cubes of S (i.e., satisfies (2.3)). This would be more attractive aesthetically than the constructions given above, and it could give rise to a more direct proof of Proposition 2.2, provided that the technical details do not make it much worse.

Concerning our proof of the fact that our sets $E(S)$ have big projections when the codimension is larger than 1, let us briefly describe a different approach which is closer to the one we used in the codimension 1 case. Recall that, when the codimension is 1, we defined a region $G = \bigcup_{Q \in S} G(Q)$, where each $G(Q)$ was the set of points in $\mathbf{R}^n$ which are not too far from Q but not too close to the d-plane $P(Q)$. [Notice that this still makes sense in codimension > 1.] The idea was that G could be naturally decomposed into two components G_+ and G_-, that every line segment going from G_+ to G_- would have to pass at a very small distance from E, and that one could find reasonably large balls $B_\pm \subset G_\pm$ so that there would be many parallel line segments which pass at a small distance from E.

In higher codimensions, one could define G similarly and use the following type of argument. For each cube $Q \in S$, one would first find a $(n-d-1)$-sphere $\mathscr{C}$ centered on $P(Q)$, contained in a $(n-d)$-plane orthogonal to $P(Q)$, of radius comparable to $\operatorname{diam} Q$, contained in G and even at distance from its complement comparable to $\operatorname{diam} Q$. To prove that $E(S)$ has big projections it would be enough to show that, for any such sphere, the $(n-d)$-ball obtained by filling in $\mathscr{C}$ (in other words, the convex hull of $\mathscr{C}$) necessarily gets very close to E. [This corresponds to what we did in codimension 1: we selected two points of G on opposite sides of E and proved that the line segment that joins them gets very close to E.]

The idea for proving that the $(n-d)$-ball must get close to E is as follows. One would construct a closed $(n-d-1)$-form α on G such that its integral over a sphere $\mathscr{C}$ as above is always $\neq 0$, so that the sphere $\mathscr{C}$ could not be contracted to a point while staying in G. There is a good candidate for α on each $G(Q)$, because $G(Q)$ looks like the complement of a d-plane, but it is not as easy to patch together the various α's coming from the various Q's and still get a closed form as it is when the codimension is 1.

2.4. The local convexity condition LCV.

There are some geometric conditions on a regular set which are similar to the bilateral weak geometric lemma and which are in appearance a little weaker, but which are nonetheless still equivalent to uniform rectifiability because they imply the BWGL. An example of this is the local symmetry condition LS of Definition I.1.79. The fact that the BWGL implies LS is trivial from the definitions, and the converse is not too difficult to prove. (See Proposition I.2.5 and its proof in §5 of [DS2].)

In this section we shall prove that another condition of a similar type—the local convexity condition—implies the BWGL (and is therefore equivalent to uniform rectifiability). Recall from Definition I.2.7 that a regular set E of dimension d in $\mathbf{R}^n$ is said to satisfy the local convexity condition (LCV) if, for each $\epsilon > 0$, the set

(2.70) $\{(x, t) \in E \times \mathbf{R}_+ :$ there exist points y, z in $E \cap B(x, t)$ such that $\operatorname{dist}\left(\frac{y+z}{2}, E\right) > \epsilon t\}$

is a Carleson set.

The main result of this section is the following.

THEOREM 2.71. *If E is a regular set of dimension d in $\mathbf{R}^n$ that satisfies the LCV, then E satisfies the BWGL.*

It is obvious that the BWGL implies the LCV, so it will follow from Theorem 2.71 and the equivalence of the BWGL with uniform rectifiability that the LCV is equivalent to uniform rectifiability.

The LCV could have been defined in slightly different (but equivalent) ways. Here is an example.

2.4. THE LOCAL CONVEXITY CONDITION

LEMMA 2.72. *Let CV be the set of compact convex subsets of $\mathbf{R}^n$. Then a d-dimensional regular set E is in the class $\mathrm{Approx}(CV)$ if and only if it satisfies the LCV.*

Recall from Definition I.2.21 that we say that $E \in \mathrm{Approx}(CV)$ if, for every $\epsilon > 0$, the set $\mathscr{B}(\epsilon)$ is a Carleson subset of $E \times \mathbf{R}_+$, where $\mathscr{B}(\epsilon)$ denotes the set of all $(x, t) \in E \times \mathbf{R}_+$ for which it is *not* possible to find a compact convex set $\mathscr{E} \subset \mathbf{R}^n$ such that

$$(2.73) \qquad \sup_{y \in E \cap \overline{B}(x,t)} \mathrm{dist}(y, \mathscr{E}) + \sup_{z \in \mathscr{E} \cap \overline{B}(x,t)} \mathrm{dist}(z, E) \leq \epsilon t.$$

The fact that every set in $\mathrm{Approx}(CV)$ satisfies the LCV is obvious. To prove the converse, it is enough to show that if $(x, 2t) \in E \times \mathbf{R}_+$ is not in the bad set (2.70) for the LCV, then there is a compact convex set $\mathscr{E}$ such that (2.73) holds, but with ϵ replaced by another constant $c(\epsilon)$, provided that $c(\epsilon)$ can be made as small as we want by taking ϵ small.

Suppose that $(x, 2t)$ is not in the bad set (2.70), and let $\mathscr{E}$ be the convex hull of $E \cap \overline{B}(x, t)$. Since $\mathscr{E}$ contains $E \cap \overline{B}(x, t)$ and since $\mathscr{E}$ is compact (which follows easily from Lemma I.2.26), we only have to prove that every point of $\mathscr{E} \cap \overline{B}(x, t)$ is within $c(\epsilon)t$ of E.

Let z be a point in $\mathscr{E}$. We know from Lemma I.2.26 that z can be written as a convex combination of $n+1$ points of $E \cap \overline{B}(x, t)$. If one starts with two of these points and then constructs new points by taking, at each step, the midpoints of pairs of points already available, then it takes less than $C \log \frac{1}{\epsilon}$ steps to get within ϵt of any point on the line segment that joins the two initial points. Iterating this, we see that our point z can be approached within ϵt using less than $Cn \log \frac{1}{\epsilon}$ steps of the same type of process, starting with points of $E \cap \overline{B}(x, t)$. Applying each time our hypothesis that $(x, 2t)$ is not in the bad set (2.70), we obtain that $\mathrm{dist}(z, E) \leq Cn \left(\log \frac{1}{\epsilon}\right) \epsilon t$ by an easy induction argument. This proves Lemma 2.72.

To prove Theorem 2.71 it will be easier to use cubes rather than working directly on $E \times \mathbf{R}_+$. So let a regular set E that satisfies the LCV be given, and let us choose a family Δ of cubes on E, as in §3.1 of Part I. For each $\epsilon > 0$ and each (large) constant $k > 1$ let

$$(2.74) \quad \mathscr{B}(k, \epsilon) = \{Q \in \Delta : \text{ there is a point } z \text{ in the convex hull of } kQ \text{ which is at distance } \geq \epsilon \operatorname{diam} Q \text{ from } E\}.$$

(The notation kQ is defined by (I.3.7).)

It follows immediately from the proof of Lemma 2.72 that our assumption that E satisfies the LCV implies that $\mathscr{B}(k, \epsilon)$ satisfies a Carleson packing condition for every choice of ϵ and k. (See Definition I.3.9 for Carleson packing conditions.)

LEMMA 2.75. *If E satisfies the LCV, then it satisfies the WGL.*

This fact is not new, and its proof is about the same as for the fact that the local symmetry condition implies the WGL. Let us sketch the argument for the convenience of the reader.

It is enough to prove that, for each $\delta > 0$, one can find an $\epsilon > 0$ such that, if Q is not in the bad set $\mathscr{B}(2, \epsilon)$, then there exists a d-plane P which satisfies $\text{dist}(x, P) \leq \delta \operatorname{diam} Q$ for all $x \in 2Q$.

To find such a d-plane P, we first observe that, because E is regular, it is possible to find $d + 1$ points $y_0, y_1, \ldots, y_d$ in Q that are affinely independent and which even satisfy the following more precise estimate: for each $j \in \{0, 1, \ldots, d-1\}$, y_{j+1} is at distance $\geq C^{-1} \operatorname{diam} Q$ from the j-plane that passes through $y_0, y_1, \ldots, y_j$. The constant C may depend on n, d, and the regularity constant for E, but not on Q. This is Lemma 5.8 on p. 28 in [DS2], and it is easy to prove using an induction on j and the fact that E cannot stay too close to a j-plane, $j < d$, because it is regular of dimension d.

We now take for P the d-plane that goes through $y_0, y_1, \ldots, y_d$. If there were a point $x \in 2Q$ such that $\text{dist}(x, P) > \delta \operatorname{diam} Q$, we would get a contradiction as follows. The convex hull $\mathscr{E}$ of the points $y_0, y_1, \ldots, y_d$ and x would have a $(d+1)$-dimensional volume $\geq C^{-1} \delta (\operatorname{diam} Q)^{d+1}$, and more precisely it would even be possible to find more than

$$C^{-1} \delta (\operatorname{diam} Q)^{d+1} (\epsilon \operatorname{diam} Q)^{-d-1} = C^{-1} \delta \epsilon^{-d-1}$$

disjoint balls of radius $2\epsilon \operatorname{diam} Q$ centered on $\mathscr{E}$. Because each point of $\mathscr{E}$ is within $\epsilon \operatorname{diam} Q$ from E (we assumed that $Q \notin \mathscr{B}(2, \epsilon)$), we would be able to find a ball centered on E and with radius $\epsilon \operatorname{diam} Q$ inside each of these disjoint balls. The total mass of the part of E which lies within $\operatorname{diam} Q$ of $\mathscr{E}$ would then have to be $\geq C^{-1} \frac{\delta}{\epsilon} (\operatorname{diam} Q)^d$, which is impossible if ϵ is chosen small enough. This concludes the proof of Lemma 2.75.

In order to prove that E satisfies the BWGL, we still have to check that E does not have too many holes, i.e., that for most cubes Q the d-plane P provided by the WGL does not have any point which is reasonably close to Q but which is not very close to E. Let us be more precise.

Denote by $\mathscr{G}$ the set of cubes $Q \in \Delta$ that do not belong to $\mathscr{B}(k, \epsilon)$, and for which there is a d-plane $P(Q)$ such that

(2.76)   every point of kQ is at distance $\leq \epsilon \operatorname{diam} Q$ from $P(Q)$.

Also let

(2.77) $$\mathscr{W} = \Delta \setminus \mathscr{G}.$$

We already know that, for each choice of ϵ and k, $\mathscr{W}$ satisfies a Carleson packing condition. Thus it is enough to prove that, for each $\delta > 0$, there is a choice of ϵ and k so that the set

(2.78) $\mathscr{H} = \{Q \in \mathscr{G} : \text{ there is a point } z \in P(Q) \text{ such that}$
$\text{dist}(z, Q) \leq \operatorname{diam} Q \text{ but } \text{dist}(z, E) \geq \delta \operatorname{diam} Q\}$

satisfies a Carleson packing condition.

2.4. THE LOCAL CONVEXITY CONDITION

So let $\delta > 0$ be given, and let $\mathscr{G}$, $\mathscr{W}$, and $\mathscr{H}$ be as above, with ϵ and k to be chosen later. Our general plan for proving that $\mathscr{H}$ satisfies a Carleson packing condition will be to associate, to each cube $Q \in \mathscr{H}$, a measurable set $E(Q) \subset Q$ such that

$$|E(Q)| \geq C^{-1}|Q|, \tag{2.79}$$

in such a way that the $E(Q)$'s have bounded overlap, i.e.,

(2.80) each $x \in E$ never belongs to more than C different sets $E(Q)$, $Q \in \mathscr{H}$.

The constant C will depend on δ through ϵ and k, but the main point is that it will not depend on x or Q. Once we have the sets $E(Q)$, the Carleson packing condition on $\mathscr{H}$ will follow at once from (2.79), (2.80) and the regularity of E.

Let us now fix a cube $Q \in \mathscr{H}$ and construct $E(Q)$. First let $c(Q)$ be a center for Q, i.e., a point of Q such that

$$\operatorname{dist}(c(Q), E \setminus Q) \geq C_1^{-1} \operatorname{diam} Q, \tag{2.81}$$

where the constant C_1 depends only on n, d, and the regularity constant for E. (See Lemma I.3.5 for the existence of such a point.)

LEMMA 2.82. *There is an $x_0 \in Q$ such that $|x_0 - c(Q)| \leq (\operatorname{diam} Q / 2C_1)$ and*

$$\text{(2.83)} \quad \text{every point in } P(Q) \cap B\left(x_0, \frac{\operatorname{diam} Q}{C_2}\right) \text{ is at a distance}$$
$$\leq 3\epsilon \operatorname{diam} Q \text{ from } E,$$

where C_2 is a geometric constant (i.e., a constant that depends only on n, d, and the regularity constant for E).

We can always require that $C_2 > 2C_1$, so that $E \cap B(x_0, (\operatorname{diam} Q)/C_2)$ will be contained in Q.

To prove this lemma, we first choose $d+1$ points $y_0, y_1, \ldots, y_d$ in Q so that their convex hull contain a d-dimensional disk of radius $C^{-1} \operatorname{diam} Q$ (i.e., the intersection of a ball of radius $C^{-1} \operatorname{diam} Q$ with some d-plane passing through the center of the ball). For instance, the points that were used in the proof of Lemma 2.75 would work here too. Then the convex hull of $c(Q)$ and the points $y_0, \ldots, y_d$ contains a d-dimensional disk D_0 which is contained in $B(c(Q), (\operatorname{diam} Q)/2C_1)$ and has a radius $R \geq (\operatorname{diam} Q)/C'$ for some geometric constant C'.

Because $Q \notin \mathscr{B}(k, \epsilon)$ (by definition of $\mathscr{H}$ and $\mathscr{G}$), every point of D_0 is at distance $\leq \epsilon \operatorname{diam} Q$ from E (see (2.74)). We take x_0 to be any point of E that is at distance $\leq \epsilon \operatorname{diam} Q$ from the center of D_0. Set $B_0 = B(x, (\operatorname{diam} Q)/C_2)$, where we set $C_2 = 2C' + 2C_1$, with C' as above.

We have to check that every point of $P(Q) \cap B_0$ is at a distance $\leq 3\epsilon \operatorname{diam} Q$ from E.

Notice that every point of D_0 is within $\epsilon \operatorname{diam} Q$ of E and, therefore, within $2\epsilon \operatorname{diam} Q$ of $P(Q)$, because (2.76) is true (since $Q \in \mathscr{H} \subset \mathscr{G}$). It then follows that every point of $P(Q) \cap B_0$ is at a distance $\leq 2\epsilon \operatorname{diam} Q$ from D_0 and, consequently, at a distance $\leq 3\epsilon \operatorname{diam} Q$ from E, as needed. [Keep in mind that we can take ϵ as small as we want.] This proves Lemma 2.82.

For each (large) integer $N > 0$, let $\chi_N(Q)$ denote the set of points of Q which belong to more than N different cubes Q' in $\mathscr{W} = \Delta \setminus \mathscr{G}$ such that $Q' \subset Q$. We know that $\mathscr{W}$ satisfies a Carleson packing condition (see the comment that follows (2.77)), and therefore

(2.84) $$|\chi_N(Q)| \leq C(\epsilon, k) N^{-1} |Q|.$$

We choose $N = N(\epsilon, k)$ sufficiently large so that this implies $|\chi_N(Q)| \leq \frac{1}{2} |E \cap (\frac{1}{2} B_0)|$, where B_0 is as above, and we set

(2.85) $$E(Q) = (E \cap \frac{1}{2} B_0) \setminus \chi_N(Q).$$

Thus $E(Q)$ is contained in Q, by the comment following the statement of Lemma 2.82, and it satisfies (2.79), so we only have to check that the $E(Q)$'s, $Q \in \mathscr{H}$, have bounded overlap. The following is the key estimate.

LEMMA 2.86. *There is a geometric constant C_3 such that the following holds. If Q and Q' are two cubes of $\mathscr{H}$ such that $\operatorname{diam} Q' \geq C_3 \operatorname{diam} Q$ and if $E(Q) \cap E(Q') \neq \varnothing$, then there is a cube $R \in \mathscr{W}$ such that $Q \subsetneq R \subsetneq Q'$.*

Before we prove the lemma, let us check that it implies the bounded overlap estimate (2.80). Let $x \in E$ be given, and suppose that $x \in E(Q_1) \cap E(Q_2) \cap \cdots \cap E(Q_L)$ for a collection $Q_1, Q_2, \ldots, Q_L$ of distinct cubes in $\mathscr{H}$. The bounded overlap estimate (2.80) will of course follow if we can prove that $L \leq (N+1)C$ for some geometric constant C. Notice that $x \in Q_i$ for $i = 1, \ldots, L$. Thus we can suppose, without loss of generality, that $Q_1 \subset Q_2 \cdots \subset Q_L$. Let us assume that $L > (N+1)C$ and prove that this leads to a contradiction. If C is large enough (depending on C_3 and the regularity constant for E), we can find $N+2$ cubes $R_1 \subset R_2 \cdots \subset R_{N+2}$, chosen among the Q_i, with the additional property that $\operatorname{diam} R_{j+1} > C_3 \operatorname{diam} R_j$ for $j = 1, \ldots, N+1$. Applying Lemma 2.86, we get cubes $S_j \in \mathscr{W}$ such that $R_j \subsetneq S_j \subsetneq R_{j+1}$ for $j = 1, \ldots, N+1$. Thus x, which belongs to all of the R_j's, lies in $N+1$ different cubes of $\mathscr{W}$ which are all contained in R_{N+2}. This contradicts the fact that $x \in E(R_{N+2})$, and thus our assumption that $L > C(N+1)$ was wrong. This proves that Theorem 2.71 follows from Lemma 2.86.

We now prove Lemma 2.86. Let Q and Q' be two cubes of $\mathscr{H}$ such that $\operatorname{diam} Q' \geq C_3 \operatorname{diam} Q$, where C_3 is large and will be specified later, and suppose that $x \in E(Q) \cap E(Q')$. Let $Q_0 = Q \subset Q_1 \subset \cdots \subset Q_l = Q'$ be the sequence of ancestors of Q that are contained in Q'. [Thus Q_{j+1} is the parent of Q_j for $j = 0, 1, \ldots, l-1$.] We want to prove that if C_3 is large enough (i.e., if l is large enough), then one of the cubes Q_j must be in the bad set $\mathscr{W}$. [Notice that Q and Q' are not in $\mathscr{W}$ because they lie in $\mathscr{H}$, so the bad cube will have to be strictly between Q and Q', as required for Lemma 2.86.]

As usual, it will be slightly easier to proceed by contradiction. So suppose that all the Q_j's lie in $\mathscr{G}$, and let us prove that this leads to a contradiction. Let m be the integer such that $Q \in \Delta_m$. [Thus $\operatorname{diam} Q \approx 2^m$.] For each $j \in \{0, 1, \ldots, l\}$, let

$$(2.87) \qquad D_j = P(Q_j) \cap B\left(x, C_4^{-1} 2^{m+j}\right),$$

where C_4 is a geometric constant that will be chosen soon. We first want to prove that for each j,

$$(2.88) \qquad \operatorname{dist}(z, E) \leq 3\epsilon \operatorname{diam} Q_j \quad \text{for all } z \in D_j.$$

We shall see later that this estimate prevents the existence of any significant hole in E near Q_0 and thus will contradict the fact that $Q_0 = Q \in \mathscr{H}$, but for the moment let us establish (2.88).

We start with the case when $j = l$. Because

$$x \in E(Q') \subset B\left(x'_0, \frac{\operatorname{diam} Q'}{2C_2}\right),$$

where x'_0 is the point of Q' that was chosen for Lemma 2.82, (2.88) for $j = l$ follows from (2.83) (applied to Q') by choosing C_4 so large that $C_4^{-1} 2^{m+l} < (\operatorname{diam} Q')/2C_2$.

We now suppose that $0 \leq j < l$ and that (2.88) holds for $j+1$, and we prove it for j. Notice that D_j is significantly smaller than D_{j+1} (which will help us), but we want a better approximation of the points in D_j by points in E than we have for D_{j+1}. We shall get this better control by noticing that all points in D_j are very close to points in the convex hull of points in E and then using our assumption that Q_j is not in the bad set $\mathscr{B}(k, \epsilon)$.

So let $z \in D_j$ be given. Choose $d+1$ points $w_0, w_1, \ldots, w_d \in \frac{3}{2} D_j$ such that z lies inside their convex hull and even such that z lies inside the convex hull of $\tilde{w}_0, \tilde{w}_1, \ldots, \tilde{w}_d$ for all choices of points $\tilde{w}_i \in P(Q_j)$ with $|\tilde{w}_i - w_i| \leq 2^{m+j}/10C_4$ for $i = 0, 1, \ldots, d$.

The approximating d-planes $P(Q_j)$ and $P(Q_{j+1})$ defined by (2.76) are very close to each other. [Recall that this can be proved by noticing that both $P(Q_j)$ and $P(Q_{j+1})$ must pass at a distance $\leq \epsilon \operatorname{diam} Q_{j+1}$ from $d+1$ given affinely independent points of Q_j.] Thus we can find points $v_i \in P(Q_{j+1})$ such that $|v_i - w_i| \leq C\epsilon \operatorname{diam} Q_j$. If ϵ is small enough, the points v_i are all in D_{j+1}, so (2.88) for $j+1$ tells us that there exist points $e_i \in E$

with $|e_i - v_i| \leq 3\epsilon \operatorname{diam} Q_{j+1}$. Let $\tilde{w}_i$ be the orthogonal projection of e_i onto $P(Q_j)$. Applying (2.76) to Q_j (with k large enough), we get that $|\tilde{w}_i - e_i| \leq \epsilon \operatorname{diam} Q_j$, and therefore $|\tilde{w}_i - w_i| \leq C\epsilon \operatorname{diam} Q_j < 2^{m+j}/10C_4$ if ϵ is small enough. Consequently, our given point $z \in D_j$ lies in the convex hull of the $\tilde{w}_i$ and hence is at distance $\leq \epsilon \operatorname{diam} Q_j$ from the convex hull of the e_i. Since we have assumed that $Q_j \in \mathscr{G}$, Q_j is not in $\mathscr{B}(k, \epsilon)$, and so $\operatorname{dist}(z, E) \leq 2\epsilon \operatorname{diam} Q_j$ (if k is large enough). This proves (2.88).

Let us now derive a contradiction from (2.88). Since Q is in the bad set $\mathscr{H}$ and $x \in E(Q) \subset Q$, there is a point $w \in P(Q)$ such that $\operatorname{dist}(w, x) \leq 2 \operatorname{diam} Q$ but $\operatorname{dist}(w, E) \geq \delta \operatorname{diam} Q$. (See the definition (2.78) of $\mathscr{H}$.) Let j_0 be the smallest integer such that $C_4^{-1} 2^{m+j_0} > 4 \operatorname{diam} Q$. Of course, j_0 is less than a constant. We now choose the constant C_3 in the statement of Lemma 2.86 so large that $j_0 \leq l$, so Q_{j_0} is among the cubes $Q_0, \ldots, Q_l$. Choosing k large enough and ϵ small enough (depending on δ and the maximum possible value of j_0), we get that the d-planes $P(Q)$ and $P(Q_{j_0})$ are so close to each other that $\operatorname{dist}(w, P(Q_{j_0})) \leq C\epsilon \operatorname{diam} Q < \frac{\delta}{2} \operatorname{diam} Q$. Hence there is a point $z \in P(Q_{j_0})$ such that $|z - w| < \frac{\delta}{2} \operatorname{diam} Q$. This point lies in D_{j_0} because of our choice of j_0, and the fact that $\operatorname{dist}(z, E) > \frac{\delta}{2} \operatorname{diam} Q$ contradicts (2.88) if ϵ is small enough. This completes our proof of Lemma 2.86. Theorem 2.71 follows.

REMARK. This proof is somewhat similar to the argument given in Chapter III.5 for proving Theorem I.2.56 when $d = n - 1$. The reader might wish to consider the exercise of rewriting the proof in a manner that more closely resembles the approach taken in §§III.5.2–4.

2.5. The weaker local convexity condition WLCV.

We are going to describe now a variation of the local convexity condition studied in the previous section.

The following notation will be convenient. For $x, y \in \mathbf{R}^n$, $I(x, y)$ will denote the closed line segment that joins x to y, and, for $0 < \rho \leq \frac{1}{2}$, we let
(2.89)
$$I_\rho(x, y) = \{u \in I(x, y) : \operatorname{dist}(u, x) \geq \rho|x-y| \text{ and } \operatorname{dist}(u, y) \geq \rho|x-y|\}.$$

Thus, when $\rho = \frac{1}{2}$, $I_\rho(x, y)$ is reduced to the midpoint $\frac{x+y}{2}$ while, when ρ is smaller, $I_\rho(x, y)$ is the interval represented in Figure 2.2.

DEFINITION 2.90. Let E be a d-dimensional regular set in $\mathbf{R}^n$. We say that E satisfies the weaker local convexity condition (WLCV) if there exists

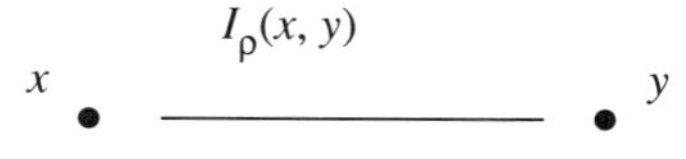

FIGURE 2.2. $I_\rho(x, y)$.

2.5. THE WEAKER LOCAL CONVEXITY CONDITION

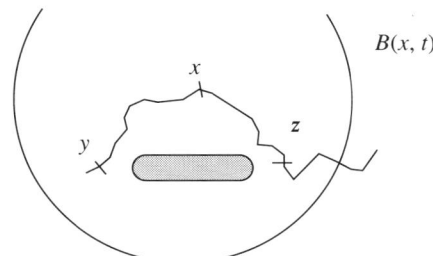

FIGURE 2.3. This (x, t) lies in $\mathscr{W}(\rho, \epsilon)$ because the tube does not meet E.

a $\rho \in (0, \frac{1}{2}]$ such that, for each $\epsilon > 0$, the set
(2.91)
$$\mathscr{W}(\rho, \epsilon) = \{(x, t) \in E \times \mathbf{R}_+ : \text{ there exist points } y, z \in E \cap B(x, t)$$
$$\text{that satisfy } \operatorname{dist}(I_\rho(y, z), E) \geq \epsilon t\}$$

is a Carleson set.

PROPOSITION 2.92. *If a d-dimensional regular set E in $\mathbf{R}^n$ satisfies the WLCV, then it satisfies the LCV (and is therefore uniformly rectifiable).*

The WCLV is clearly weaker than the LCV (which corresponds to the case $\rho = \frac{1}{2}$). When $d = 1$, the WLCV is still stronger (in appearance!) than the weak connectedness condition; thus Proposition 2.92 is of interest only when $d \geq 2$.

To prove the proposition, let a (small) constant $\delta > 0$ and a ρ, $0 < \rho \leq \frac{1}{2}$, be given. We shall prove that there is an $\epsilon > 0$ with the property that if (x, t) is not in the bad set $\mathscr{W}(\rho, \epsilon)$, then $\operatorname{dist}\left(\frac{y+z}{2}, E\right) \leq \delta t$ for all $y, z \in E \cap B(x, t)$. Proposition 2.92 will follow at once (compare with Definition I.2.7). Notice, incidentally, that in order to prove that E is uniformly rectifiable, it is enough to show that the set (2.70) is a Carleson set for a single value of ϵ (that may be computed in terms of n, d, and the regularity constant for E). It follows from this and the proof given below that for each ρ, $0 < \rho \leq \frac{1}{2}$, there is an $\epsilon > 0$ (depending on n, d, ρ, and the regularity constant for E) such that E is uniformly rectifiable as soon as $\mathscr{W}(\rho, \epsilon)$ is a Carleson set.

The idea of the proof is quite simple. Given $(x, t) \in E \times \mathbf{R}_+ \setminus \mathscr{W}(\rho, \epsilon)$ and y, z in $E \cap B(x, t)$, we shall be able to find a sequence of points in E that get closer and closer to $\frac{y+z}{2}$, by applying repeatedly our hypothesis that $(x, t) \notin \mathscr{W}(\rho, \epsilon)$. If we were able to take $\epsilon = 0$, we could start from the interval $I(y, z)$ and then construct a nested sequence of intervals $I_0 = I(y, z) \supset I_1 \supset \cdots \supset I_m \supset \cdots$ by replacing each time one of the end points of I_m by another point of $E \cap I_m$, in such a way that $\frac{y+z}{2}$ would still belong to I_{m+1}, and also $|I_{m+1}| \leq (1 - \rho)|I_m|$. The fact that we cannot take $\epsilon = 0$ will force us to introduce errors of size about ϵt at each step of the construction. This will not cause a significant problem because we shall be able to take ϵ as small as we wish and because the total number of steps required to get within δt of $\frac{y+z}{2}$ will be controlled.

Let us now describe the construction more precisely.

Let $(x, t) \in \mathscr{W}(\rho, \epsilon)$ be given, and fix $y, z \in E \cap B(x, t)$. We want to prove that the midpoint $w = \frac{y+z}{2}$ satisfies $\text{dist}(w, E) \leq \delta t$ (if we take ϵ small enough, depending on δ and ρ). To do this, we shall construct two sequences $\{y_m\}$ and $\{z_m\}$ of points of $E \cap B(x, t)$, with the hope that they will eventually be at distance $\leq \delta t$ from w. If we ever have that

$$(2.93) \qquad |y_m - w| \leq \delta t \text{ or } |z_m - w| \leq \delta t,$$

we shall stop the construction of the sequences immediately and be happy. Otherwise, we shall continue the construction (until (2.93) eventually occurs). Our two sequences will also have the following additional features:

(2.94) $(\frac{\rho}{2})^m |y - z| \leq |y_m - z_m| \leq (1 - \frac{\rho}{2})^m |y - z|$;

(2.95) $\text{Max}\{\text{dist}(y_m, I(y, z)), \text{dist}(z_m, I(y, z))\} \leq m\epsilon t$;

(2.96) if Π denotes the orthogonal projection onto the line through y and z, then $w \in I(\Pi(y_m), \Pi(z_m))$.

Let us now construct the two sequences. We start with $y_0 = y$ and $z_0 = z$. These two points obviously satisfy (2.94)–(2.96). Now assume that we have already constructed our sequences up to y_m and z_m, with the properties above, and let us choose y_{m+1} and z_{m+1}. As we explained before, we assume that (2.93) has not occurred yet.

From (2.96) we have that $\text{dist}(w, y_m) \leq |y_m - \Pi(y_m)| + |\Pi(y_m) - \Pi(z_m)|$, so

$$(2.97) \qquad \text{dist}(w, y_m) \leq m\epsilon t + |y_m - z_m| \leq m\epsilon t + \left(1 - \frac{\rho}{2}\right)^m |y - z|,$$

because of (2.95) and (2.96).

Since (2.93) is not true, we obtain

$$\delta t < \text{dist}(w, y_m) \leq m\epsilon t + 2\left(1 - \frac{\rho}{2}\right)^m t.$$

In fact, (2.93) has not occurred yet, so $\delta < j\epsilon + 2(1 - \frac{\rho}{2})^j$ for all $0 \leq j \leq m$. In particular, if j_0 is the smallest integer such that $(1 - \frac{\rho}{2})^{j_0} < \frac{\delta}{4}$ and if we choose ϵ smaller than $\delta/2j_0$, we see that m must be $< j_0$.

Since y_m and z_m both belong to $E \cap B(x, t)$ by induction hypothesis, the fact that $(x, t) \notin \mathscr{W}(\rho, \epsilon)$ implies there is a point $u_m \in E$ such that

$$(2.98) \qquad \text{dist}(u_m, I_\rho(y_m, z_m)) \leq \epsilon t.$$

We want to take $y_{m+1} = u_m$, so we should check that $u_m \in B(x, t)$. Notice first that $|y - z| \geq \delta t$, since otherwise (2.93) holds when $m = 0$.

FIGURE 2.4

Combining this with (2.94) and our bound on m we have that $|y_m - z_m| \geq C(\rho, \delta)^{-1} t$. The strict convexity of $B(x, t)$ now implies that

$$\text{dist}\left(I_\rho(y_m, z_m), \mathbf{R}^n \setminus B(x, t)\right) \geq C'(\rho, \delta)^{-1} t,$$

since $y_m, z_m \in B(x, t)$. Therefore, u_m does lie in $B(x, t)$ if ϵ is small enough.

Since $|y_m - z_m| \geq C(\rho, \delta)^{-1} t$, we also have that $|\Pi(y_m) - \Pi(z_m)| \geq C''(\rho, \delta)^{-1} t$, at least if ϵ is small enough, because of (2.95) and our upper bound on m. Hence

$$\Pi(u_m) \in I(\Pi(y_m), \Pi(z_m))$$

if ϵ is small enough, because of (2.98). (See Figure 2.4.)

Set $y_{m+1} = u_m$, and take z_{m+1} to be y_m if $w \in I(\Pi(y_m), \Pi(u_m))$ and take z_{m+1} to be z_m otherwise. Thus y_{m+1} and z_{m+1} lie in $E \cap B(x, t)$ and satisfy (2.96) by construction, and it is easy to check that (2.94) and (2.95) are also satisfied (with m replaced by $m+1$) using (2.98).

This finishes our recursive construction of the y_m's and the z_m's. We have seen that this process must stop in a finite number of steps, so it produces an element of E at distance $\leq \delta t$ from w. This completes the proof of Proposition 2.92.

2.6. Weak starlikeness.

DEFINITION 2.99. Let E be a d-dimensional regular set in $\mathbf{R}^n$. We say that E satisfies the central starlikeness condition (CSL) if for each $\epsilon > 0$ we have that $E \times \mathbf{R}_+ \setminus \mathscr{G}_s(\epsilon)$ is a Carleson set, where

$$\mathscr{G}_s(\epsilon) = \{(x, t) \in E \times \mathbf{R}_+ : \text{ if } y \in E \cap \overline{B}(x, t) \text{ and } z \text{ lies on the}$$
$$\text{line segment which joins } x \text{ to } y, \text{ then } \text{dist}(z, E) \leq \epsilon t\}.$$

In other words, E satisfies the CSL if $E \cap \overline{B}(x, t)$ is approximately starlike with respect to x for most $(x, t) \in E \times \mathbf{R}_+$. It is an easy exercise to check that E satisfies the CSL if and only if it lies in $\text{Approx}(\mathscr{S})$, where $\mathscr{S}$ denotes the collection of closed subsets of $\mathbf{R}^n$ which are starlike about the origin, and $\text{Approx}(\cdot)$ is as in Definition I.2.21.

PROPOSITION 2.100. *The CSL is equivalent to the LCV (and hence uniform rectifiability also).*

This is fairly trivial. The LCV implies the CSL because of Lemma 2.72, for instance. Let us prove the converse. Let E be a regular set which satisfies the CSL, and set

$$\mathscr{G}_s^*(\epsilon) = \{(x, t) \in E \times \mathbf{R}_+ : (y, 2t) \in \mathscr{G}_s(\epsilon/2) \text{ whenever } y \in E \cap \overline{B}(x, t)\}.$$

It is not hard to show that $E \times \mathbf{R}_+ \setminus \mathscr{G}_s^*(\epsilon)$ must be a Carleson set for each $\epsilon > 0$, using the easy fact that $(y, 2t) \in \mathscr{G}_s(\epsilon/2)$ as soon as $(z, r) \in \mathscr{G}_s(\epsilon/10)$ for any $(z, r) \in E \times \mathbf{R}_+$ such that $|z - y| \leq \epsilon t/10$ and $3t \leq r \leq 4t$. [To put it differently, we are saying that if $(x, t) \notin \mathscr{G}_s^*(\epsilon)$, then there must be a substantial set of $(z, r) \in E \times \mathbf{R}_+$ near (x, t) such that $(z, r) \notin \mathscr{G}_s(\epsilon/10)$.] On the other hand, if (x, t) lies in $\mathscr{G}_s^*(\epsilon)$, then (x, t) will not lie in the bad set (2.70) for the LCV. This proves the proposition.

The CSL is slightly amusing because it appears to be less rigid than the LCV. Of course, this is not really true, as the preceding argument shows.

A more subtle condition arises if we drop the "central" from CSL. For example, we could consider the class of regular sets E which lie in Approx($\mathscr{S}_r$), where $\mathscr{S}_r$ denotes the collection of closed subsets of $\mathbf{R}^n$ which are starlike with respect to some point in $\overline{B}(0, r)$ (instead of the origin). We do not know if the resulting class of sets must be uniformly rectifiable, except when $d = 1$, which is covered by the WCC, and when $r > 0$ is sufficiently small, how small depending on d, n, and the regularity constant for the given set E. This last observation follows from the fact that uniform rectifiability is implied by the weaker version of the CSL in which $E \times \mathbf{R}_+ \setminus \mathscr{G}_s(\epsilon)$ is required to be a Carleson set for only one very small ϵ which depends on d, n, and the regularity constant for E. This fact is in turn a consequence of the corresponding statements for the BWGL and the LCV. (See Remark 2.5.)

Note that there is a substantial difference between $r < 1$ and $r \geq 1$ in this noncentral version of the CSL.

For the record let us state the condition which has the same relationship with the CSL that the WLCV has with the LCV.

DEFINITION 2.101. A d-dimensional regular set E in $\mathbf{R}^n$ satisfies the weaker central starlikeness condition (WCSL) if there is a $\rho \in (0, \frac{1}{2}]$ such that $E \times \mathbf{R}_+ \setminus \mathscr{G}_s(\epsilon, \rho)$ is a Carleson set for every $\epsilon > 0$, where

$$\mathscr{G}_s(\epsilon, \rho) = \{(x, t) \in E \times \mathbf{R}_+ : \text{dist}(I_\rho(y, x), E) \leq \epsilon t$$
$$\text{for all } y \in E \cap \overline{B}(x, t)\}.$$

Here $I_\rho(x, y)$ is as in (2.89).

PROPOSITION 2.102. *The WCSL is equivalent to the WLCV (and hence uniform rectifiability too).*

We leave the proof of this as an exercise.

2.7. Some questions about variants of the LCV and the LS.

To conclude this chapter, let us mention that there are many weaker versions of the local symmetry condition and the local convexity condition for which we do not know whether they imply the BWGL. Here are some examples.

Let E be a regular set of dimension d in $\mathbf{R}^n$. For each τ, $0 < \tau < \frac{1}{2}$, let $\mathscr{G}(\tau)$ be the set of pairs $(x, t) \in E \times \mathbf{R}_+$ for which, whenever $y, z \in E \cap B(x, t)$ are such that $|y - z| \geq \frac{t}{10}$, there is an element of E in the ball $\overline{B}\left(\frac{y+z}{2}, \tau|y - z|\right)$. (See Figure 2.5.) For which values of τ is it true that if the complement of $\mathscr{G}(\tau)$ is a Carleson set, then E is uniformly rectifiable?

We know that, when $d = 1$, any value of $\tau < \frac{1}{2}$ will work, because the condition above implies the weak connectedness condition WCC of Chapter 1. Also, for any dimensions d and n, we know that for each $C_0 > 0$ there is a $\tau = \tau(C_0, d, n)$ such that if E is regular with constant $\leq C_0$ and the complement of $\mathscr{G}(\tau)$ is a Carleson set, then E is uniformly rectifiable. This can be derived from our proof of the fact that the local convexity condition implies uniform rectifiability.

On the other hand, we cannot expect that all $0 < \tau < \frac{1}{2}$ will work for all integer values of the dimensions n and d. One way to produce counterexamples is to construct, for any $d > 1$, a curve $\Gamma \subset \mathbf{R}^n$ (where the dimension n depends on d and might be quite large) which is Ahlfors-regular of dimension d, and for which all $(x, t) \in \Gamma \times \mathbf{R}_+$ lie in $\mathscr{G}(\tau)$ for some $\tau = \tau(d) < \frac{1}{2}$. Rather than saying any more about this approach let us describe a different way to generate counterexamples.

These counterexamples will be obtained by taking Cartesian products of sufficiently many quasicircles with dimensions close to 1. First observe that, for each $0 < \tau < \frac{1}{2}$, there is a dimension $d(\tau) > 1$ such that, for any $\lambda \in (1, d(\tau))$, one can find a quasicircle $\Gamma_\lambda \subset \mathbf{R}^2$ which is Ahlfors-regular of dimension λ and for which all $(x, t) \in \Gamma_\lambda \times \mathbf{R}_+$ belong to $\mathscr{G}(\tau/2)$, say. [Use the standard snowflake construction but with flatter snowflakes.] For each integer $d > 1$ we can construct a set E by first choosing an integer m so that $d^{1/m} < d(\tau)$ and then taking E to be the m-fold product $\Gamma_\lambda \times \cdots \times \Gamma_\lambda \subset \mathbf{R}^{2m}$, with $\lambda = d^{1/m}$. Then E is regular with dimension d, and the fact that all $(x, t) \in E \times \mathbf{R}_+$ belong to $\mathscr{G}(\tau)$ follows from the same property for Γ_λ. The

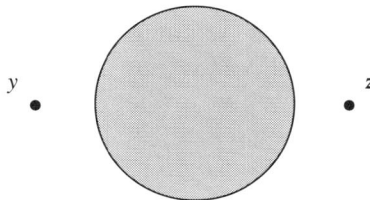

FIGURE 2.5. There should be a point of E in this ball if $(x, t) \in \mathscr{G}(\tau)$.

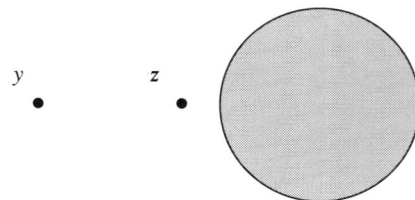

FIGURE 2.6. There should be a point of E in this ball if $(x, t) \in \mathscr{G}'(\tau)$.

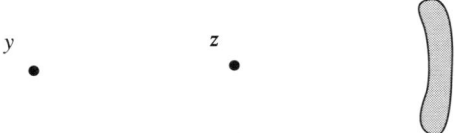

FIGURE 2.7. A variant of the condition defining $\mathscr{G}'(\tau)$.

set E is a counterexample because it is not even rectifiable.

Unlike the local convexity and the local symmetry conditions, the requirement that $E \times \mathbf{R}_+ \setminus \mathscr{G}(\tau)$ be a Carleson set (for a fixed τ) can be satisfied by Ahlfors-regular sets of dimension d even when d is not an integer. We do not know the precise set of triples (τ, d, n) for which this is possible.

The local symmetry condition also has weaker variants that do not seem to be too easy to work with. For instance, let $\mathscr{G}'(\tau)$ be the set of $(x, t) \in E \times \mathbf{R}_+$ for which, whenever y, z are two points of $E \cap B(x, t)$ such that $|y - z| \geq \frac{t}{10}$, there is a point of E in the ball $B(2z - y, \tau|y - z|)$. (See Figure 2.6.)

For which values of $0 < \tau < 1$ is it true that if the complement of $\mathscr{G}'(\tau)$ is a Carleson set, then E is uniformly rectifiable?

This time we do not even know the answer when $d = 1$. As before, it follows from the proof of the fact that the local symmetry condition implies uniform rectifiability that for each choice of n, d, and the regularity constant C_0 there is a $\tau = \tau(n, d, C_0)$ that works. Also, the same examples as for $\mathscr{G}(\tau)$ show that for all values of d and τ there is a regular set of E of dimension d in some $\mathbf{R}^n$ that is totally unrectifiable and for which $\mathscr{G}'(\tau)$ is the whole $E \times \mathbf{R}_+$.

We could also ask the same question with the ball $B(2z - y, \tau|y - z|)$ in the definition of $\mathscr{G}'(\tau)$ replaced by its intersection with a thin neighborhood of the sphere centered at z and with radius $|y - z|$. (See Figure 2.7.) Here again, we do not know the answer even when $d = 1$.

These conditions should be compared with the weaker local symmetry condition (see Definition 1.20), which we do understand when $d = 1$, anyway.

One can also look for variants of the LCV which are more topological. When $d = 1$ we already have the WCC, and it would be hard to beat that, but $d > 1$ is an entirely different matter. We would like to have sufficient

2.7. VARIANTS OF THE LCV AND THE LS

conditions for uniform rectifiability which are given in terms of our regular set E being well approximated by sets which have no holes, in much the same way that the LCV is about approximating E by compact convex sets. However, one can produce examples (like the ones above) which show that this is not so easy to do. The simple fact that there are 2-dimensional regular sets which are quasicircles is already an indication of trouble.

If we toss in the WGL, then it is probably a different story. The WGL, which does not imply uniform rectifiability by itself, requires that the given set E be well approximated by subsets of d-planes at most scales and locations; if we demanded also that these subsets be contractible, then perhaps we would get a sufficient condition for uniform rectifiability. It is not so clear what happens even if we simply combine the WGL with the WCC.

The preceding section provides a good starting place for looking at these issues, since we understand the CSL but not its noncentral variants. Chapter 4 is also relevant here.

CHAPTER 3

The WHIP and Related Conditions

3.1. The WHIP, the WTP, and uniform rectifiability.
Let E be, as usual, a d-dimensional regular set in $\mathbf{R}^n$. We are going to define two new "weak conditions"—the WHIP and the WTP—which together imply that E is uniformly rectifiable. We begin with the definition of the WHIP, which is by far the more substantial of the two.

We first need to define a good set $\mathscr{G}(\alpha, A, N) \subset E \times \mathbf{R}_+$ for all choices of a (small) constant α, $0 < \alpha < \frac{1}{10}$, a (large) constant $A \geq 1$, and a (large) integer $N > 0$.

Let $\mathscr{G}(\alpha, A, N)$ be the set of pairs $(x, t) \in E \times \mathbf{R}_+$ such that, for all d-planes P, the following is true. Let Π denote the orthogonal projection onto P and let B be any (open) ball of radius αt inside P which is contained in $\Pi(B(x, \frac{t}{2}))$. If

(3.1) $B \cap \Pi(E \cap B(x, t)) = \varnothing$ and $\partial B \cap \Pi(E \cap \overline{B}(x, \frac{t}{2})) \neq \varnothing$,

then (we require that)

(3.2)   there exist N points $x_1, x_2, \ldots, x_N$ in $E \cap B(x, At)$ such that $|x_i - x_j| \geq \alpha t$ whenever $i \neq j$, and $\Pi(x_i) \in AB$ for all i.

Here ∂B denotes the boundary of B (in P), and AB is the ball in P with the same center as B and radius multiplied by A.

The existence of a ball B satisfying (3.1) can be interpreted as the existence of a hole in the projection of $E \cap B(x, t)$; the pairs (x, t) in $\mathscr{G}(\alpha, A, N)$ are those for which the existence of such a hole implies the existence of points as in (3.2), i.e., many points $x_i \in E \cap B(x, At)$ whose projections are all close to B. (We shall call this a post.) See Figure 3.1 for a symbolic picture of a hole; the apparent "hole" in Figure 3.2 is not considered a hole for the purposes of the definition of the WHIP because ∂B does not meet $\Pi(E \cap \overline{B}(x, \frac{t}{2}))$. The exclusion of this case makes the WHIP a little more stable and easy to use.

To make our definition even more stable, we define a smeared-up version of $\mathscr{G}(\alpha, A, N)$ by

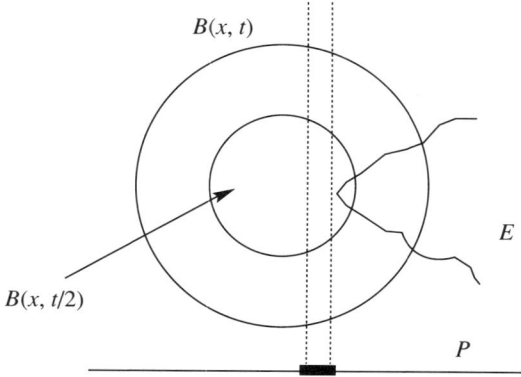

FIGURE 3.1. A hole.

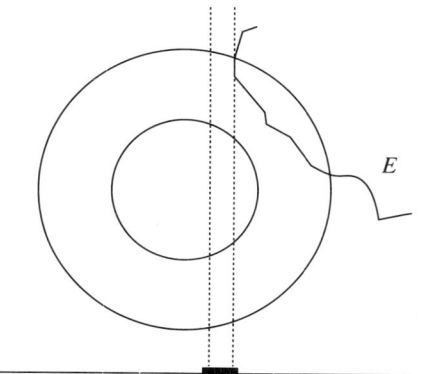

FIGURE 3.2. Not a hole.

(3.3)
$$\mathscr{G}^*(\alpha, A, N) = \{(x, t) \in E \times \mathbf{R}_+ : \text{ all pairs } (y, s) \in E \times \mathbf{R}_+ \text{ such that}$$
$$|y - x| \leq t \text{ and } \tfrac{t}{2} \leq s < t$$
$$\text{lie in } \mathscr{G}(\alpha, A, N)\}.$$

DEFINITION 3.4. We say that a d-dimensional regular set E in $\mathbf{R}^n$ satisfies the WHIP (weak "holes imply posts" condition) if there is an $A \geq 1$ so that, for every (large) integer $N > 0$, one can find an α, $0 < \alpha < \tfrac{1}{10}$, so that $E \times \mathbf{R}_+ \setminus \mathscr{G}^*(\alpha, A, N)$ is a Carleson set.

(See Definition I.1.69 for Carleson sets.)

There are some minor variations of this definition that are a priori weaker but which are nonetheless equivalent. The formulation which we have given here seems, however, to be the most convenient. The definition of the WHIP is a little painful, but there are some simpler and more geometrically significant conditions that imply it. We shall discuss some examples in the next three sections, but for the moment let us state the main theorem of this chapter.

We would like to say that the WHIP implies uniform rectifiability, but at present we need an auxiliary condition (the WTP) which is fairly mild and perhaps even automatic.

Given $\alpha > 0$ and $a > 0$ (both small), let $\mathscr{H}(\alpha, a)$ be the set of all $(x, t) \in E \times \mathbf{R}_+$ for which there is a d-plane P such that

$$(3.5) \qquad |\{p \in P : \text{dist}(p, \Pi(E \cap B(x, t))) < \alpha t\}| \geq at^d,$$

where Π denotes the orthogonal projection onto P and $|\cdot|$ stands for the Lebesgue measure on P. In other words, $(x, t) \in \mathscr{H}(\alpha, a)$ if, for some P, the αt-neighborhood of $\Pi(E \cap B(x, t))$ has reasonably large mass.

DEFINITION 3.6. The d-dimensional regular set E in $\mathbf{R}^n$ satisfies the WTP (weak "thick projections" condition) if there is an $a > 1$ such that, for every $\alpha > 0$, $E \times \mathbf{R}_+ \setminus \mathscr{H}(\alpha, a)$ is a Carleson set.

The WTP is somewhat similar to the big projections condition in Definition I.1.74 (which clearly implies the WTP). However, the fact that we may look at a given level of precision (specified by α) and throw away a Carleson set of pairs (x, t) before checking (3.5) makes the WTP much weaker and easier to verify. In particular, the BWGL implies the WTP trivially, but even more is true:

$$(3.7) \qquad \text{the WGL implies the WTP.}$$

Let us sketch the easy proof. Let $\alpha > 0$ be given, and suppose that $(x, t) \in E \times \mathbf{R}_+$ is such that $E \cap B(x, t)$ stays at distance $\leq \frac{\alpha t}{10}$ from some d-plane P. [We know that if E satisfies the WGL, this will be the case for all but a Carleson set of (x, t)'s.] Next, consider a maximal set of points of $E \cap B(x, t)$ that are at mutual distances $\geq 3\alpha t$; because E is regular, there are more than $C^{-1}\alpha^{-d}$ points in this set. On the other hand, the projections onto P of the balls centered on these points and with radius αt are all disjoint and contained in the set on the left side of (3.5). This proves (3.5) with a constant a that depends only on the regularity constant for E, and (3.7) follows.

REMARK 3.8. Define $\mathscr{H}^*(\alpha, a)$ in terms of $\mathscr{H}(\alpha, a)$ in the same way that $\mathscr{G}^*$ was defined in terms of $\mathscr{G}$ in (3.3). It is easy to check that E satisfies the WTP if and only if there is an $a > 0$ so that $E \times \mathbf{R}_+ \setminus \mathscr{H}^*(\alpha, a)$ is a Carleson set for every $\alpha > 0$.

THEOREM 3.9. *Let E be a d-dimensional regular set in $\mathbf{R}^n$. If E satisfies the WHIP and the WTP, then E is uniformly rectifiable.*

We shall see in the next section that the BWGL implies the WHIP trivially. Hence the converse to Theorem 3.9 is true (but not too interesting).

REMARK 3.10. Theorem 3.9 is still true with slightly weaker hypotheses. The WHIP and the WTP can be replaced by the weaker conditions that there exist A, $a > 0$, and $\alpha_0 \in (0, \frac{1}{10})$ such that the complements of $\mathscr{G}^*(\alpha_0, A, N)$ and $\mathscr{H}(\alpha_1, a)$ are Carleson sets, where N is a large integer which can be computed in terms of n, d, A, a, and the regularity constant for E, and α_1 is a small positive number which depends on these quantities and also the choice of α_0. A second observation (which will follow from the proof) is that precise estimates for the uniform rectifiability of E (such as the Carleson measure norm in the geometric lemma (I.1.59), or the constants θ and C_1 in the BPBI condition of Definition I.1.33) can be computed in terms of the Carleson norms of $E \times \mathbf{R}_+ \setminus \mathscr{G}^*(\alpha, A, N)$ and $E \times \mathbf{R}_+ \setminus \mathscr{H}(\alpha, a)$ and the various constants. This is of course similar to what happened in Chapter 2 with the BWGL and its variants.

Theorem 3.9 will be proved in §§5 and 6 of this chapter. Before doing that, it is probably a good idea to describe the relationships between the WHIP and other conditions and to try to explain why Theorem 3.9 can be useful. We do this in the next three sections.

Let us conclude this section with an "amusing" exercise: try to show directly that Garnett's counterexample (the set K described in § 1.1 of Part I) does not satisfy the WHIP. We know that it cannot, because it has big projections (and hence satisfies the WTP) and because it is not even rectifiable.

3.2. The WHIP and weaker versions of the BWGL.

The definition of the WHIP is not extremely friendly, but it seems to have some (slightly hidden) geometric significance, because Theorem 3.9 will allow us to prove that a few simpler (and geometrically more natural) conditions on E are sufficient to imply that E is uniformly rectifiable. In this section we shall restrict our attention to some weaker versions of the BWGL.

Let us start with the BWGL itself (see Definition I.2.2) and check that

(3.11) the BWGL implies the WHIP.

Let E be a regular set that satisfies the BWGL. To prove that E satisfies the WHIP, we take $A = 2$ and, for each integer $N > 0$, choose $\alpha = (6N)^{-1}$. It is enough to prove that if $(x, t) \in E \times \mathbf{R}_+$ has the property that there exists a d-plane P_0 for which

$$(3.12) \qquad \sup_{y \in E \cap B(x, 2t)} \operatorname{dist}(y, P_0) + \sup_{z \in P_0 \cap B(x, 2t)} \operatorname{dist}(z, E) \leq \frac{\alpha t}{10},$$

then $(x, t) \in \mathscr{G}^*(\alpha, 2, N)$.

Let us check first that $(x, t) \in \mathscr{G}(\alpha, 2, N)$. Let P be any d-plane and B be a ball of radius αt in P which is contained in $\Pi\left(B\left(x, \frac{t}{2}\right)\right)$ and satisfies (3.1). Let y be a point in $E \cap \overline{B}\left(x, \frac{t}{2}\right)$ such that $\Pi(y) \in \partial B$. Let q be the point of P_0 which is closest to y. Thus $|q - y| \leq \frac{\alpha t}{10}$, so $\operatorname{dist}(\Pi(q), B) \leq \frac{\alpha t}{10}$. On the other hand, $\Pi\left(P_0 \cap B\left(q, \frac{t}{3}\right)\right)$ does not meet $\frac{9B}{10}$, because of (3.1) and (3.12). Therefore, P_0 is almost perpendicular to P. More precisely, there is a unit vector v parallel to P_0 such that $|\Pi(v)| \leq \frac{3}{5}\alpha$. It is now easy to find a post. For each positive integer j such that $j \leq \frac{1}{6\alpha} = N$, we choose a point $x_j \in E$ such that $\text{dist}(x_j, q + 2\alpha j t v) \leq \frac{\alpha t}{10}$. The points x_j obviously satisfy all the requirements of (3.2). Thus $(x, t) \in \mathscr{G}(\alpha, 2, N)$. The fact that $(x, t) \in \mathscr{G}^*(\alpha, 2, N)$ is proved in the same manner, and (3.11) follows.

Another condition that implies the WHIP and the WTP is the "other unilateral weak geometric lemma", or OUWGL, which we define now.

Given $\epsilon > 0$ (small), let $\mathscr{G}_o(\epsilon)$ be the set of $(x, t) \in E \times \mathbf{R}_+$ for which there is a d-plane P containing x such that every point of $P \cap \overline{B}(x, t)$ is at distance $\leq \epsilon t$ from E.

DEFINITION 3.13. We say that the d-dimensional set E in $\mathbf{R}^n$ satisfies the OUWGL (other unilateral weak geometric lemma) if, for every $\epsilon > 0$, $E \times \mathbf{R}_+ \setminus \mathscr{G}_o(\epsilon)$ is a Carleson set.

We can also think in terms of the functorial notation of §I.2.2. Then the regular set E satisfies the OUWGL if and only if it is in the class $\text{Approx}(\mathscr{E})$, where $\mathscr{E}$ is the collection of closed subsets of $\mathbf{R}^n$ that contain a d-plane through the origin.

We want to prove that the OUWGL implies the WHIP and the WTP. To do this, it will be convenient to establish first the fact that the OUWGL is equivalent to the following variant of the BWGL. For each $\epsilon > 0$, let $\mathscr{G}_u(\epsilon)$ be the set of all $(x, t) \in E \times \mathbf{R}_+$ such that there exists a family $\mathscr{F}$ of d-planes with the property that every point of $E \cap \overline{B}(x, t)$ is at distance $\leq \epsilon t$ from $\bigcup_{P \in \mathscr{F}} P$ and every point of $(\bigcup_{P \in \mathscr{F}} P) \cap \overline{B}(x, t)$ is at distance $\leq \epsilon t$ from E.

DEFINITION 3.14. The regular set E satisfies the condition of bilateral approximation by unions of d-planes (in short, the BAUP) if, for every $\epsilon > 0$, $E \times \mathbf{R}_+ \setminus \mathscr{G}_u(\epsilon)$ is a Carleson set.

Before we get back to the OUWGL, we want to digress briefly with some remarks about the BAUP.

In more functorial terms, E satisfies the BAUP if and only if it lies in $\text{Approx}(\mathscr{E}')$, where $\mathscr{E}'$ is the class of closures of unions of d-planes. It is not hard to check that this is also equivalent to $E \in \text{Approx}(\mathscr{E}'(C_0))$, where $\mathscr{E}'(C_0)$ is the collection of unions of less than C_0 d-planes and C_0 is a constant that depends on n, d, and the regularity constant for E. Since we are not directly interested in this remark, we shall content ourselves with only a moderately detailed sketch of the proof.

The desired result will follow if we can prove that, whenever $(x, t) \in \mathscr{G}_u(\epsilon)$ is given and $\mathscr{F}$ is a family of d-planes as in the definition of $\mathscr{G}_u(\epsilon)$, it is possible to find a subset $\mathscr{F}'$ of $\mathscr{F}$ with less than C_0 elements such that all points $y \in E \cap B\left(x, \frac{t}{3}\right)$ still lie at distance $\leq C_1 \epsilon t$ from the union of the d-planes $P \in \mathscr{F}'$. Here C_0 is as above, and C_1 is another constant that may depend on n, d, and C_0.

To choose $\mathscr{F}'$, we proceed as follows. We first select a d-plane $P_1 \in \mathscr{F}$. If all points of $E \cap B\left(x, \frac{t}{3}\right)$ are within $C_1 \epsilon t$ of P_1, we can stop and take $\mathscr{F}' = \{P_1\}$. Otherwise, we choose a point $x_2 \in E \cap B\left(x, \frac{t}{3}\right)$ at distance

$> C_1 \epsilon t$ from P_1 and add to P_1 a new d-plane $P_2 \in \mathcal{F}$ that passes at distance $\leq \epsilon t$ from x_2. We continue with the same strategy: if $P_1, \ldots, P_k$ were already selected and if all points of $E \cap B(x, \frac{t}{3})$ are within $C_1 \epsilon t$ from $P_1 \cup \cdots \cup P_k$, we stop and take $\mathcal{F}' = \{P_1, \ldots, P_k\}$. Otherwise, we select a new d-plane P_{k+1} in $\mathcal{F}$ that passes at distance $\leq \epsilon t$ from some point $x_{k+1} \in E \cap B(x, \frac{t}{3})$ which is at distance $> C_1 \epsilon t$ from $P_1 \cup \cdots \cup P_k$. We have to prove that the process stops before k becomes larger than C_0.

To do this, we shall show that if C_1 is large enough (depending on C_0) and if E_k is the set of points $y \in E \cap B(x, t)$ such that $\text{dist}(y, P_k) \leq 2\epsilon t$ but $\text{dist}(y, P_1 \cup \cdots \cup P_{k-1}) > 2\epsilon t$, then

$$(3.15) \qquad |E_k| \geq \frac{t^d}{C} \quad \text{for all } k \leq C_0,$$

where C is a constant that may depend on the regularity constant for E but not on C_0. This will give the desired result because the E_k's are disjoint and E is regular.

For each $k \leq C_0$, let F_k be the set of $w \in P_k \cap B(x, \frac{2t}{3})$ with the property that $\text{dist}(w, P_1 \cup P_2 \cdots \cup P_{k-1}) \geq 4\epsilon t$. If we prove that

$$(3.16) \qquad |F_k| \geq \frac{t^d}{C'},$$

then (3.15) will follow. [Indeed, if (3.16) holds, then we can find more than $(C''\epsilon)^{-d}$ disjoint balls B_i of radius $2\epsilon t$ and centered on F_k; each $\frac{1}{2}B_i$ meets E because $P_k \in \mathcal{F}$, so (3.15) follows because the $B_i \cap B$ are disjoint and contained in E_k.]

To prove (3.16), consider the sets $A_{kl} = \{w \in P_k \cap B(x, \frac{2t}{3}) : \text{dist}(w, P_l) < 4\epsilon t\}$ for $l = 1, \ldots, k-1$. By definition of P_k, there is a point of $P_k \cap B(x, \frac{t}{2})$ which is at a distance $> (C_1 - 1)\epsilon t$ from P_l, so $|A_{k,l}| \leq \mu(C_1) t^d$ for some constant $\mu(C_1)$ which tends to 0 as C_1 tends to $+\infty$. This is not hard to prove, but it is easier to omit the proof. [The main point is that we can get an estimate that does not depend on ϵ. The fact that $\mu(C_1)$ tends to 0 is a reflection of the fact that $P_k \cap P_l$ is a $(d-1)$-plane.] Now $|F_k| \geq t^d/C - (k-1)\mu(C_1)t^d \geq t^d/C - C_0 \mu(C_1)t^d$, which implies (3.16) if we choose C_1 large enough. This completes the sketch of the proof.

Let us come back to the discussion about the OUWGL.

PROPOSITION 3.17. *The OUWGL is equivalent to the BAUP.*

The fact that the BAUP implies the OUWGL is essentially obvious. We only need to observe that if E is well approximated in $\overline{B}(x, t)$ by a union of d-planes, then one of them must pass at a small distance from x (and hence could be replaced by a d-plane through x).

Now let us prove that the OUWGL implies the BAUP. To do this, we first check that if E satisfies the OUWGL, then $E \times \mathbf{R}_+ \setminus \mathcal{G}_o^*(\epsilon)$ is a Carleson set for each $\epsilon > 0$, where

$$\mathscr{G}_o^*(\epsilon) = \{(x, t) \in E \times \mathbf{R}_+ : (y, s) \in \mathscr{G}_o(\epsilon) \text{ whenever } (y, s) \in E \times \mathbf{R}_+$$
$$\text{satisfies } t/2 \leq s \leq t \text{ and } |y - x| \leq t\}.$$

This is not hard to verify, using the simple observation that $(y, s) \in \mathscr{G}_o(\epsilon)$ as soon as there is a $(z, r) \in \mathscr{G}_o(\epsilon/10)$ such that $2s \leq r \leq 3s$ and $|z-y| \leq \epsilon s/2$.

To prove that the OUWGL implies the BAUP, it now suffices to show that

$$\mathscr{G}_u(\epsilon) \supset \{(x, t) \in E \times \mathbf{R}_+ : (x, 2t) \in \mathscr{G}_o^*(\epsilon/2)\}$$

for any (small) $\epsilon > 0$. Let ϵ be given, and let (x, t) be a given element of the right-hand side of the above inclusion. Then for each $y \in E \cap \overline{B}(x, t)$ we have that $(y, 2t) \in \mathscr{G}_o(\epsilon/2)$, so that there is a d-plane $P(y)$ which contains y and satisfies $\text{dist}(z, E) \leq \epsilon t$ for all $z \in P \cap \overline{B}(x, t)$. From here it follows easily that $(x, t) \in \mathscr{G}_u(\epsilon)$. This proves Proposition 3.17.

PROPOSITION 3.18. *The BAUP implies the WHIP and the WTP (and hence, by Theorem 3.9, that E is uniformly rectifiable).*

The fact that the BAUP (or the OUWGL) implies the WTP is trivial. The proof of the fact that the BAUP implies the WHIP is the same as for the BWGL. [In the proof of (3.11), we only used the fact that there was a d-plane P_0 that was very close to y and stayed very close to E inside $B(x, 2t)$.]

The conditions that we have discussed in this section have the minor defect that they are very rigid; for instance, the fact that the OUWGL is invariant under bilipschitz mappings is not at all obvious a priori. This rigidity is obviously connected to the rigidity of the WHIP and the WTP, and the various conditions studied in the next sections will have the same sort of defect.

Let us conclude this section with a question about the OUWGL. Suppose that, in the definition of the OUWGL, we had only required, for all but a Carleson set of (x, t) in $E \times \mathbf{R}_+$, the existence of a d-plane P that intersects $B\left(x, \frac{t}{2}\right)$ (instead of containing x) and which satisfies $\text{dist}(y, E) \leq \epsilon t$ for all $y \in P \cap B(x, t)$. Does the resulting (weaker) variant of the OUWGL also imply uniform rectifiability?

3.3. The weak exterior convexity condition and the GWEC.

Let us begin with a review of the weak exterior convexity condition WEC. The WEC is a condition on regular sets of codimension 1, and it will be used in Part III in connection with square function estimates for the Cauchy kernel. Recall from Definition I.2.17 that a d-dimensional regular set E in $\mathbf{R}^{d+1}$ satisfies the WEC if, for each $\epsilon > 0$, the complement in $E \times \mathbf{R}_+$ of the set $\mathscr{G}_e(\epsilon)$ is a Carleson set. Here $\mathscr{G}_e(\epsilon)$ denotes the set of all $(x, t) \in E \times \mathbf{R}_+$ for which, whenever u, v are two points of $B(x, t) \setminus E$ and γ is a path in $B(x, t) \setminus E$ that joins u to v and satisfies

(3.19) $$\text{dist}(\gamma, E) \geq \epsilon t,$$

we have that the line segment $L(u, v)$ that joins u to v lies in $\mathbf{R}^{d+1} \setminus E$.

The WEC also has a simple functorial description: Lemma I.2.23 says that E satisfies the WEC if and only if it belongs to Approx($\mathscr{E}_0$), where $\mathscr{E}_0$ is the class of all closed subsets F of $\mathbf{R}^{d+1}$ for which every connected component of $\mathbf{R}^{d+1} \setminus F$ is convex. Notice that the WEC is weaker than the BWGL and even the BAUP, because closures of unions of d-planes belong to $\mathscr{E}_0$. However, the WEC implies uniform rectifiability because of Theorem 3.9 and the following result.

PROPOSITION 3.20. *The WEC implies the WHIP and the WTP.*

To prove that the WEC implies the WHIP, notice first that if E satisfies the WEC, then for each $\epsilon > 0$ the complement in $E \times \mathbf{R}_+$ of $\mathscr{G}_e^*(\epsilon)$ is a Carleson set. Here $\mathscr{G}_e^*(\epsilon)$ is defined in terms of $\mathscr{G}_e(\epsilon)$ in the same way that $\mathscr{G}^*(\alpha, A, N)$ was defined in terms of $\mathscr{G}(\alpha, A, N)$ in (3.3).

We want to prove that, for any integer $N > 0$, we can find an α, $0 < \alpha < \frac{1}{10}$, such that

$$(3.21) \qquad \mathscr{G}_e\left(\frac{\alpha}{10}\right) \subset \mathscr{G}(\alpha, 2, N).$$

This will be enough to prove that the WEC implies the WHIP because, if E satisfies the WEC, then (3.21) and the previous remark imply that for all N there is a choice of α such that the complement of $\mathscr{G}^*(\alpha, 2, N)$ is a Carleson set.

To prove (3.21), let E, N, and $(x, t) \in E \times \mathbf{R}_+$ be given. Suppose that $(x, t) \in \mathscr{G}_e\left(\frac{\alpha}{10}\right)$, where α will be chosen soon. Let P be a d-plane, and denote by Π the orthogonal projection onto P. Suppose that $B \subset P$ is a ball of radius αt which is contained in $\Pi\left(B\left(x, \frac{t}{2}\right)\right)$ and satisfies (3.1). We want to find a post (i.e., a bunch of points $x_1, \ldots, x_N$ as in (3.2)).

Let y be a point in $E \cap \overline{B}\left(x, \frac{t}{2}\right)$ such that $\Pi(y) \in \partial B$, and let $y_1, \ldots, y_M$ be a maximal family of points in $\mathbf{R}^{d+1}$ such that

(3.22) $\qquad |y_i - y_j| > 3\alpha t \quad \text{for } i \neq j,$

(3.23) $\qquad \Pi(y_i) = \Pi(y) \quad \text{and} \quad \frac{t}{4} \leq |y_i - y| \leq \frac{t}{3}$

for all i and also such that the last coordinate of each y_i is larger than the last coordinate of y. [By "last coordinate" we mean the orthogonal projection onto some oriented line perpendicular to P. As we shall see later, we do not really use the fact that E is of codimension 1 here; the main point of this restriction on the last coordinates is to make sure that y_i is also far away from the reflections of the y_j's about y.]

We take $\alpha = (CN)^{-1}$, where C is chosen so large that $M \geq N$. If, for each j, E meets $B_j = B(y_j, \alpha t)$ or its reflection B_j' about y, then we obtain the desired post by picking a point $x_j \in E \cap (B_j \cap B_j')$ for each j. (See Figure 3.3.) Otherwise, choose a j such that $B_j \cup B_j'$ does not meet E.

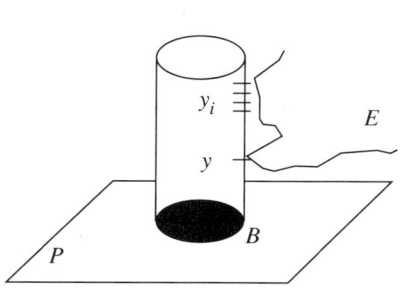
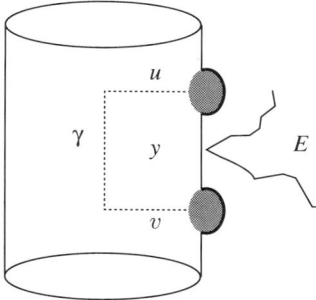

FIGURE 3.3   FIGURE 3.4

Let $u = y_j$, and let v be its reflection about y. We connect u and v by a path γ as follows. Let $\tilde{u}$ and $\tilde{v}$ be the points which have the same last coordinate as u and v, respectively, and which lie on the line perpendicular to P that passes through the center of B. We form the path γ by taking the union of the three line segments which join u to $\tilde{u}$, $\tilde{u}$ to $\tilde{v}$, and $\tilde{v}$ to v. (See Figure 3.4.)

The path γ is contained in $B(x, t)$ because $y \in \overline{B}(x, \frac{t}{2})$ and $|u - y| = |v - y| \leq \frac{t}{3}$ by (3.23). Also, $\mathrm{dist}(\gamma, E) \geq \frac{\alpha t}{10}$ because B satisfies (3.1) and u and v lie at distance $\geq \alpha t$ from E. We now use our hypothesis that $(x, t) \in \mathscr{G}_e\left(\frac{\alpha}{10}\right)$ to get that the line segment that joins u and v does not meet E. This is impossible because y lies on that line segment. This contradiction shows that our second case cannot occur. Thus we have found the desired post, and $(x, t) \in \mathscr{G}(\alpha, 2, N)$. This proves (3.21), and the fact that the WEC implies the WHIP follows immediately.

Let us now prove that the WEC implies the WTP. We want to show that there is an $a > 0$ such that, for every $\alpha > 0$, we can find $\epsilon > 0$ so that $\mathscr{G}_e(\epsilon) \subset \mathscr{H}(\alpha, a)$. Let $(x, t) \in \mathscr{G}_e(\epsilon)$ be given. We first select two balls B_1 and B_2 with the following properties: B_1 and B_2 are contained in $B\left(x, \frac{t}{2}\right)$, they have the same radius $\frac{t}{C}$, they do not intersect E, and B_2 is the reflection of B_1 about x. Such a pair of balls exists if we choose C large enough, because E is a regular set with dimension $d < d + 1$.

Let u and v be the respective centers of B_1 and B_2. Since the line segment between u and v meets E (at the point x) and $(x, t) \in \mathscr{G}_e(\epsilon)$, it is impossible to connect u and v by a path $\gamma \subset B(x, t)$ that stays at a distance $\geq \epsilon t$ from E. Notice that every point of $\frac{1}{2}B_1$ can be connected to u by a short line segment which stays at distance $\geq \epsilon t$ from E (if ϵ is small enough). The same thing is true for the ball $\frac{1}{2}B_2$ and v, and consequently it is impossible to connect any point of $\frac{1}{2}B_1$ to any point of $\frac{1}{2}B_2$ by a path $\gamma \subset B(x, t)$ such that $\mathrm{dist}(\gamma, E) \geq \epsilon t$.

Choose a d-plane P that is perpendicular to the line L through u, v, and x, and let Π be the orthogonal projection onto P. Since every line segment parallel to L that joins $\frac{1}{2}B_1$ to $\frac{1}{2}B_2$ must pass at a distance $\leq \epsilon t$ from E, the set $\{p \in P : \mathrm{dist}(p, \Pi(E \cap B(x, t))) \leq \epsilon t\}$ contains a ball of

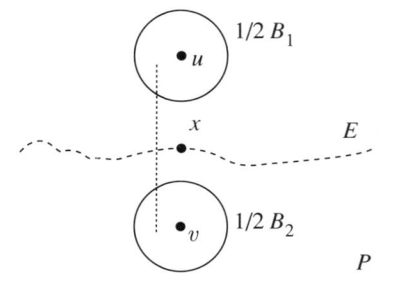

FIGURE 3.5

radius $\frac{t}{2C}$, where C is as above. [See Figure 3.5.] Hence, the estimate (3.5) is satisfied for some constant a that depends only on d and the regularity constant for E, provided that we take $\epsilon < \alpha$. Thus $(x, t) \in \mathcal{H}(\alpha, a)$. This completes the proof of Proposition 3.20.

For an application of the WEC, we refer the reader to Part III, Chapters 2 and 3, where it is used to prove that the usual square function estimates for the Cauchy kernel (or its Clifford generalization to higher dimensions) implies uniform rectifiability.

The WEC has a generalization to higher codimensions, the GWEC, which we now define.

In the following discussion, we shall use the expression "k-sphere" to mean the intersection of a sphere in $\mathbf{R}^n$ with some $(k+1)$-dimensional plane.

Let a d-dimensional regular set $E \subset \mathbf{R}^n$ be given. For each (small) $\epsilon > 0$, let $\mathcal{B}_g(\epsilon)$ be the (bad) set of all $(x, t) \in E \times \mathbf{R}_+$ for which there exists a $(n - d - 1)$-sphere S with the following properties:

(3.24) $S \subset B(x, t)$ and $\text{dist}(S, E) > \epsilon t$;
(3.25) S can be contracted to a point inside $\{z \in B(x, t) : \text{dist}(z, E) > \epsilon t\}$;
(3.26) the flat $(n - d)$-dimensional ball of which S is the boundary (i.e., the convex hull of S) intersects E.

See Figure 3.6 for a symbolic picture that describes a pair $(x, t) \in \mathcal{B}_g(\epsilon)$ for $d = 1$ and $n = 3$.

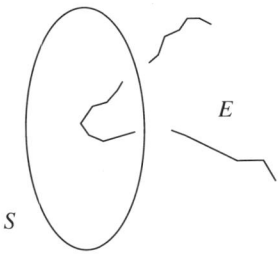

FIGURE 3.6

3.3. WEAK EXTERIOR CONVEXITY

DEFINITION 3.27. We say that a d-dimensional regular set $E \subset \mathbf{R}^n$ satisfies the GWEC (generalized weak exterior convexity condition) if $\mathscr{B}_g(\epsilon)$ is a Carleson set for each $\epsilon > 0$.

Notice that, when $d = n - 1$, the GWEC is the same thing as the WEC. [The $(n-d-1)$-sphere S is reduced to two points u, v, and the contraction of S defines a path from u to v.]

THEOREM 3.28. *A d-dimensional regular set E in $\mathbf{R}^n$ satisfies the GWEC if and only if it is uniformly rectifiable.*

The "if" part is easy; the GWEC can be derived from the BWGL using elementary topology. [More generally, the BAUP also implies the GWEC rather easily.] The "only if" part will be proved using Theorem 3.9.

We postpone the verification that the GWEC implies the WHIP until the next section, where we shall prove that a slightly more general condition than the GWEC, the WNR, implies the WHIP. [See Proposition 3.50 and (3.51).]

Let us prove now that the GWEC implies the WTP. Let E be a regular set; it is enough to prove that there is a constant $a > 0$, that depends only on n, d, and the regularity constant for E, such that for every choice of $\alpha > 0$ there is an $\epsilon > 0$ (that depends only on α and the previous constants) so that

$$(3.29) \qquad E \times \mathbf{R}_+ \setminus \mathscr{B}_g(\epsilon) \subset \mathscr{H}(\alpha, a)$$

(the good set for the WTP). Let $(x, t) \in E \times \mathbf{R}_+ \setminus \mathscr{B}_g(\epsilon)$ be given, and let us try to find a d-plane P such that (3.5) holds. We need the following.

LEMMA 3.30. *Let E be a d-dimensional regular set in $\mathbf{R}^n$. Then there exists a constant $a > 0$, which depends only on n, d, and the regularity constant for E, with the following property. For all $x \in E$ and $t > 0$ there is a $(n - d - 1)$-sphere S, whose radius r satisfies $\frac{t}{20} \leq r \leq \frac{t}{10}$, which is centered at x and satisfies $\mathrm{dist}(S, E) \geq 2\sqrt{d}\,at$.*

Let us assume the lemma for the moment and use it to find the d-plane P. Let E and (x, t) be given as before, and let S be the $(n-d-1)$-sphere given by Lemma 3.30. Let Q be the $(n-d)$-plane that contains S, and let P be the d-plane through x which is orthogonal to Q. Let $A = \{p \in P : \mathrm{dist}(p, \Pi(E \cap B(x, t))) < \alpha t\}$ be the set in the left-hand side of (3.5). (As usual, Π denotes the orthogonal projection onto P.) In order to prove (3.5), it suffices to show that if ϵ is small enough (depending on α), then

$$(3.31) \qquad A \supset P \cap B(x, \sqrt{d}\,at).$$

Suppose that (3.31) is not true, and let $p \in P \cap B(x, \sqrt{d}\,at)$ be such that

$$\mathrm{dist}(p, \Pi(E \cap B(x, t))) \geq \alpha t.$$

Let $\widetilde{S}$ be the $(n-d-1)$-sphere with center p and radius r which lies in the $(n-d)$-plane $\widetilde{Q}$ through p that is parallel to Q. Notice that $\widetilde{S} \subset B\left(x, \frac{2t}{10}\right)$

if we were cautious enough to take $a < (10\sqrt{d})^{-1}$. Hence the convex hull of $\widetilde{S}$ is at a distance $\geq \alpha t$ from E. Consider the contraction of S to a point obtained by first translating S to $\widetilde{S}$ and then contracting $\widetilde{S}$ to the point p in the natural way. This deformation takes place in the union of $\{z \in \mathbf{R}^n : \mathrm{dist}(z, S) \leq \sqrt{d}at\}$ and the convex hull of $\widetilde{S}$, which is contained in $\{z \in B(x, t) : \mathrm{dist}(z, E) > \epsilon t\}$ if we choose $\epsilon < \inf(\sqrt{d}a, \alpha)$. Therefore, S satisfies (3.25). Since it also satisfies (3.24) and (3.26), we obtain that $(x, t) \in \mathscr{B}_g(\epsilon)$. This contradicts our assumption that $(x, t) \notin \mathscr{B}_g(\epsilon)$, and so we conclude that (3.31), and hence (3.5), are true.

It remains to prove Lemma 3.30. We may as well assume that $x = 0$ and $t = 1$. Let Σ denote the set of $(n - d - 1)$-spheres in $\mathbf{R}^n$ with center 0 and radius $r \in [\frac{1}{20}, \frac{1}{10}]$. We identify Σ with $G(n - d, n) \times [\frac{1}{20}, \frac{1}{10}]$, where $G(n - d, n)$ denotes the Grassmann manifold of $(n - d)$-planes through the origin in $\mathbf{R}^n$, and we equip Σ with the product of the usual invariant probability measure on $G(n - d, n)$ with Lebesgue measure on $[\frac{1}{20}, \frac{1}{10}]$.

Set $E_a = \{z \in \mathbf{R}^n : \mathrm{dist}(z, E) \leq 2\sqrt{d}a\}$. Because E is regular, we can cover $E_a \cap B(0, 1)$ by a family $\{B_i\}$ of less than Ca^{-d} balls of radius a. For each i, set $\Sigma(B_i) = \{S \in \Sigma : S \cap B_i \neq \varnothing\}$.

LEMMA 3.32. $|\Sigma(B_i)| \leq C_0 a^{d+1}$ *for some constant* C_0 *that depends only on* n *and* d.

Let us first explain why Lemma 3.30 follows from this. We have that

$$|\{S \in \Sigma : S \cap E_a \neq \varnothing\}| \leq \sum_i |\Sigma(B_i)| \leq CC_0 a.$$

If we choose a small enough, this is less than the total measure of Σ, and hence there is an element S of Σ which does not meet E_a. This $(n-d-1)$-sphere S satisfies all the requirements of Lemma 3.30.

Let us now prove Lemma 3.32. It suffices to show that $\Sigma(y) = \{S \in \Sigma : y \in S\}$ is a codimension $d + 1$ submanifold in Σ for each $y \in \overline{B}\left(0, \frac{1}{10}\right) \setminus B\left(0, \frac{1}{20}\right)$. [Indeed, $\Sigma(B_i)$ is contained in a Ca-neighborhood of $\Sigma(p_i)$ (for the natural metric on Σ), where p_i is the center of B_i; the rotation invariance of the relevant objects is helpful in making the estimates even more trivial.]

The problem reduces immediately to proving that

(3.33) $$\{Q \in G(n - d, n) : L \subset Q\}$$

is a codimension d submanifold in $G(n - d, n)$ for all lines $L \subset \mathbf{R}^n$ that pass through the origin.

Fix such a line L and also an $(n - d)$-plane $Q_0 \in G(n - d, n)$ such that $L \subset Q_0$. Let P_0 denote the orthogonal complement of Q_0, and let $\mathscr{L}(Q_0, P_0)$ be the set of linear mappings from Q_0 to P_0. The mapping from $\mathscr{L}(Q_0, P_0)$ to $G(n-d, n)$ that associates to each $A \in \mathscr{L}(Q_0, P_0)$ the $(n-d)$-plane $Q = \{v + A(v), v \in Q_0\}$ (i.e., the graph of A in $\mathbf{R}^n \simeq Q_0 + P_0$)

provides a parameterization of a neighborhood of Q_0 in $G(n-d, n)$ by a neighborhood of 0 in $\mathscr{L}(Q_0, P_0)$. In these coordinates, the condition $L \subset Q$ corresponds to the requirement that $A|_L = 0$. This last equation defines a subspace of $\mathscr{L}(Q_0, P_0)$ with codimension equal to $\dim P_0 = d$, and hence (3.33) is a codimension d submanifold in $G(n-d, n)$.

This completes the proof of the fact that the GWEC implies the WTP.

3.4. The weak-no-mugs, weak-no-boxes, and weak-no-reels conditions.

The weak-no-mugs condition (WNM), which we are going to describe now, is a close relative of the GWEC, at least when $d = 1$. We first need to define the "standard ϵ-mug" $M(\epsilon)$, where $0 < \epsilon < 1$ is a small constant. It is the subset of $\mathbf{R}^n$ defined by

(3.34) $$M(\epsilon) = B \cup S,$$

where

(3.35) $$B = \{(x_1, \ldots, x_n) \in \mathbf{R}^n : 0 \leq x_n \leq \epsilon \text{ and } x_1^2 + \cdots + x_{n-1}^2 \leq 1\}$$

is a flat cylinder (the bottom of the mug) and

(3.36) $$S = \{(x_1, \ldots, x_n) \in \mathbf{R}^n : 0 \leq x_n \leq 10\epsilon \text{ and } 1 - \epsilon \leq x_1^2 + \cdots + x_{n-1}^2 \leq 1\}$$

is the side of the mug. (See Figure 3.7.)

By the "content" of the mug $M(\epsilon)$ we mean the set

(3.37) $$C = \{(x_1, \ldots, x_n) \in \mathbf{R}^n : \epsilon < x_n < 2\epsilon \text{ and } x_1^2 + \cdots + x_{n-1}^2 < 1 - \epsilon\}.$$

Note that, since ϵ will often be quite small, our mugs look more like flat halves of camembert boxes than real coffee mugs.

We call a subset of $\mathbf{R}^n$ an ϵ-mug if it can be written as $\phi(M(\epsilon))$, where ϕ is obtained by composing a dilatation with an isometry of $\mathbf{R}^n$. The content of the mug is then $\phi(C)$.

The good set (for the weak-no-mug condition) $\mathscr{G}_m(\epsilon)$ is the set of pairs $(x, t) \in E \times \mathbf{R}_+$ for which it is impossible to find an ϵ-mug $M \subset B(x, t)$ such that

(3.38) $\operatorname{diam} M \geq \epsilon t$, and

(3.39) $M \cap E = \varnothing$, but E intersects the content of M.

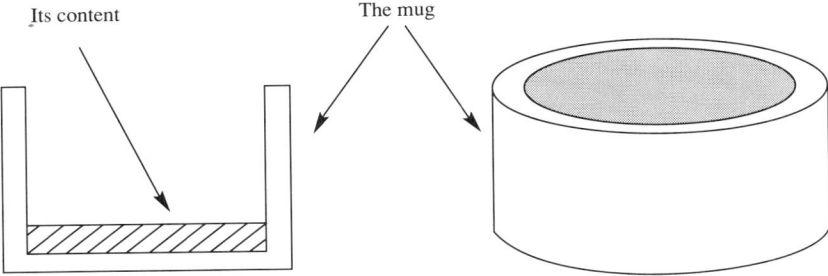

FIGURE 3.7. $M(\epsilon)$ in dimensions 2 and 3.

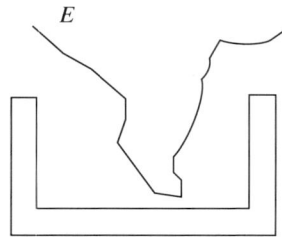

FIGURE 3.8

[Figure 3.8 is a symbolic picture of a bad pair $(x, t) \notin \mathscr{G}_m(\epsilon)$.]

DEFINITION 3.40. A d-dimensional regular set E in $\mathbf{R}^n$ satisfies the weak-no-mug property (WNM) if, for every $\epsilon > 0$, $E \times \mathbf{R}_+ \setminus \mathscr{G}_m(\epsilon)$ is a Carleson set.

The WNM is a reasonably natural condition for all values of n and d. Unfortunately, though, we can derive interesting information from the WNM only when $d = 1$.

PROPOSITION 3.41. *If a one-dimensional regular set $E \subset \mathbf{R}^n$ satisfies the WNM, then it satisfies the WHIP and the WTP (and hence is uniform rectifiable).*

The fact that the WNM implies the WHIP is a special case of Proposition 3.50, so we refer the reader to that result. The WTP can be deduced from the WNM in exactly the same way as it was derived from the GWEC (when $d = 1$) in the previous section. Indeed, the WNM is really just a minor variant of the GWEC when $d = 1$; basically, the difference is that for the WNM we replace (3.25) by a much more restrictive kind of contraction to a point. However, when we proved that the GWEC implies the WTP, we only needed this more restrictive version of (3.25).

Next we want to describe a minor variant of the WNM, the weak-no-boxes condition.

The weak-no-boxes condition WNB is defined in the same way as the WNM was, except that we replace the standard ϵ-mug $M(\epsilon)$ by the "standard ϵ-box" $M'(\epsilon) = B' \cup S'$, where

$$(3.42) \quad B' = \left\{(x_1, \ldots, x_n) \in \mathbf{R}^n : 0 \leq x_n \leq \epsilon \text{ and } \sup_{1 \leq i \leq n-1} |x_i| \leq 1\right\},$$

and
(3.43)
$$S' = \left\{(x_1, \ldots, x_n) \in \mathbf{R}^n : 0 \leq x_n \leq 10\epsilon \text{ and } 1 - \epsilon \leq \sup_{1 \leq i \leq n-1} |x_i| \leq 1\right\}.$$

[Thus we have simply replaced the euclidian norm on $\mathbf{R}^{n-1}$ with the sup norm; see Figure 3.9 for a picture of $M'(\epsilon)$.] The rest of the definition of

3.4. NO-MUGS, NO-BOXES, NO-REELS

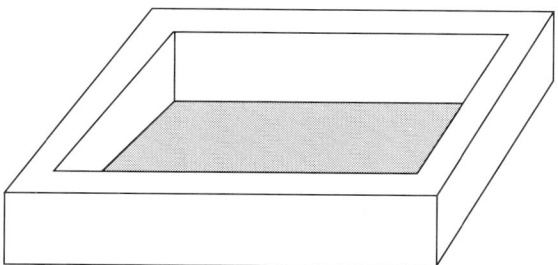

FIGURE 3.9. The standard ϵ-box $M'(\epsilon)$.

the WNB follows the same lines as for the WNM. We let

$$C' = \left\{ (x_1, \ldots, x_n) \in \mathbf{R}^n : \epsilon < x_n < 2\epsilon \text{ and } \sup_{1 \leq i \leq n-1} |x_i| < 1 - \epsilon \right\}$$

be the content of $M'(\epsilon)$. The general ϵ-box and its contents are then obtained from $M'(\epsilon)$ and C' by translations, rotations, and dilations. The good set $\mathscr{G}_b(\epsilon)$ is the set of $(x, t) \in E \times \mathbf{R}_+$ for which it is impossible to find an ϵ-box $M' \subset B(x, t)$ with diameter $\geq \epsilon t$ such that M' does not meet E but such that its content does intersect E.

DEFINITION 3.44. We say that a regular set E of dimension d in $\mathbf{R}^n$ satisfies the WNB (weak-no-box condition) if $E \times \mathbf{R}_+ \setminus \mathscr{G}_b(\epsilon)$ is a Carleson set for each $\epsilon > 0$.

Notice that the WNB is the same as the WNM when $n = 2$.

PROPOSITION 3.45. *For regular sets of dimension 1 in $\mathbf{R}^n$, the WNB implies the WHIP and the WTP (and hence uniform rectifiability).*

We shall use this result in Chapter 4 of Part III to show that the condition of weak approximation of Lipschitz functions by affine functions (i.e., the WALA) implies uniform rectifiability for 1-dimensional regular sets in $\mathbf{R}^n$. We could just as well have used Proposition 3.41, but for reasons which are more embarrassing than significant we decided to work with the WNB instead of the WNM.

The proof of the part about the WHIP is again the same as what will be done for Proposition 3.50, so we omit it. To take care of the WTP, we shall use an argument which is more direct than the approach that we took for the GWEC and the WNM.

Let us prove that if $(x, t) \in \mathscr{G}_b\left(\frac{\alpha}{10}\right)$, then the estimate (3.5) holds with $a = 10^{-n}$ and some choice of line P. Let $(x, t) \in \mathscr{G}_b\left(\frac{\alpha}{10}\right)$ be given. For each i, $1 \leq i \leq n$, let us try $P = L_i$, where L_i is the ith coordinate axis. If (3.5) is not satisfied with this choice of P, then there is a point $y_i \in \mathbf{R}$ such that

(3.46) $x_i - \frac{t}{3n} \leq y_i \leq x_i - \frac{t}{6n}$ (where x_i is the ith coordinate of x), and

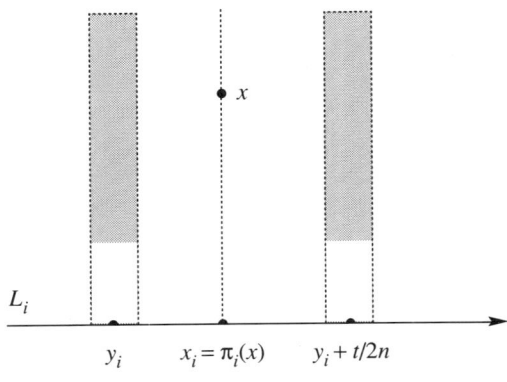

FIGURE 3.10

(3.47) both y_i and $y_i + \frac{t}{2n}$ are at distance $\geq \alpha t$ from $\Pi_i(E \cap B(x, t))$, where Π_i denotes the standard projection onto L_i. In (3.47) we are identifying L_i with the real line in the obvious way. (See Figure 3.10.)

Assume that (3.5) is not true for any choice of line L_i, and let us derive a contradiction. Consider the ball β for the l^∞-norm on $\mathbf{R}^n$ with center $z = (y_1 + \frac{t}{4n}, y_2 + \frac{t}{4n}, \ldots, y_n + \frac{t}{4n})$ and radius $\frac{t}{4n}$. We know that $x \in \beta$, because of (3.46), and $\partial \beta$ is at distance $\geq \alpha t$ from E. It is now easy to find an $\frac{\alpha}{10}$-box M' such that diam $M' \geq \frac{\alpha t}{10}$, $M' \cap E = \varnothing$, but the content of M' meets E: remove the top of the set of points w whose l^∞-distance to $\partial \beta$ is $\leq \alpha t$ and lift its bottom until you almost touch E; the set you obtain contains the desired $\frac{\alpha}{10}$-box. This contradicts our assumption that $(x, t) \in \mathscr{G}_b\left(\frac{\alpha}{10}\right)$. This proves that E does satisfy the WTP.

Incidentally, notice that for the purposes of Proposition 3.45 we could have replaced the WNB by the slightly weaker condition where we only consider boxes with sides parallel to the axes. Indeed, for the proof of the WTP this makes no difference, and a close look at the proof of Theorem 3.9 shows that the WHIP is always applied to d-planes parallel to one of the d-planes given by the WTP (in our case, one of the coordinate axes). Also, the argument for proving that the WNB implies the WHIP (when $d = 1$) still works if we restrict ourselves to boxes with sides parallel to coordinate hyperplanes and to projections onto lines parallel to the coordinate axes.

Unfortunately, the analogues of Propositions 3.41 and 3.45 for regular sets of dimension $d > 1$ are not true. A counterexample of dimension 2 in $\mathbf{R}^3$ can be obtained by taking the product of the Cantor set $K \subset \mathbf{R}^2$ described at the beginning of §I.1.1 (Garnett's counterexample) with the remaining axis. Let $E_0 = K \times \mathbf{R} \subset \mathbf{R}^2 \times \mathbf{R}$ be the resulting set. (See Figure 3.11 for a picture of a first approximation of a piece of E_0.) If you want a connected regular set, you can add to E_0 the plane $\mathbf{R}^2 \times \{0\}$. This gives a set E which is not rectifiable but which satisfies the WNM and the WNB. Notice that E has big projections (and hence satisfies the WTP) because K has the same

3.4. NO-MUGS, NO-BOXES, NO-REELS

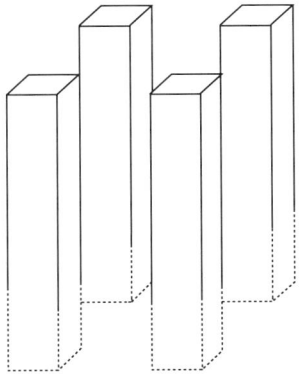

FIGURE 3.11

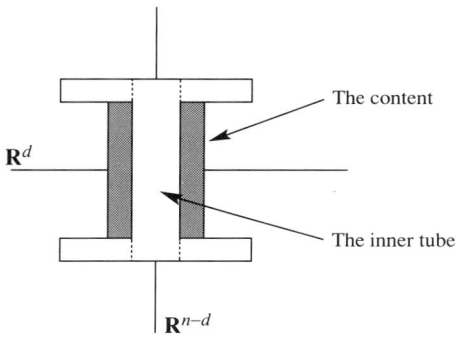

FIGURE 3.12. The standard ϵ-reel. Imagine a rotation invariance in $\mathbf{R}^d$ and in $\mathbf{R}^{n-d}$.

property with $d = 1$. Consequently, E does not even satisfy the WHIP (since the WHIP and the WTP imply uniform rectifiability). The same sort of counterexamples work in all dimensions n and d, $d \geq 2$. This is a little sad, because it prevents us from being able to show that the WALA implies uniform rectifiability when $d > 1$, at least for the present. See §III.4.2.

Let us now describe another generalization of the WNM which works when $d > 1$. This time we shall replace the mugs with "reels" and keep most of the rest of the definition of the WNM unchanged.

Let the integers $0 < d < n$ be fixed, and let us first define the standard ϵ-reel. [See Figure 3.12.] It will be convenient to write points of $\mathbf{R}^n$ as $x = (x', x'')$, where $x' = (x_1, \ldots, x_d) \in \mathbf{R}^d$ and $x'' = (x_{d+1}, \ldots, x_n) \in \mathbf{R}^{n-d}$. For each $0 < \epsilon < \frac{1}{10}$, let

(3.48)
$$R(\epsilon) = \{x \in \mathbf{R}^n : |x'| \leq \epsilon \text{ and } |x''| \leq 1\}$$
$$\cup \{x \in \mathbf{R}^n : |x'| < 10\epsilon \text{ and } (1-\epsilon) \leq |x''| \leq 1\}$$

be the standard ϵ-reel.

The "content" of $R(\epsilon)$ is the set $C = \{x \in \mathbf{R}^n : \epsilon < |x'| < 2\epsilon$ and $|x''| < 1 - \epsilon\}$; the "inner tube" of $R(\epsilon)$ is $\{x \in \mathbf{R}^n : |x'| \leq \epsilon$ and $|x''| \leq 1\}$.

Note that, when $d = 1$, the standard reel looks like two halves of camember boxes put back to back. More precisely, it is the union of the standard mug with its reflection about the origin (modulo a change in the order of the variables).

A closed set in $\mathbf{R}^n$ will be called an ϵ-reel if it is of the form $R = \phi(R(\epsilon))$ for some composition ϕ of an isometry of $\mathbf{R}^n$ and a dilation. The content and the inner tube of R will then be the images under ϕ of the content and the inner tube of $R(\epsilon)$.

We take $\mathscr{G}_r(\epsilon)$ to be the good set of all pairs $(x, t) \in E \times \mathbf{R}_+$ for which it is impossible to find an ϵ-reel $R \subset B(x, t)$ with $\operatorname{diam} R \geq \epsilon t$ such that R does not intersect E but the content of R does.

DEFINITION 3.49. Let E be a regular set of dimension d in $\mathbf{R}^n$. We say that E satisfies the WNR (weak-no-reels condition) if $E \times \mathbf{R}_+ \setminus \mathscr{G}_r(\epsilon)$ is a Carleson set for every $\epsilon \in (0, \frac{1}{10})$.

When $d = 1$, it is easy to see that the WNR is equivalent to the WNM. This is no longer true when $d \geq 2$; in this case, the WNR seems to be the correct notion to consider.

PROPOSITION 3.50. *The WNR implies the WHIP and the WTP and hence uniform rectifiability.*

The fact that the WNR implies the WTP can be proved in the same way as for the GWEC. Just as in the case of the WNM (when $d = 1$), the main difference between the WNR and the GWEC is that (3.25) is replaced by a more special kind of contraction for the WNR. For the purposes of proving the WTP this difference does not really matter.

Notice, incidentally, that the WNR actually implies the WEC when $d = n - 1$. This follows easily from (the simple) Lemma III.2.92 and its obvious generalization to higher dimensions.

It remains to show that the WNR implies the WHIP. The proof below is a fairly straightforward unravelling of the definitions, but it might appear to be more complicated than that, because of all the notation.

Let E be a regular set, and let $N > 0$ be given. It is enough to prove that there exist constants $0 < \alpha < \frac{1}{10}$ and $0 < \epsilon < \frac{1}{100}$ so that $(x, t) \in \mathscr{G}^*(\alpha, 2, N)$ as soon as $(x, 10t) \in \mathscr{G}_r(\epsilon)$.

So let $(x, 10t) \in \mathscr{G}_r(\epsilon)$ be given, and let us first prove that, if α and ϵ are chosen correctly, (x, t) lies in $\mathscr{G}(\alpha, 2, N)$. Let P be a d-plane, Π be the orthogonal projection onto P, and B be a ball of radius αt inside P which is contained in $\Pi(B(x, \frac{t}{2}))$. We assume that (3.1) holds, and we want to find a post.

Let y be a point in $E \cap \overline{B}(x, \frac{t}{2})$ such that $\Pi(y) \in \partial B$. If we choose $\alpha = (100N)^{-1}$, say, then we can find radii $r_1, r_2, \ldots, r_N$ such that $\frac{t}{5} \leq r_i \leq \frac{t}{4}$ for all i and $|r_i - r_j| > 3\alpha t$ for $i \neq j$. For each i, let C_i be the

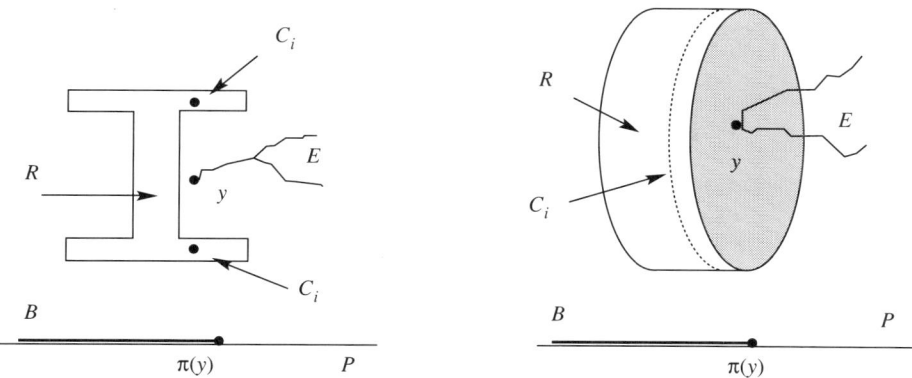

FIGURE 3.13. A picture of R in codimensions 1 and 2.

$(n-d-1)$-sphere given by $C_i = \{w \in \mathbf{R}^n : \Pi(w) = \Pi(y) \text{ and } |w-y| = r_i\}$. If $\operatorname{dist}(C_i, E) < \alpha t$ for each i, then we can select points $x_i \in E$ such that $\operatorname{dist}(x, C_i) < \alpha t$; the x_i, $1 \le i \le N$, provide the desired post (i.e., they satisfy (3.2)).

So suppose that there is an i such that C_i is at a distance $\ge \alpha t$ from E. We are going to show that this is impossible by constructing a reel R whose existence contradicts our assumption that $(x, 10t) \in \mathscr{G}_r(\epsilon)$.

Let ϕ be a composition of a dilation with an isometry of $\mathbf{R}^n$ which has the following properties. First, $\phi^{-1}(P)$ is a d-plane parallel to $\mathbf{R}^d$. Next, $\phi^{-1}(\Pi(y))$ is the point in $\phi^{-1}(\partial B)$ with the largest possible first coordinate (which is easily arranged by composing with a rotation of $\mathbf{R}^d$), and $\phi^{-1}(C_i)$ is a $(n-d-1)$-sphere of radius $(1-\frac{\epsilon}{2})$ (just compose with a dilation). Finally, $\phi^{-1}(y) = (\frac{3\epsilon}{2}, 0, \ldots, 0)$ (compose with a translation). Let R be the ϵ-reel which is the image under ϕ of the standard reel $R(\epsilon)$. [See Figure 3.13.] It is clear that $R \subset B(x, t)$ (since $y \in \overline{B}(x, \frac{t}{2})$) and that $\operatorname{diam} R \ge \frac{2t}{5}$ (because $\frac{t}{5} \le r_i \le \frac{t}{4}$). The content of R meets E because it contains y. The inner tube of R does not meet E because it is contained in $B(x, t)$ and its projection is contained in B, provided that we choose $\epsilon < \frac{\alpha}{100}$, say. Finally, the rest of R does not meet E either, because it stays at a distance $\le 12\epsilon \operatorname{diam} R \le \frac{\alpha t}{2}$ from C_i, which is itself at a distance $\ge \alpha t$ from E. Since $\operatorname{diam} R \ge \frac{t}{4} \ge 10\epsilon t$, the existence of R contradicts our assumption that $(x, 10t) \in \mathscr{G}_r(\epsilon)$.

We have now proved that if $(x, 10t) \in \mathscr{G}_r(\epsilon)$, then $(x, t) \in \mathscr{G}(\alpha, 2, N)$. The same argument also gives us that $(z, s) \in \mathscr{G}(\alpha, 2, N)$ for all pairs $(z, s) \in E \times \mathbf{R}_+$ such that $|z - x| \le t$ and $\frac{t}{2} \le s \le t$, and therefore $(x, t) \in \mathscr{G}^*(\alpha, 2, N)$, under the same hypothesis. This completes our proof of Proposition 3.50.

As we announced before, the same argument (or a minor variation of it) works to show that if E is a regular set of dimension 1 that satisfies the WNM or the WNB, then it satisfies the WHIP.

To conclude this section, let us check that

(3.51)   the GWEC implies the WNR.

This point was left over from the proof of Theorem 3.28 in the preceding section. We have of course implicitly asserted before that the GWEC is stronger than the WNR and that this is basically trivial, but it seems worthwhile to write down some details at least once.

Let a regular set E and $\epsilon > 0$ be given, and assume that (x, t) in $E \times \mathbf{R}_+ \setminus \mathscr{G}_r(\epsilon)$ is bad for the WNR. We want to associate to (x, t) some bad points for the WEC. Let R be an ϵ-reel which is contained in $B(x, t)$, has a diameter $\geq \epsilon t$, does not meet E, and whose content contains a point $z \in E$. Map R to the standard reel $R(\epsilon)$ by a composition ϕ^{-1} of a dilation and an isometry of $\mathbf{R}^n$. Set $z_0 = \phi^{-1}(z)$, let $\Pi(z_0)$ be the orthogonal projection of z_0 onto $\mathbf{R}^d$, and let $\mathscr{C}_0$ be the $(n-d-1)$-sphere of radius $(1 - \frac{\epsilon}{2})$ which is centered at $\Pi(z_0)$ and contained in the $(n-d)$-plane through $\Pi(z_0)$ that is perpendicular to $\mathbf{R}^d$. There is a contraction of $\mathscr{C}_0$ to a point that takes place inside the reel $R(\epsilon)$ and even at distance $\geq \frac{\epsilon}{2}$ from the complement of $R(\epsilon)$. Mapping this back by ϕ, we get a deformation of the $(n-d-1)$-sphere $\mathscr{C} = \phi(\mathscr{C}_0)$ to a point which takes place well inside R and, hence, in $\{w \in B(z, \operatorname{diam} R) : \operatorname{dist}(w, E) \geq \frac{\epsilon \operatorname{diam} R}{4}\}$. Since the convex hull of $\mathscr{C}$ meets E at the point z, we get that all the pairs $(y, s) \in E \times \mathbf{R}_+$ such that $|y - z| \leq \operatorname{diam} R$ and $2 \operatorname{diam} R \leq s \leq 3 \operatorname{diam} R$ lie in the bad set $\mathscr{B}_g(\frac{\epsilon}{12})$ for the GWEC (described just before Definition 3.27). Remember that $\epsilon t \leq \operatorname{diam} R \leq 2t$; (3.51) now follows from the usual argument.

3.5. The proof of Theorem 3.9 (part 1).

The overall organization of the proof of Theorem 3.9 is the same as for the fact that the bilateral weak geometric lemma (BWGL) implies uniform rectifiability. The idea will again be to apply Theorem I.3.42 on generalized corona decompositions. To do this, we shall first choose a coronization $(\mathscr{B}, \mathscr{G}, \mathscr{F})$ of E where the good cubes $Q \in \mathscr{G}$ are sufficiently good for both the WHIP and the WTP (with the precise values of the various constants involved to be chosen near the end). Then, for each stopping-time region $S \in \mathscr{F}$, we shall construct an approximating regular set $E(S)$. (This part of the argument will be the same as in Chapter 2.) These initial steps follow a general pattern that has essentially nothing to do with the WHIP or the WTP.

The most complicated part of the argument will be to prove that, if the good set $\mathscr{G}$ is chosen correctly, then the sets $E(S)$ have BPLG (big pieces of Lipschitz graphs) with uniform estimates. This will be done in two main steps. First, we shall prove that the $E(S)$'s have (uniformly) big projections and also that they satisfy an additional property that says, in rather vague

3.5. THE PROOF OF THEOREM 3.9. (PART 1)

terms, that if Π is the projection onto some d-plane and if the projection of $E(S) \cap B(x_0, CR)$ does not cover most of $B(\Pi(x_0), R)$, then there is a large pile of cubes not too far from x_0 whose projections all lie very close to $\Pi(x_0)$. This additional property will permit us to pass from the existence of big projections to the BPLG, using a stopping-time argument. The stopping-time argument will not use any of the properties of E related to the WHIP or the WTP other than the big projections and the additional property just mentioned, and it will be described in the next section.

Let us now begin the proof of Theorem 3.9. Let E be a d-dimensional regular set in $\mathbf{R}^n$ that satisfies the WHIP and the WTP, and let Δ be a family of cubes on E, as in §I.3.1. These assumptions will be in force throughout this section and the next.

We first define a good set of cubes by

(3.52)
$$\mathscr{G}_0 = \{Q \in \Delta : \ (x,t) \in \mathscr{G}(\alpha, A, N) \cap \mathscr{H}\left(\frac{\alpha}{k_0}, a\right) \ \text{whenever} \ x \in k_1 Q$$
$$\text{and} \ k_1^{-1} \operatorname{diam} Q \leq t \leq k_1 \operatorname{diam} Q\},$$

where the notation $k_1 Q$ is the same as in (I.3.7) and the constants α, A, N, a, k_0, and k_1 will be chosen as follows. The constants A and a are given to us by the definitions (3.4) and (3.6) of the WHIP and the WTP; the constant $k_0 > 1$ and the large integer N will be chosen later, and $\alpha \in (0, \frac{1}{10})$ will be chosen according to the definition of the WHIP and the value of N; k_1 will be chosen last. Our choices of A, a, and then α ensure that the complement of $\mathscr{G}_0$ in Δ satisfies a Carleson packing condition. [This follows easily from the fact that the complements in $E \times \mathbf{R}_+$ of $\mathscr{G}^*(\alpha, A, N)$ and $\mathscr{H}^*(\alpha/k_0, a)$ are Carleson sets. Note that we pay for k_0 and k_1 being large by having large Carleson constants, but we do not mind.]

To be completely honest, we should admit that the final choice of N does not take place until the next section, so k_1 is not chosen until then either. However, k_1 will not play an active role after this section, nor will k_0.

Because of Lemma I.3.22, there is a coronization $(\mathscr{B}, \mathscr{G}, \mathscr{F})$ of E such that $\mathscr{G} \subset \mathscr{G}_0$. We now associate to each of the stopping-time regions $S \in \mathscr{F}$ a regular set $E(S)$, as in §1 of Chapter 2. Let us review the construction of $E(S)$, and its main properties, for the convenience of the reader.

Let $S \in \mathscr{F}$ be given. Define the associated "distance" function $d(x)$ by

$$d(x) = \inf_{Q \in S} \{\operatorname{dist}(x, Q) + \operatorname{diam} Q\},$$

and set $Z = \{x \in E : d(x) = 0\}$. The set $E(S)$ is defined by

(3.53)
$$E(S) = Z \cup \left\{ \bigcup_{Q \in m(S)} \Sigma(Q) \right\} \cup P_0,$$

where P_0 is any d-plane such that $\operatorname{dist}(P_0, Q(S)) = \operatorname{diam} Q(S)$ (where $Q(S)$ denotes the top cube of S, as usual), $m(S)$ is the set of minimal cubes of S, and $\Sigma(Q)$, $Q \in m(S)$, is a star-shaped set which is constructed as follows.

We start from the "unit star" Σ_0, which is the intersection of the closed unit ball in $\mathbf{R}^n$ with a finite union of d-planes through the origin. The finite union of d-planes is chosen so that the projection of Σ_0 onto any d-plane P through the origin contains $P \cap \overline{B}\left(0, \frac{1}{2}\right)$ (see (2.8)). We then choose for each $Q \in m(Q)$ a center b_Q (i.e., a point of Q that satisfies (2.9)), and we take $\Sigma(Q)$ to be the image of Σ_0 under the composition of the translation that takes 0 to b_Q with the "dilation" that fixes b_Q and stretches lengths by the factor

$$\tfrac{1}{3}\operatorname{Min}\{\operatorname{diam} Q, \ \operatorname{dist}(b_Q, E \setminus Q)\}.$$

It was proved in Chapter 2 that the sets $E(S)$, $S \in \mathscr{F}$, are regular (with uniform estimates) and that they satisfy

(3.54) $\operatorname{dist}(x, E(S)) \leq 4 \operatorname{diam} Q$ whenever $x \in 2Q$ for some $Q \in S$.

(See Lemma 2.14 and (2.3).) Thus E has a generalized corona decomposition in terms of the $E(S)$'s, and Theorem 3.9 will follow from Theorem I.3.42 as soon as we prove that the $E(S)$'s have BPLG with uniform estimates. [Up to now, the argument is exactly the same as for the BWGL.]

The fact that the sets $E(S)$ have BPLG will be derived from the following lemma.

LEMMA 3.55. *There is a constant $a_1 > 0$ so that, for any $S \in \mathscr{F}$ and each cube $Q_0 \in S$, there is a d-plane $P = P(Q_0)$, a closed set $Z_0 \subset Z \cap Q_0$, and a (possibly empty) collection $\mathscr{C} = \mathscr{C}(Q_0)$ of cubes $T \in m(S)$ which are contained in Q_0, with the properties that*

(3.56) $$\left| Z_0 \cup \left(\bigcup_{T \in \mathscr{C}} T \right) \right| \geq a_1 |Q_0|$$

and

(3.57) $|\Pi(x) - \Pi(y)| \geq a_1 |x - y|$
        whenever x and y belong to $Z_0 \cup \{b_T : T \in \mathscr{C}\}$.

Here $\Pi : \mathbf{R}^n \to P$ denotes the orthogonal projection on P.

The proof of this lemma will occupy us for most of the rest of the chapter. Let us first check that Lemma 3.55 implies Theorem 3.9. As we said above, we only need to prove that each $E(S)$ has big pieces of Lipschitz graphs, with estimates that do not depend on S.

To do this, let $S \in \mathscr{F}$, $x \in E(S)$, and $R > 0$ be given. We want to find a big piece of Lipschitz graph in $E(S) \cap B(x, R)$. If x lies in the d-plane P_0 from (3.53) or if $R \geq 2 \operatorname{diam} Q(S)$, then $P_0 \cap B(x, R)$ has a large enough piece of a Lipschitz graph contained in $E(S) \cap B(x, R)$, and there is nothing

3.5. THE PROOF OF THEOREM 3.9. (PART 1)

to prove. The case when x lies in some $\Sigma(Q)$, where $Q \in m(S)$ and $R \leq 10 \operatorname{diam} Q$, is also trivial. [This time we take a piece of $\Sigma(Q)$.] Otherwise, x belongs to Z, or some $\Sigma(Q)$ with $\operatorname{diam} Q < \frac{R}{10}$. In either case there is a cube $Q_0 \in S$ such that $\operatorname{diam} Q_0 < \frac{R}{10}$ and $\operatorname{dist}(x, Q_0) \leq \frac{1}{3} \operatorname{diam} Q_0$. Replacing Q_0 by one of its ancestors if necessary, we can even assume that $\operatorname{diam} Q_0 \geq \frac{R}{C}$.

We now apply Lemma 3.55 to Q_0, and we get a d-plane $P = P(Q_0)$, a set Z_0, and a collection $\mathscr{C}$ of minimal cubes of S with the properties (3.56) and (3.57). If $|Z_0| \geq \frac{1}{2} a_1 |Q_0|$, then Z_0 is the piece of Lipschitz graph for which we are looking. Indeed, $Z_0 \subset Q_0 \subset B(x, R)$ by our choice of Q_0, and (3.57) implies that Z_0 is contained in an a_1^{-1}-Lipschitz graph over P.

Otherwise, (3.56) tells us that $\sum_{T \in \mathscr{C}} |T| \geq \frac{1}{2} a_1 |Q_0|$. Set $\mathscr{L} = \bigcup_{T \in \mathscr{C}} \mathscr{L}(T)$, where $\mathscr{L}(T) = \overline{B}(b_T, a_1 r_T / 10) \cap P(T)$, $r_T = \min(\operatorname{dist}(b_T, E \setminus T), \operatorname{diam} T)$, and $P(T)$ is one of the d-planes that was used to construct $\Sigma(T)$. Because of the way Σ_0 was defined, we can choose $P(T)$ so that it makes an angle $\leq \theta_0$ with the d-plane P, where $0 < \theta_0 < \frac{\pi}{2}$ depends on our choice of Σ_0. [A reasonable choice of Σ_0 gives $\theta_0 \leq \frac{\pi}{4}$.] The fact that $\mathscr{L}$ is contained in $E(S) \cap B(x, R)$ follows immediately from our choice of Q_0; the fact that it is contained in some Lipschitz graph over P and that it has measure $\geq C^{-1} a_1^{-d-1} |Q_0|$ follows from (3.57) and (2.10).

Thus Theorem 3.9 does indeed follow from Lemma 3.55.

We are left with the task of verifying Lemma 3.55. This verification will have three main ingredients: Lemmas 3.58 and 3.65 below and then a stopping-time argument. The two lemmas will contain all the relevant information that we get from the assumption that E satisfies the WHIP and the WTP, whereas the stopping-time argument will permit us to derive Lemma 3.55 from these two lemmas through a general procedure which is essentially the same as the one used in [D3].

Before we state the two lemmas, let us announce in advance the order in which the various constants that will be used in the proof will be chosen. (This might help to convince the reader that we are not cheating.) The constants A and a are already given to us by the definition of the WHIP and the WTP. The other constants will be chosen in the following order: ξ (small), k_0 (large), a_2 (small), C_1 (large), A_1 (large), N (large), α (small), C_2 (large), k_1 (large), and finally a_1 (very small).

LEMMA 3.58. *There is a constant $a_2 > 0$ that depends on A, a, the regularity constant for E, and the dimensions n and d, but not on N or α, such that for each $Q_0 \in S$ there is a d-plane $P = P(Q_0)$ which satisfies*

$$(3.59) \qquad \left| \Pi \left[(Z_0 \cap Q_0) \cup \left(\bigcup_{\substack{T \in m(S) \\ T \subset Q_0}} \beta(T) \right) \right] \right| \geq a_2 |Q_0|,$$

where Π is the orthogonal projection onto P and where
$$(3.60) \qquad \beta(T) = B(b_T, \tfrac{1}{10} \operatorname{dist}(b_T, E \setminus T)).$$

We need some additional notation to state the next lemma. Let C_2 be some large constant, which will arise in the lemma. Given any cube Q, define $W(Q)$ to be the union of all the cubes T that are of the same generation as Q (i.e., that belong to the same Δ_m) and which satisfy
$$(3.61) \qquad \operatorname{dist}(T, Q) \leq C_2 \operatorname{diam} Q.$$

Also, for each cube $Q \in S$, set
$$(3.62) \qquad \gamma(Q) = \overline{B}(b_Q, \operatorname{diam} Q)$$
and then
$$(3.63) \qquad \widehat{Q} = (Q \cap Z) \cup \left(\bigcup_{\substack{T \in m(S) \\ T \subset Q_0}} 2\gamma(T) \right).$$

More generally, if $\mathscr{I}$ is a finite union of cubes $Q \in S$, we define $\widehat{\mathscr{I}}$ to be the union of the $\widehat{Q}$ or, more directly,
$$(3.64) \qquad \widehat{\mathscr{I}} = (\mathscr{I} \cap Z) \cup \left(\bigcup_{\substack{T \in m(S) \\ T \subset \mathscr{I}}} 2\gamma(T) \right).$$

LEMMA 3.65. *For each (large) constant $C_1 > 1$, there is a constant A_1 (independent of N) such that, for every choice of N (and hence of α, through Definition 3.4), there is a constant C_2 such that the following holds.*

Let $Q \in S$ be given. Suppose that all of the cubes in $W(Q)$ which are of the same generation as Q lie in S (so that $\widehat{W}(Q)$ is defined, in particular). Let P be any d-plane, and let Π be the orthogonal projection onto P. Then at least one of the following alternatives occurs:

$(3.66) \quad |[B(\Pi(b_Q), C_1 \operatorname{diam} Q) \cap P] \setminus \Pi(\widehat{W}(Q))| \leq N^{-1/2} |Q|$; *or*

$(3.67) \quad$ *one can find at least $N^{1/2}/A_1$ cubes T of the same generation as Q, all contained in $W(Q)$, such that each $\Pi(T)$ intersects $P \cap B(\Pi(b_Q), 10AC_1 \operatorname{diam} Q)$.*

[This last "A" really is an A—the constant from the definition of the WHIP — and not an A_1.]

Note that the statement of Lemma 3.65 is somewhat similar to the statement of the WHIP; it says that if the projection $\Pi(\widehat{W}(Q))$ does not cover a large ball in P well enough, then there is a large pile of cubes which have nearly the same projections.

Before we prove the two lemmas, let us make a few comments about how they will be used in the stopping-time argument. Lemma 3.58 will be used first, to choose a d-plane P such that (roughly speaking) $\Pi(E(S))$ is large. Then the stopping-time construction will be applied to remove a part of $E(S)$ each time that there are points in $E(S)$ which are faraway from each other but whose projections are close together. Lemma 3.65 will then be used, in a technically complicated way, to show that whenever we have to remove a cube Q because its projection is too close to the projection of a faraway cube, we can manage to increase the ratio of the measure of the projection of the remaining set to its own measure. That will be good.

The constant C_1 in Lemma 3.65 will be chosen some time in the next section, but it is not dangerous; it will only depend on n, d, and the regularity constant for E. The choice of N will also be made in the next section. (It is a little more complicated than the choice of C_1.) This will give the value of α, through the definition of the WHIP.

Let us now prove Lemma 3.58. Let $Q_0 \in S$ be given. Set

$$x_0 = b_{Q_0} \quad \text{and} \quad t_0 = \tfrac{1}{2}\operatorname{dist}(x_0, E \setminus Q_0).$$

By definition of the coronization, $Q_0 \in \mathcal{G} \subset \mathcal{G}_0$, so $(x_0, t_0) \in \mathcal{H}(\alpha/k_0, a)$ (if k_1 is large enough). Therefore, we can choose a d-plane $P = P(Q_0)$ such that (3.5) holds for the pair (x_0, t_0). Set

$$(3.68) \qquad U = \{p \in P : \operatorname{dist}(p, \Pi(E \cap B(x_0, t_0))) < \alpha t_0/k_0\},$$

so that we have

$$(3.69) \qquad |U| \geq at_0^d.$$

Rather than working directly with $B(x_0, t_0)$, we first choose a comparable ball B such that a small spherical shell near ∂B will meet as little of E as possible. More precisely, let $\xi > 0$ be a small constant (to be chosen soon). We choose a radius $r_0 \in (t_0, \tfrac{3}{2}t_0)$ such that

$$(3.70) \qquad |E \cap B(x_0, (1+2\xi a)r_0) \setminus B(x_0, (1-2\xi a)r_0)| \leq C\xi ar_0^d.$$

Such a choice is possible if we choose a large enough constant C, because we can try approximately $\tfrac{1}{\xi a}$ choices of r_0 for which the corresponding shells are disjoint, and the total mass of those shells is $\leq Cr_0^d$. Note that C depends only on the regularity constant for E.

The interest of this choice of r_0 is that, if ξ is small enough, we shall not have to consider the points of U which come from the projection of the spherical shell in (3.70).

To prove (3.59), it suffices to show that there are constants K_1 and K_2, independent of N and α, such that, if

$$(3.71) \qquad V_1 = (Z \cap Q_0) \cup \left\{ \bigcup_{\substack{T \in m(S) \\ T \subset Q_0}} K_1 \beta(T) \right\},$$

then
(3.72) $$|\Pi(V_1)| \geq K_2^{-1}|Q_0|.$$

Indeed, if
$$V_0 = (Z \cap Q_0) \cup \left\{ \bigcup_{\substack{T \in m(S) \\ T \subset Q_0}} \beta(T) \right\}$$

is the set occurring in (3.59), then we have that $|\Pi(V_1)| \leq C(K_1)|\Pi(V_0)|$, as is easily seen using an argument based on the Vitali covering lemma for balls in P. (The same kind of argument was given just after (2.39).)

To prove (3.72), we intend to prove that most of the set U in (3.68) is contained in $\Pi(V_1)$. In order to control $|U \setminus \Pi(V_1)|$ we are going to use the WHIP to show that a certain maximal function is large on $U \setminus \Pi(V_1)$.

Let p be an element of $U \setminus \Pi(V_1)$. Set $\delta = \operatorname{dist}(p, \Pi(E \cap B(x_0, r_0)))$. By definition of U we have that

(3.73) $$\delta < \frac{\alpha}{k_0} t_0.$$

Assume first that $\delta < \operatorname{dist}(p, \Pi(E \cap \mathscr{A}))$, where $\mathscr{A} = \overline{B}(x_0, (1+\xi a)r_0) \setminus B(x_0, (1-\xi a)r_0)$ is a slightly smaller shell than the one that occurs in (3.70). Pick a point $x \in E \cap \overline{B}(x_0, r_0)$ such that $|p - \Pi(x)| = \delta$. Since
$$\operatorname{dist}(p, \Pi(E \cap \mathscr{A})) > \delta,$$
we have that $x \in \overline{B}(x_0, r_0) \setminus \mathscr{A} = B(x_0, (1-\xi a)r_0)$.

Notice that $x \in Q_0$, because of the definition of t_0 and the way we chose r_0.

We want to apply the condition in the definition of the WHIP to the ball $B(x, \alpha^{-1}\delta)$, so we need to check that $(x, \alpha^{-1}\delta) \in \mathscr{G}(\alpha, A, N)$. Notice first that

(3.74) $$\alpha^{-1}\delta \leq \frac{t_0}{k_0} \leq \frac{C \operatorname{diam} Q_0}{k_0},$$

where C is a geometric constant.

LEMMA 3.75. *We have that*
$$\alpha^{-1}\delta > \rho, \quad \text{where } \rho = \inf\{\operatorname{diam} Q : Q \in S, \ Q \subset Q_0, \ \text{and} \ x \in Q\}.$$

The lemma is trivial when $\rho = 0$. Otherwise, there is a cube $T \in m(S)$ such that $x \in T \subset Q_0$. If the inequality of the lemma were false, we would have $\delta \leq \alpha\rho = \alpha \operatorname{diam} T < \operatorname{diam} T$, and hence $p \in \Pi(B(x, \operatorname{diam} T))$ by definition of x. We choose the constant K_1 in (3.71) so large that $B(x, \operatorname{diam} T) \subset K_1 \beta(T)$ (which is itself contained in V_1, by definition). Thus we would have that $p \in \Pi(V_1)$, which contradicts the definition of p. This proves Lemma 3.75.

From Lemma 3.75 it follows that there is a cube $Q \in S$ such that $x \in Q$, $Q \subset Q_0$, and $\operatorname{diam} Q \leq \alpha^{-1}\delta$. The largest ancestor of Q that satisfies this inequality is still contained in Q_0 because of (3.74) (assuming that k_0 is sufficiently large), so we may assume that $\alpha^{-1}\delta \leq C \operatorname{diam} Q$ as well. By definition of our coronization $(\mathscr{B}, \mathscr{G}, \mathscr{F})$, Q is in the good set $\mathscr{G}_0$ defined by (3.52), so $(x, \alpha^{-1}\delta) \in \mathscr{G}(\alpha, A, N)$, provided that we choose k_1 large enough. Thus we are allowed to apply the condition in the definition of $\mathscr{G}(\alpha, A, N)$ to the ball $B(x, \alpha^{-1}\delta)$.

Let B be the ball in P with center p and radius δ. Because $\alpha < \frac{1}{10}$, B is contained in $\Pi(B(x, \frac{1}{2}\alpha^{-1}\delta))$. Clearly ∂B intersects $\Pi(E \cap B(x, \frac{1}{2}\alpha^{-1}\delta))$, because $\Pi(x)$ lies in both sets. We also know that $B \cap \Pi(E \cap B(x_0, r_0)) = \varnothing$, by definition of δ, so the fact that $B \cap \Pi(E \cap B(x, \alpha^{-1}\delta))$ is empty will follow from the inclusion $B(x, \alpha^{-1}\delta) \subset B(x_0, r_0)$. This inclusion comes from the fact that $x \in B(x_0, r_0) \setminus \mathscr{A} = B(x_0, (1 - \xi a)r_0)$ and from the inequalities $\alpha^{-1}\delta \leq t_0/k_0$ (by (3.74)) $\leq r_0/k_0$. For this to work, we have to choose $k_0 \geq (\xi a)^{-1}$. [This will not create any problem, because ξ will depend on a and geometric constants but not on k_0.]

We have just checked that B satisfies all the requirements in the definition of $\mathscr{G}(\alpha, A, N)$. With a crack of the WHIP we get a post, i.e., N points $x_1, \ldots, x_N$ in $E \cap B(x, A\alpha^{-1}\delta)$ which are at mutual distances $\geq \alpha(\alpha^{-1}\delta) = \delta$ and which satisfy $\Pi(x_j) \in AB$ for each j. (Compare with (3.2).) We want to use these points to make a certain maximal function large at the point p, in order to show that the set of p's is not too large.

Define a measure μ on P by

$$(3.76) \qquad \mu(F) = |\Pi^{-1}(F) \cap E \cap B(x_0, 10Ar_0)|$$

for all Borel sets $F \subset P$. Let $\mu^*(z)$ denote the value at $z \in P$ of the Hardy-Littlewood maximal function of μ on P, so that

$$(3.77) \qquad |\{z \in P : \mu^*(z) > \lambda\}| \leq \frac{CA^d r_0^d}{\lambda}.$$

The existence of the post implies that $\mu^*(p)$ is rather large:

$$\mu^*(p) \geq \frac{\mu(2AB)}{|2AB|} \geq C^{-1}A^{-d}\delta^{-d} \sum_{i=1}^{N} \left| E \cap B\left(x_i, \frac{\delta}{2}\right) \right|$$
$$\geq C^{-1}A^{-d}\delta^{-d} N\delta^d = C^{-1}A^{-d}N.$$

With this estimate, we can control the set of points $p \in U \setminus \Pi(V_1)$ which satisfy our assumption that $\delta < \operatorname{dist}(p, \Pi(E \cap \mathscr{A}))$; if

$$U' = \{p \in U \setminus \Pi(V_1) : \operatorname{dist}(p, \Pi(E \cap B(x_0, r))) < \operatorname{dist}(p, \Pi(E \cap \mathscr{A}))\},$$

then we have that

$$|U'| \leq |\{z \in P : \mu^*(z) > C^{-1}A^{-d}N\}| \leq CA^{2d}N^{-1}r_0^d \leq \tfrac{1}{3}|U|,$$

at least if we choose N large enough depending on A and a. (See (3.69).)

It remains to show that the set
$$U'' = \{p \in U \setminus \Pi(V_1) : \operatorname{dist}(p, \Pi(E \cap B(x_0, r_0))) \geq \operatorname{dist}(p, \Pi(E \cap \mathscr{A}))\}$$
satisfies

(3.78) $\qquad\qquad\qquad |U''| \leq \tfrac{1}{3}|U|,$

for then (3.72) will follow at once with $K_2 = Ca^{-1}$ for some geometric constant C.

By definition of U (see (3.68)), we see that U'' is contained in the set $U''' = \{p \in P : \operatorname{dist}(p, \Pi(E \cap \mathscr{A})) \leq \alpha t/k_0\}$. Let B_i, $i \in I$, be a maximal set of disjoint balls with centers in $E \cap \mathscr{A}$ and radii $\alpha t_0/k_0$. Thus $E \cap \mathscr{A} \subset \bigcup_{i \in I} 2B_i$, so $U'' \subset \bigcup_{i \in I} \Pi(3B_i)$.

On the other hand, each B_i is contained in the slightly larger annulus
$$\widehat{\mathscr{A}} = B(x_0, (1 + 2\xi a)r_0) \setminus B(x_0, (1 - 2\xi a)r_0),$$
provided that we choose k_0 large enough (depending on a and ξ). Notice that $\widehat{\mathscr{A}}$ is the same annulus as in (3.70), so we get

$$\begin{aligned}|U''| &\leq \sum_{i \in I} |\Pi(3B_i)| \leq C \sum_{i \in I} \left(\frac{\alpha}{k_0} t_0\right)^d \\ &\leq C \sum_{i \in I} |E \cap B_i| \leq C|E \cap \widehat{\mathscr{A}}| \leq C\xi a r_0^d.\end{aligned}$$

We now choose ξ small enough — depending only on a and the regularity constant for E — to get (3.78) by comparing with (3.69); (3.72) and Lemma 3.58 follow. Notice that, as promised, the constant K_2, and then a_2, do not depend on N. However, a sufficiently large choice of N, depending on A and a, is required.

REMARK 3.79. Lemma 3.58 is the counterpart here to Lemma 2.33 in the proof that the BWGL implies uniform rectifiability. Because the BWGL implies the WHIP and the WTP trivially, the proof of Lemma 3.58 that was just given provides a third proof of Lemma 2.20 (i.e., in essence, that the approximating sets $E(S)$ that were associated to a regular set E which satisfies the BWGL have big projections with uniform estimates). (See the comments in the first paragraph of § 2.2.)

Let us now prove Lemma 3.65. The reader might find it helpful to review some of the notation that was set immediately before Lemma 3.65, as well as the statement of the lemma. Note that the proof of this lemma will use only the WHIP and not the WTP. The argument will be somewhat similar to the previous one.

Let C_1 be given, let C_2 be large (to be chosen later), and let $Q \in S$ and P be given, as in the statement of the lemma. Assume that (3.67) is not true, and let us prove (3.66).

3.5. THE PROOF OF THEOREM 3.9. (PART 1)

Let $\mathscr{A}_j$ denote the annular shell
(3.80)
$$\mathscr{A}_j = \overline{B}\left(b_Q, 3\alpha^{-1}(j+2)C_1 \operatorname{diam} Q\right) \setminus B\left(b_Q, 3\alpha^{-1}(j-2)C_1 \operatorname{diam} Q\right).$$

[Recall that b_Q is a center for Q, as in (2.9). Of course, the geometric constant in (2.9) has nothing to do with the C_1 that is currently in play, even though it was given the same name.] We shall take C_2 large enough to ensure that

(3.81) $$\overline{B}\left(b_Q, 3\alpha^{-1}N^{1/2}C_1 \operatorname{diam} Q\right) \cap E \subset W(Q).$$

We claim that there is an integer j_0 such that $20 \leq j_0 \leq 20 + N^{1/2}/10$ and

(3.82) $$\Pi(E \cap \mathscr{A}_{j_0}) \cap B(\Pi(b_Q), 10AC_1 \operatorname{diam} Q) = \varnothing.$$

Indeed, if we could not find such a j_0, then there would exist more than $N^{1/2}/50$ points in $W(Q)$ which are at mutual distances $\geq 3\alpha^{-1}C_1 \operatorname{diam} Q$ from each other and whose projections all lie in $B(\Pi(b_Q), 10AC_1 \operatorname{diam} Q)$, and then (3.67) would hold if we take A_1 large enough, contradicting our assumption that (3.67) is not true. Let us now choose such a j_0 and set $\mathscr{A} = \mathscr{A}_{j_0}$.

Before we proceed further we should dispense with a minor detail.

LEMMA 3.83. $|\Pi(\overline{\widehat{W}(Q)}) \setminus \Pi(\widehat{W}(Q))| = 0$.

Since Π does not increase Hausdorff measure, it is enough to check that $\overline{R} \setminus \widehat{R}$ has measure 0 for each of the (finitely many) cubes R that make up $W(Q)$. Set

$$H = \bigcup_{\substack{T \in m(S) \\ T \subset R}} \{2\gamma(T)\}.$$

Since each $\gamma(T)$ is closed by definition (3.62), every point $x \in \overline{R} \setminus H$ must be the limit of a sequence of points x_k such that $\lim_{k \to \infty} d(x_k) = 0$. Any such x must lie in $\overline{R} \cap Z$, and the lemma follows because $\overline{R} \setminus R$ has measure 0 (since our cubes have "small boundaries").

Set

(3.84) $$\Omega = \{P \cap B(\Pi(b_Q), C_1 \operatorname{diam} Q)\} \setminus \Pi(\overline{\widehat{W}(Q)}).$$

We want to prove that $|\Omega| \leq N^{-1/2}|Q|$. Assume that Ω is not empty, and let p be an element of Ω. Set

(3.85) $$\delta = \operatorname{dist}(p, \Pi[W(Q) \cap B(b_Q, 3\alpha^{-1}j_0 C_1 \operatorname{diam} Q)]).$$

Note that

(3.86) $$0 < \delta < C_1 \operatorname{diam} Q.$$

[The second inequality is trivial because $\delta \leq |p - \Pi(b_Q)|$, and the first one follows from the definition of p and the fact that $W(Q) \subset \widehat{W}(Q)$.] Choose a point x which lies in $\overline{W(Q)} \cap \overline{B}(b_Q, 3\alpha^{-1} j_0 C_1 \operatorname{diam} Q)$ such that $|p - \Pi(x)| = \delta$. Set

(3.87) $\qquad \rho = \inf\{ \operatorname{diam} T : T \text{ is a cube in } S$
$\qquad\qquad\qquad$ such that $\overline{T}$ contains x and $T \subset W(Q)\}$.

We claim that

(3.88) $\qquad\qquad\qquad\qquad \delta > \rho.$

This is certainly true if $\rho = 0$. Otherwise, let T be a cube that realizes the infimum. Then T is a minimal cube of S and $\rho = \operatorname{diam} T$. Remember that $|p - \Pi(x)| = \delta$; so, if $\delta \leq \rho$, $p \in \Pi(\overline{B}(x, \operatorname{diam} T)) \subset \Pi(\overline{B}(b_T, 2\operatorname{diam} T)) = \Pi(2\gamma(T)) \subset \Pi(\widehat{W}(Q))$. This is impossible because $p \in \Omega$ (which is defined by (3.84)). This proves (3.88).

We are going to apply the condition in the definition of the WHIP to (x, t), where x is as above and $t = \alpha^{-1} \delta$. Let us first check that $(x, t) \in \mathscr{G}(\alpha, A, N)$. Since $\rho < \delta < C_1 \operatorname{diam} Q$, we see that there is a cube $T \subset S$ such that $x \in \overline{T}$ and $\operatorname{diam} T \leq \delta \leq CC_1 \operatorname{diam} T$. By definition of our coronization, T is in the class $\mathscr{G}_0$ defined by (3.52), and we get that (x, t) lies in $\mathscr{G}(\alpha, A, N)$ if we choose k_1 large enough. [Our choice of k_1 may depend on C_1 and on α (and therefore on N), but this does not cause a problem.]

To crack the WHIP, we use our d-plane P and take $B = P \cap B(p, \delta)$. Clearly $B \subset \Pi\left(B\left(x, \frac{t}{2}\right)\right)$, and ∂B contains $\Pi(x)$ and therefore intersects $\Pi\left(E \cap \overline{B}\left(x, \frac{t}{2}\right)\right)$. The fact that $B \cap \Pi(E \cap B(x, t)) = \varnothing$ will follow from

(3.89) $\qquad\qquad B(x, t) \subset \mathscr{A} \cup B(b_Q, 3\alpha^{-1} j_0 C_1 \operatorname{diam} Q),$

(3.90) $\qquad\qquad B \cap \Pi(E \cap \mathscr{A}) = \varnothing,$ and

(3.91) $\qquad\qquad B \cap \Pi[E \cap B(b_Q, 3\alpha^{-1} j_0 C_1 \operatorname{diam} Q)] = \varnothing.$

[Recall that $\mathscr{A} = \mathscr{A}_{j_0}$, where $\mathscr{A}_{j_0}$ is defined by (3.80).]

Now, (3.89) holds because $x \in \overline{B}(b_Q, 3\alpha^{-1} j_0 C_1 \operatorname{diam} Q)$ by definition, and $t = \alpha^{-1} \delta \leq \alpha^{-1} C_1 \operatorname{diam} Q$ by (3.86). Next, (3.90) follows because

$$B \subset B(p, C_1 \operatorname{diam} Q) \subset B(\Pi(b_Q), 2C_1 \operatorname{diam} Q)$$

(since $p \in \Omega$), which is contained in $B(\Pi(b_Q), 10AC_1 \operatorname{diam} Q)$, and is therefore disjoint from $\Pi(E \cap \mathscr{A})$ by our choice of j_0 (see (3.82)). Finally, (3.91) follows from (3.81) and the definition (3.85) of δ.

Thus B satisfies (3.1), and because $(x, t) \in \mathscr{G}(\alpha, A, N)$, there exists a post, i.e., points $x_1, \ldots, x_N$ as in (3.2).

Define a measure μ on P by

(3.92) $$\mu(F) = \left| E \cap \Pi^{-1}(F) \cap \left[\bigcup_T T \right] \right|,$$

where the union is taken over all cubes $T \in S$ such that $T \subset W(Q)$ and $\Pi(T)$ meets $B(\Pi(b_Q), 10AC_1 \operatorname{diam} Q)$. [Of course, we can restrict ourselves to cubes T of the same generation as Q.]

Let μ^* denote the Hardy-Littlewood maximal function of μ on P. To estimate $\mu^*(p)$ from below, it is helpful to observe first that the balls $B_i = B(x_i, \frac{\delta}{2})$ are disjoint (by definition of a post and because $\delta = \alpha t$). They also satisfy

(3.93) $$E \cap B_i \subset W(Q).$$

Indeed, because of (3.81), we need only check that each B_i is contained in the ball centered at b_Q with radius $R = 3\alpha^{-1} N^{1/2} C_1 \operatorname{diam} Q$. The radius of B_i is $\frac{\delta}{2} < (C_1 \operatorname{diam} Q)/2$ (by (3.86)), which is much less than $R/3$. Similarly,

$$|x_i - x| \leq At \leq A\alpha^{-1}\delta \leq A\alpha^{-1} C_1 \operatorname{diam} Q < \frac{R}{3}$$

(if we choose N large enough, depending on A). Finally, $|x - b_Q| \leq 3\alpha^{-1} j_0 C_1 \operatorname{diam} Q < \frac{R}{3}$, this time by definition of x and our choice of j_0. This proves (3.93).

The last property of the B_i we shall need is that

$$\Pi(B_i) \subset P \cap B\left(p, \left(A + \frac{1}{2}\right)\delta\right) \subset P \cap B(\Pi(b_Q), 10AC_1 \operatorname{diam} Q).$$

From all of this we deduce that

$$\mu^*(p) \geq \frac{\mu(2AB)}{|2AB|} \geq \frac{(2A)^{-d}}{|B|} \sum_{i=1}^N |E \cap B_i| \geq C^{-1} A^{-d} N.$$

Altogether, we have proved that

(3.94) $$\Omega \subset \{ p \in P : \mu^*(p) \geq C^{-1} A^{-d} N \},$$

where Ω is the set defined by (3.84). To control $|\Omega|$, we need to control the total mass of μ. Since we are assuming that (3.67) fails, there are at most $N^{1/2}/A_1$ cubes T of the same generation as Q that can arise in the union in (3.92). Hence $\mu(P) \leq CN^{1/2}|Q|/A_1$. From this and (3.94) we get $|\Omega| \leq CN^{-1/2}|Q|/A_1$ (where C may depend on A) by the maximal theorem. Taking A_1 large enough gives $|\Omega| \leq N^{-1/2}|Q|$, which implies (3.66) because of Lemma 3.83. This completes the proof of Lemma 3.65.

3.6. Part 2 of the proof: The stopping-time argument.

To finish the proof of Theorem 3.9, we still need to derive Lemma 3.55 from Lemmas 3.58 and 3.65. As we said before, the proof will no longer use the WHIP or the WTP directly but only the fact that E is regular (and, mostly, the existence of cubes). The argument is a minor variation of the one used in [**D3**], in which, unfortunately, the statements were not written in enough generality to allow a direct application here. The present implementation of this method will also benefit from a couple of simplifications.

In order to help the reader check the details, let us announce already the order in which the various constants in the argument will eventually be chosen. We are given constants A, a, and a_2 coming from the definitions of the WHIP, the WTP, and Lemma 3.58. We shall choose C_1 first (a geometric constant); a choice of A_1 will follow (by Lemma 3.65). Then the integer N will be chosen, and a choice of C_2 will follow (by Lemma 3.65). Next we shall choose C_3 (just a little larger than C_2), η, η_1, η_2, η_0 (smaller and smaller), and then ϵ, θ, and finally a constant a_1 for Lemma 3.55.

(a) *The description of the stopping-time procedure.* Rather than working with cubes of all generations, we shall find it more convenient to use only cubes of generations that are multiples of some very large integer. So let ω be a very large integer, and set $\epsilon = 2^{-\omega}$. Thus ω will be determined by our eventual choice of ϵ.

The first thing to do is to modify, if necessary, our coronization $(\mathscr{B}, \mathscr{G}, \mathscr{F})$ of E, in such a way that for all regions $S \in \mathscr{F}$, the top cube $Q(S)$ and the minimal cubes T, $T \in m(S)$, all belong to $\widetilde{\Delta} = \bigcup_{m \in \mathbf{Z}} \Delta_{m\omega}$ (i.e., are of generations which are multiples of ω). This can be arranged as follows. Let $S \in \mathscr{F}$ be given. For each cube Q in Δ let $\alpha(Q)$ denote the smallest cube in $\widetilde{\Delta}$ that contains Q. If Q is a cube in S such that $\alpha(Q) \notin S$ (so that Q is in the top layers of S), then we remove Q from S and add it to $\mathscr{B}$. We also remove from S and add to $\mathscr{B}$ any cube $Q \in S$ such that there is a minimal cube T in S and a cube R in $\widetilde{\Delta}$ with the properties that $T \notin \widetilde{\Delta}$, $R \subsetneq Q \subsetneq \alpha(T)$, and $R \in \Delta_{(m-1)\omega}$ where m satisfies $\alpha(T) \in \Delta_{m\omega}$. The remaining part of S — call it S' — is no longer coherent, but it decomposes naturally into the disjoint union of maximal coherent subsets, each of which has its top cube and all of its minimal cubes lying in $\widetilde{\Delta}$. [This is not hard to check. Notice that whenever we removed a cube from S we also removed either its parent or all of its siblings.] This new enlarged version of $\mathscr{B}$, as well as the collection of top cubes of the new stopping-time regions, satisfies a Carleson packing condition. The Carleson constants will be much worse than they were before, and they will depend on ϵ in particular, but we do not mind, because this does not prevent us from concluding that E has a generalized corona decomposition or from applying Theorem I.3.42; nor does it matter for the purposes of applying Lemmas 3.58 and 3.65.

We may also require that the cube Q_0 in the statement of Lemma 3.55 lies in $\widetilde{\Delta}$. This time we simply have to replace any other Q_0 by a smaller cube in $S \cap \widetilde{\Delta}$ and replace a_1 by a (much) smaller constant.

Let us now begin the description of the stopping-time procedure. Let a cube Q_0 be given, as in Lemma 3.55. We may as well require that Q_0 not be a minimal cube of S, since otherwise the lemma is trivial. For simplicity, we shall assume that Q_0 is an element of Δ_0, so that $\operatorname{diam} Q_0 \approx 1$. We start by choosing a d-plane P such that (3.59) holds. This is the only place where we use Lemma 3.58. The d-plane P will remain fixed throughout the entire argument; the point of the construction is to remove from Q_0 just enough subcubes to make the orthogonal projection Π onto P essentially bilipschitz on the remaining set.

The main idea behind the argument is the following. We shall not have too much control a priori on the number of cubes we have to remove from Q_0. Essentially, whenever we have two cubes of the same size which are far away from each other but whose images under Π are close to each other, we shall have to remove at least one of them. We shall make sure that we do not have to remove too many cubes by showing that, each time we remove a cube, some quantity associated to the piece of Q_0 that is left increases rapidly enough and that that quantity never gets too large. The quantity in question will be, grosso modo, the measure of the projection by Π of the remaining set, divided by its total mass.

Let us be more precise. If F is a finite union of cubes $Q \in S$, we set $p(F) = \Pi(\widehat{F})$, where $\widehat{F}$ is defined by (3.64) and (3.62). Thus

$$(3.95) \qquad p(F) = \Pi(F \cap Z) \cup \left\{ \bigcup_{\substack{T \in m(S) \\ T \subset F}} \left[P \cap \overline{B}(\Pi(b_Q), 2\operatorname{diam} Q) \right] \right\}.$$

The quantity we are interested in is

$$(3.96) \qquad \partial(F) = \frac{|p(F)|}{|F|}.$$

Let us give a name to the initial value of $\partial(F)$: set $\delta = \partial(Q_0)$. Note that

$$|p(Q_0)| \geq \left| \Pi(Q_0 \cap Z) \cup \left\{ \bigcup_{\substack{T \in m(S) \\ T \subset Q_0}} \Pi(\beta(T)) \right\} \right|,$$

where $\beta(T)$ is defined by (3.60). So, by our choice of P satisfying (3.59), we get

$$(3.97) \qquad a_2 \leq \delta \leq C,$$

where C is some innocuous geometric constant.

We shall find it convenient to adopt the following conventions concerning cubes. As was announced earlier, we shall only work with cubes in $\widetilde{\Delta} = \bigcup_{m \in \mathbb{Z}} \Delta_{-m\omega}$. The cubes Q in $\Delta_{-m\omega}$ will now be referred to as cubes of the mth generation or even as m-cubes. The "size" of such a cube Q will be ϵ^m (where again m is the integer such that $Q \in \Delta_{-m\omega}$). Note that ϵ^m is comparable to the diameter of Q (with a constant that does not depend on ϵ).

We shall now construct a decreasing sequence $E_0 = Q_0 \supset E_1 \supset \cdots \supset E_k \supset \cdots$ of subsets of Q_0, where each E_k will be a finite union of cubes of S. Each E_{k+1} will be obtained from E_k by removing one cube or sometimes a few cubes of the same generation. We shall of course start by removing cubes of the first generation, then cubes of the second generation, and so on. The part of this process in which we remove cubes of the first generation will be called the first generation of the construction, and similarly for the later generations.

We now describe the mth generation of the construction, assuming that we have just finished the $(m-1)$st generation. Let $k(m-1)$ be the largest value of k such that E_k was obtained from E_{k-1} during the $(m-1)$st generation of the construction, and call $E(m-1) = E_{k(m-1)}$ the last set we obtained during that generation of the construction. Since we only removed cubes of generations $\leq m-1$ so far, $E(m-1)$ is a finite union of $(m-1)$-cubes. However, we decide to view it now as a union of m-cubes.

The process for completing the $(m-1)$st generation of the construction will also produce a pair of sets $G(m-1)$ and $\mathscr{C}(m-1)$ contained in $E(m-1)$. For reasons that will be explained later, we shall not allow ourselves to make any changes to the cubes of which these sets are composed, so all the modifications that will be made to the sets E_k during the mth generation of the construction will only affect the set $E(m-1) \setminus \{G(m-1) \cup \mathscr{C}(m-1)\}$.

The mth generation of the construction will be performed in six stages. The order of the stages has some importance, because we want to remove first the cubes for which we know that removing them will increase $\partial(E_k)$ substantially. As we said before, each E_{k+1} will be obtained by removing from E_k one or a finite number of m-cubes which are still contained in

$$\widetilde{E}_k = E_k \setminus \{G(m-1) \cup \mathscr{C}(m-1)\}.$$

Stage 1 (removing the piles). During this stage, we look for $[N^{1/3}]$ (= the greatest integer part of $N^{1/3}$) m-cubes Q_j, $j \in J$, which are still contained in the current $\widetilde{E}_k$ and which have the property that the $p(Q_j)$'s all intersect some ball of radius $C_1 A \epsilon^m$ in P. If we find such a collection of cubes, we remove them. More precisely, we set $\mathscr{D}_k = \bigcup_{j \in J} Q_j$ and let $E_{k+1} = E_k \setminus \mathscr{D}_k$. We then look for another pile of cubes like $\{Q_j\}$, and so on. After some time, we shall no longer be able to find new families of cubes $\{Q_j\}$ as above, and at that point we proceed to Stage 2 of the construction.

Stage 2 (good cubes). We now look at all the m-cubes R which are still contained in $\widetilde{E}_k$ at the end of Stage 1 and which satisfy

(3.98) $$\partial(R) \geq (1 + \eta_0)\delta.$$

3.6. THE STOPPING-TIME ARGUMENT

We pay special attention to these cubes, and we call them good. Let us explain why we like them so much. The main goal of the present argument is to show that we can reduce to the special case of cubes Q_0 which satisfy $\partial(Q_0) \geq (1+\eta_0)\delta$. By iterating this argument enough times we shall be able to reduce to the case where $\partial(Q_0) \geq (1+\eta_0)^M \delta$; the case when M is sufficiently large is trivial, because we know from the second half of (3.97) that this case cannot occur. The precise induction procedure will be spelled out later, but for the moment we shall content ourselves with keeping as many of the cubes satisfying (3.98) as we can, while taking comfort in the knowledge that finding sets Z_0 and $\mathscr{C} \subset m(S)$ for each such cube R will be easier.

Select a maximal collection R_i, $i \in g(m)$, of cubes of generation m which satisfy $R_i \subset \widetilde{E}_k$, (3.98), and

$$(3.99) \qquad \operatorname{dist}(R_i, R_j) \geq \epsilon^m \quad \text{and} \quad \operatorname{dist}(p(R_i), p(R_j)) \geq \epsilon^m$$

for all $i \neq j$ in the set $g(m)$. [We should perhaps emphasize that $g(m)$ is merely some set of labels which may be chosen capriciously.]

Once this selection is made, we can augment the current good set by setting

$$(3.100) \qquad G(m) = G(m-1) \cup \left\{ \bigcup_{i \in g(m)} R_i \right\}.$$

We now eliminate all the cubes which interact too closely with our new good cubes, as follows. We equip $g(m)$ with an (arbitrary) linear ordering, and we perform, for each $i \in g(m)$ (taken in order), the following manipulation. Let k be the current index (k will increase by 1). We let $\mathscr{D}_k$ be the union of all the m-cubes $Q \subset \widetilde{E}_k$, $Q \neq R_i$, such that

$$(3.101) \qquad \operatorname{dist}(Q, R_i) < \epsilon^m \quad \text{or} \quad \operatorname{dist}(p(Q), p(R_i)) < \epsilon^m,$$

and then set $E_{k+1} = E_k \setminus \mathscr{D}_k$. Once this has been done for each $i \in g(m)$, we proceed to Stage 3.

Notice that the maximality of the collection of R_i's ensures that any good m-cube which is not among the R_i's is eliminated at this stage of the construction.

Stage 3 (minimal cubes). We never want to decompose any of the minimal cubes of S, because it would be useless, and even dangerous, since the children of such a cube are no longer in S. So we also want to keep such cubes in a separate collection and not touch them again. We choose a maximal set R_i, $i \in c(m)$, of m-cubes which are still contained in $E_k \setminus \{G(m) \cup \mathscr{C}(m-1)\}$ at the end of Stage 2, such that every R_i lies in $m(S)$ and such that (3.99) holds for $i \neq j \in c(m)$. We are then ready to define the new set of minimal cubes by

$$(3.102) \qquad \mathscr{C}(m) = \mathscr{C}(m-1) \cup \left\{ \bigcup_{i \in c(m)} R_i \right\}.$$

Next, exactly as we did for the good cubes, we linearly order $c(m)$, and then we take the cubes R_i one after the other and remove from the current set E_k the union $\mathscr{D}_k$ of all the m-cubes $Q \neq R_i$ which are still contained in $E_k \setminus \{G(m) \cup \mathscr{C}(m-1)\}$ and which satisfy (3.101). Notice that all the minimal cubes of this generation which are not among the R_i's are eliminated here, because of the maximality of the collection of R_i's.

Note that, during Stages 2 and 3, $\mathscr{D}_k$ is never composed of more than $C(1 + N^{1/3})$ m-cubes. This is because we made sure, during Stage 1, to remove all the piles of more than $N^{1/3}$ cubes. Since, during Stage 1, each $\mathscr{D}_k$ was composed of exactly $[N^{1/3}]$ cubes, we have the estimate

$$(3.103) \qquad |\mathscr{D}_k| \leq C(1 + N^{1/3})\epsilon^{md}$$

on the mass of $\mathscr{D}_k$ in each case.

For each $R \in G(m) \cup \mathscr{C}(m)$, denote by $k(R)$ the value of k where all the cubes that satisfy (3.101) with $R_i = R$ were removed from E_k. (When $R \in G(m-1) \cup \mathscr{C}(m-1)$, the m in (3.101) should be replaced by the generation of R.) We claim that

$$(3.104) \qquad \operatorname{dist}(R, E_{k(R)+1} \setminus R) \geq \epsilon^m$$

and

$$(3.105) \qquad \operatorname{dist}(p(R), p\left(E_{k(R)+1} \setminus R\right)) \geq \epsilon^m$$

for every $R \in G(m) \cup \mathscr{C}(m)$.

Let us verify these inequalities. When $R \in G(m-1) \cup \mathscr{C}(m-1)$, (3.104) and (3.105) follow from their counterparts for the previous generation (i.e., "induction hypothesis"). Now suppose that

$$R \in [G(m) \cup \mathscr{C}(m)] \setminus \{G(m-1) \cup \mathscr{C}(m-1)\}.$$

If we replace the $E_{k(R)+1}$ in (3.104) and (3.105) with

$$E_{k(R)+1} \setminus \{G(m-1) \cup \mathscr{C}(m-1)\},$$

then the resulting inequalities follow immediately from the fact that we removed all of the cubes which satisfy (3.101) during Stages 2 and 3. We also have that

$$\operatorname{dist}(R, G(m-1) \cup \mathscr{C}(m-1)) \geq \epsilon^{m-1},$$

because we already removed in the earlier generations any cubes that got too close to the cubes that make up $G(m-1) \cup \mathscr{C}(m-1)$. Similarly,

$$\operatorname{dist}(p(R), p(G(m-1) \cup \mathscr{C}(m-1))) \geq \epsilon^{m-1}$$

for the same reason. This proves (3.104) and (3.105).

Stage 4 (cubes with small projections). At this point it is convenient to get rid of all the cubes whose projection is too small. It should not be necessary to do this, but it does not hurt anyway because removing these cubes will help increase $\partial(E_k)$. Also, it will be pleasant to know, in the future, that all the cubes left have essentially the same value of ∂. Stage 4 consists in removing, one by one, all of the cubes Q of the mth generation which are still contained in $E_k \setminus \{G(m) \cup \mathscr{E}(m)\}$ and which satisfy

(3.106) $$\partial(Q) \leq (1-\eta)\delta.$$

Each time we find such a cube Q, we set $\mathscr{D}_k = Q$ and $E_{k+1} = E_k \setminus Q$. When there are no such cubes left, we go on to Stage 5.

Stage 5 (large intersections). We now look for ordered pairs (Q, T) of distinct m-cubes which are still contained in $E_k \setminus \{G(m) \cup \mathscr{E}(m)\}$ and which satisfy

(3.107) $$|p(Q) \cap p(T)| \geq \eta_1 \delta |Q|.$$

As long as we can find such a pair, we remove Q from E_k (i.e., we set $\mathscr{D}_k = Q$ and $E_{k+1} = E_k \setminus Q$). After some time, no such pairs can be found, and we proceed to the last stage.

Stage 6 (what has to be done). We now remove, one by one, any m-cube $Q \subset E_k \setminus \{G(m) \cup \mathscr{E}(m)\}$ for which there is another cube R contained in $E_k \setminus \{G(m) \cup \mathscr{E}(m)\}$ such that

(3.108) $$\mathrm{dist}(Q, R) \geq C_3 \epsilon^m \quad \text{but} \quad \mathrm{dist}(p(Q), p(R)) < \epsilon^m.$$

[The constant C_3 will be chosen soon. It is just a little larger than C_2.]

Once all of this is over, we are finished with the mth generation of our construction. We have already defined the sets $G(m)$ and $\mathscr{E}(m)$, and we set $E(m) = E_{k(m)}$, where $k(m)$ is the current value of k. We can then proceed to the next generation of the construction.

We must now say when we stop the construction. We stop as soon as any one of the following events takes place:

(3.109) $$|G(m)| \geq \theta |Q_0|;$$

(3.110) $$|\mathscr{E}(m)| \geq \theta |Q_0|;$$

(3.111) $$\partial(E_k) \geq (1 + 2\eta_0)\delta;$$

(3.112) $$|E_k| \leq \tfrac{1}{2}|E_0|;$$

(3.113) $$\partial(E_k) \leq (1 - \eta_2)\delta.$$

We should emphasize that (3.109) and (3.110) are only tested for the values of m for which $G(m)$ and $\mathscr{E}(m)$ is defined. For example, at the end of the second stage of the mth generation of the construction, $G(m)$ and $\mathscr{E}(m-1)$ are defined, but $\mathscr{E}(m)$ is not.

We shall see later how to proceed when none of the stopping-time conditions occur, or when (3.109), (3.110), or (3.111) happens. In each of these cases, it will be rather easy to find large sets on which Π is bilipschitz (using, in particular, the fact that (3.112) did not occur).

(b) *The events* (3.112) *and* (3.113) *never happen*. Let k_0 be a fixed positive integer, and assume that none of the stopping-time conditions have occurred for any $k \leq k_0$. In this subsection we are going to prove that (3.112) and (3.113) are still not true for k_0+1. Note, by the way, that the large number of stopping-time conditions provides us with a large number of false inequalities to play with (which will be useful).

As we said earlier, the point of the argument will be that $\partial(E_k)$ tends to increase essentially as fast as $|E_k|$ decreases, so that, if η_0 is small enough with respect to the other constants, (3.111) should happen before (3.112), and (3.113) should never happen.

Our first task is to produce suitable lower bounds for the ratio

$$(3.114) \qquad \mu(k) = \partial(E_{k+1})\partial(E_k)^{-1},$$

where k is any integer $\geq k_0$. We shall try to prove that $\mu(k)$ is significantly larger than 1 as often as we can.

To simplify the statement of these estimates we need some more notation. First, we shall continue to denote by $\mathscr{D}_k$ the cube or the family of cubes that was removed from E_k to get E_{k+1}. [In short, $\mathscr{D}_k = E_k \setminus E_{k+1}$.] Also, for $j = 1, \ldots, 6$, we shall denote by $J(j)$ the set of integers k such that $\mathscr{D}_k$ was removed from E_k during Stage j of some generation of the construction. We summarize our lower bounds for $\mu(k)$ in the following lemma.

LEMMA 3.115. *There is a geometric constant C such that the following estimates hold for all $k \leq k_0$ if N is large enough (depending on C_1, A, and a_2):*

$$(3.116) \qquad \mu(k) \geq 1 + C^{-1}|\mathscr{D}_k| \quad \text{whenever } k \in J(1);$$

$$(3.117) \qquad \mu(k) \geq 1 - C\delta^{-1}|\mathscr{D}_k| \quad \text{when } k \in J(2) \cup J(3);$$

$$(3.118) \qquad \mu(k) \geq 1 + C^{-1}\eta|\mathscr{D}_k| \quad \text{whenever } k \in J(4);$$

$$(3.119) \qquad \mu(k) \geq 1 + C^{-1}\eta_1|\mathscr{D}_k| \quad \text{whenever } k \in J(5); \text{ and}$$

$$(3.120) \qquad \mu(k) \geq 1 - C\eta_2|\mathscr{D}_k| \quad \text{when } k \in J(6).$$

Let us make some preliminary observations before we begin the proof of this. Note that $\frac{1}{2}|E_0| \leq |E_k| \leq |E_0|$, because (3.112) has not happened yet. Also, if k corresponds to the generation m, we have the inequality (3.103). [We stated it only for Stages 1, 2, and 3, but it is trivial for the other stages, because $\mathscr{D}_k$ then consists of a single m-cube.] Consequently, $|\mathscr{D}_k|/|E_k|$ is as small as we wish, provided that we take ϵ small enough (depending on N). (Remember that $m \geq 1$.) Thus all the right-hand sides of the inequalities in Lemma 3.115 will be as close to 1 as we want. This will make the computations of products of $\mu(k)$ simpler. Since we also have that $(1-\eta_2)\delta \leq \partial(E_k) \leq (1+2\eta_0)\delta$ (because (3.113) and (3.111) are false), we get that

$$(3.121) \qquad C^{-1}\delta \leq |E_{k+1}|\partial(E_k) \leq C\delta$$

for some geometric constant C.

3.6. THE STOPPING-TIME ARGUMENT

To prove (3.116), we notice that when $k \in J(1)$, $|\mathscr{D}_k| \geq C^{-1} N^{1/3} \epsilon^{md}$ (we shall systematically use the implicit convention that m is the generation associated to k). On the other hand, $p(\mathscr{D}_k)$ is contained in a ball of radius $CC_1 A \epsilon^m$, so that

$$|p(E_{k+1})| \geq |p(E_k)| - |p(\mathscr{D}_k)| \geq \partial(E_k)|E_k| - C(C_1 A \epsilon^m)^d$$
$$= \partial(E_k)|E_{k+1}| + \partial(E_k)|\mathscr{D}_k| - C(C_1 A \epsilon^m)^d.$$

Now $\partial(E_k)|\mathscr{D}_k| \geq (1 - \eta_2)\delta C^{-1} N^{1/3} \epsilon^{md}$, which is much larger than $C(C_1 A \epsilon^m)^d$ if we choose N large enough (depending on C_1, A, and δ (or rather a_2, through (3.97))). This yields

$$|p(E_{k+1})| \geq \partial(E_k)|E_{k+1}| + \frac{C^{-1}}{2} \delta N^{1/3} \epsilon^{md} \geq \partial(E_k)|E_{k+1}| + (C')^{-1}\delta|\mathscr{D}_k|.$$

We now divide both sides of the inequality by $\partial(E_k)|E_{k+1}|$ to obtain $\mu(k)$ on the left-hand side; (3.116) follows from (3.121).

To prove (3.117), we simply write

$$|p(E_{k+1})| \geq |p(E_k)| - |p(\mathscr{D}_k)| \geq \partial(E_k)|E_k| - C|\mathscr{D}_k| \geq \partial(E_k)|E_{k+1}| - C|\mathscr{D}_k|$$

and then divide by $\partial(E_k)|E_{k+1}|$ and use (3.121) to get (3.117). Notice that (3.117) is true in all cases, but it will be useful only when we do not have a better estimate, i.e., when $k \in J(2) \cup J(3)$.

For (3.118), we use the estimate (3.106):

$$|p(E_{k+1})| \geq |p(E_k)| - |p(\mathscr{D}_k)| \geq \partial(E_k)\left[|E_{k+1}| + |\mathscr{D}_k|\right] - (1 - \eta)\delta|\mathscr{D}_k|$$
$$\geq \partial(E_k)|E_{k+1}| + (1 - \eta_2)\delta|\mathscr{D}_k| - (1 - \eta)\delta|\mathscr{D}_k|$$

(because (3.113) is still false).

We decide to choose $\eta_2 < \frac{1}{2}\eta$ so that we get $|p(E_{k+1})| \geq \partial(E_k)|E_{k+1}| + \frac{1}{2}\eta\delta|\mathscr{D}_k|$, from which (3.118) follows in the usual way.

For (3.119) we shall use the fact that $\mathscr{D}_k$ is not a good cube (we never remove any good cube after the end of Stage 2) and also the observation that $|p(\mathscr{D}_k) \cap p(E_{k+1})| \geq \eta_1 \delta |\mathscr{D}_k|$, which comes from (3.107). We get

$$|p(E_{k+1})| = |p(E_k)| - |p(\mathscr{D}_k)| + |p(\mathscr{D}_k) \cap p(E_{k+1})|$$
$$\geq \partial(E_k)\left[|E_{k+1}| + |\mathscr{D}_k|\right] - (1 + \eta_0)\delta|\mathscr{D}_k| + \eta_1 \delta |\mathscr{D}_k|$$
$$\geq \partial(E_k)|E_{k+1}| + \delta|\mathscr{D}_k|\{(1 - \eta_2) - (1 + \eta_0) + \eta_1\}$$

(we have used again the fact that (3.113) fails). We decide to choose η_2 and η_0 small enough with respect to η_1 so that we get $|p(E_{k+1})| \geq \partial(E_k)|E_{k+1}| + (\eta_1/2)\delta|\mathscr{D}_k|$, from which (3.119) follows as usual.

The estimate for (3.120) is the same, except that we do not have any estimate on $|p(E_{k+1}) \cap p(\mathscr{D}_k)|$, so we lose the additional $\eta_1 \delta |\mathscr{D}_k|$ in the estimate. We only obtain

$$|p(E_{k+1})| \geq \partial(E_k)|E_{k+1}| + \delta|\mathscr{D}_k|\{(1 - \eta_2) - (1 + \eta_0)\},$$

which yields (3.120) by the usual division, provided that we take $\eta_0 \leq \eta_2$. This completes the proof of Lemma 3.115.

Let us now say a few words about our strategy. We would like to say that each time we remove a $\mathscr{D}_k$ (i.e., that $|E_k|$ decreases), we increase the value of $\partial(E_k)$ about as fast as $|E_k|$ decreases. This is what happens when we can use the estimates (3.116), (3.118), or (3.119). If we only had to use those, we would take the product of our estimates for $\mu(k)$ and deduce that (3.111) must occur before (3.112) does (if η_0 is small enough). The problem is that we shall have to use (3.117) and (3.120) from time to time. For (3.117), we do not have a serious problem, because the fact that (3.109) and (3.110) have not happened yet gives a bound on the number of times that we have to apply (3.117), which is as small as we want (by taking θ small enough). Thus we shall apply (3.117) so seldomly that our estimates on $\partial(E_k)$ will not be significantly disturbed. Our only serious problem is with (3.120). We are going to show that, each time we have to use (3.120), we have already used one of the other estimates, and that the fact that we now have to use (3.120) only destroys a small portion of the prior favorable estimate.

In order to do the accounting, we shall associate to each value of $k \in J(6)$ a prior value of k where we had a favorable estimate. We shall also have to check that, for each favorable k, there are not too many $k' \in J(6)$ which are associated to it in this way.

We are implicitly assuming here that $k \leq k_0$, and so none of the stopping-time conditions have yet obtained. We shall remind the reader of this when it becomes important.

So let $k \in J(6)$ be given. Let $Q = \mathscr{D}_k$ be the m-cube that was removed from E_k to get E_{k+1}. Let $W(Q)$ be the large neighborhood of Q that was defined a little before Lemma 3.65 (between (3.60) and (3.61)).

Let us check that
$$W(Q) \cap (G(m-1) \cup \mathscr{E}(m-1)) = \varnothing.$$
If this were not the case, then we would have that
$$\text{dist}(Q, G(m-1) \cup \mathscr{E}(m-1)) \leq CC_2 \epsilon^m < \epsilon^{m-1}$$
if ϵ is small enough (depending on C_2, which is O.K.). This last would mean that an ancestor of Q was removed during Stage 2 or 3 of an earlier generation, which is impossible, since Q is still around.

Now we want to consider some cases separately. Assume first that $W(Q) \subseteq E(m-1)$ so that
$$W(Q) \subseteq E(m-1) \setminus (G(m-1) \cup \mathscr{E}(m-1)).$$
This implies that each of the m-cubes which make up $W(Q)$ lies in S. [Otherwise, $W(Q)$ intersects a minimal $(m-1)$-cube contained in $E(m-1)$, and all of those are contained in $\mathscr{E}(m-1)$, by construction.] Therefore, we can apply Lemma 3.65 and further distinguish the cases where (3.66) or

(3.67) holds. We start with the case when (3.67) is satisfied. Note that if T is one of the cubes given by (3.67), then $p(T)$ is contained in a ball of radius $CC_1 A \operatorname{diam} Q$ centered on $p(Q)$. If we choose N large enough, then we can find more than $N^{1/3}$ cubes T, chosen among the $N^{1/2}/A_1$ cubes which are given by (3.67), such that the "projections" $p(T)$ all intersect a single ball of radius $C_1 A \epsilon^m$. [Of course, we can do this without letting N depend on ϵ.] Since $W(Q) \subseteq E(m-1)$ and $W(Q)$ is disjoint from $G(m-1) \cup \mathscr{E}(m-1)$, all these cubes were available at the beginning of Stage 1, and, since they constitute a pile, they cannot all be present at the end of Stage 1. Hence there is an integer $\phi(k)$, corresponding to Stage 1 of the mth generation, such that $\mathscr{D}_{\phi(k)}$ contains one of the cubes T given by (3.67). There may be more than one such integer $\phi(k)$, but we do not care, we simply pick one.

Note that the whole $p(\mathscr{D}_{\phi(k)})$ remains at a distance $\leq CC_1 A \epsilon^n$ from $p(Q)$. This implies in particular that, for a given $l \in J(1)$, there are no more than $CN^{1/3}$ integers $k' \in J(6)$ such that $\phi(k') = l$. Indeed, for any such k' we would have to have that $p(\mathscr{D}_{k'})$ lies within $CC_1 A \epsilon^m$ of $p(\mathscr{D}_l)$ (whose diameter is $\leq CC_1 A \epsilon^m$), but all of these k''s belong to Stage 6 of the generation m of the construction, and we know that, after the end of Stage 1, there are never more than $N^{1/3}$ m-cubes $Q' \subseteq E_{\tilde{k}} \setminus \{G(m-1) \cup \mathscr{E}(m-1)\}$ such that $p(Q')$ intersects a given ball of radius $C_1 A \epsilon^m$. (Here $\tilde{k}$ is the index that corresponds to the end of Stage 1 of the mth generation.)

Our second case is when $W(Q)$ is still contained in $E(m-1)$, but (3.67) does not hold, so that (3.66) must be true. Because Q was removed during Stage 6, there is another cube $R \subset E_{k+1}$ such that (3.108) holds. Because R is still around at time k, it does not satisfy (3.106), so $|p(R)| \geq \frac{\delta}{2}|R|$. Because of (3.108), we know that $p(R)$ is contained in $B(\Pi(b_Q), C_1 \operatorname{diam} Q) \cap P$, at least if we take C_1 large enough. This is actually the place where we choose C_1; as promised, it is a geometric constant. Using (3.66) and the fact that $|p(R)| \geq \frac{\delta}{2}|R|$, we get

$$(3.122) \qquad |p(R) \cap \Pi(\widehat{W}(Q))| \geq \frac{\delta}{4}|R|$$

if we take N large enough. This is in fact the last time we shall put restrictions on the value of N, so we can now consider N to be officially chosen.

Now remember that $\widehat{W}(Q)$ is the union of all the $\widehat{T}$, where T runs among the m-cubes composing $W(Q)$ (as discussed before Lemma 3.65), and $\pi(\widehat{T}) = p(T)$ by definition of p (given just before (3.95)). Because there are no more than CC_2^d m-cubes in $W(Q)$, we can find an m-cube $T \subset W(Q)$ such that

$$(3.123) \qquad |p(R) \cap p(T)| \geq \frac{\delta}{CC_2^d}|R|.$$

We now choose the constant C_3 of (3.108) so large (depending on C_2) that (3.108) implies that R is not part of $W(Q)$. Hence T is different from R. We also require η_1 to be so small that (3.123) implies that (R, T) satisfies (3.107). Since R is still around after the end of Stage 5 of the mth generation of the construction, we conclude that T was removed before the end of Stage 5. This time we set $\phi(k) = l$, where l is the integer such that $T \subset \mathscr{D}_l$. Thus l corresponds to one of the first five stages of the mth generation of the construction.

As in the first case, the total number of values of k' such that $\phi(k')$ is a given l satisfies the estimate

$$\#\{\phi^{-1}(l)\} \leq C(C_2, N). \tag{3.124}$$

This time the estimate comes from the fact that $T \subset W(Q)$, so that Q is not too far from T, and the fact there are no more than $CN^{1/3}$ cubes T in a given $\mathscr{D}_l$ (see (3.103)).

We now switch to the case when $W(Q)$ is not entirely contained in $E(m-1)$, and again we look for a favorable event to associate to k. One reason that $W(Q)$ might not be contained in $E(m-1)$ is that $W(Q)$ might not be contained in Q_0. In this case we define an integer $\psi(k)$ by $\psi(k) = -1$. [We choose a different letter from ϕ to help distinguish between the two cases.] Otherwise, there is an integer $l \geq 0$ such that $\mathscr{D}_l$ contains a piece of $W(Q)$ and l belongs to a generation $\leq m - 1$ of the construction. We set $\psi(k) = l$ in this case.

Notice that Q is not contained in $\mathscr{D}_l$ (because it is still in E_k), and thus the distance from Q to the boundary of $\mathscr{D}_l$ is $\leq CC_2\epsilon^m$, which is much smaller than the diameter of the cubes that make up $\mathscr{D}_l$ (because the cubes that make up $\mathscr{D}_l$ have diameters $\geq C^{-1}\epsilon^{m-1}$, and we shall choose ϵ to be much smaller than C_2^{-1}). We now use the fact that our cubes have "small boundaries" (see (3.4) in Part I) to get

$$\sum_{k\,:\,\psi(k)=l} |\mathscr{D}_k| \leq \tau(\epsilon)|\mathscr{D}_l|, \tag{3.125}$$

where $\tau(\epsilon)$ might depend on C_2 and N but tends to 0 as ϵ tends to 0. [The sum is of course naturally restricted to the $k \in J(6)$ for which $\psi(k)$ is defined.]

The same argument also gives

$$\sum_{k\,:\,\psi(k)=-1} |\mathscr{D}_k| \leq \tau(\epsilon). \tag{3.126}$$

We are now ready to associate, to our original $k \in J(6)$, an integer $\rho(k)$ such that $-1 \leq \rho(k) < k$ and $\rho(k) \notin J(6)$. In the first two cases (i.e., when $W(Q) \subset E(m-1)$) we simply take $\rho(k) = \phi(k)$. In the other cases, the choice of $\rho(k)$ will depend on the value of $\psi(k)$. When $\psi(k) \notin J(6)$ we set $\rho(k) = \psi(k)$. Otherwise, we can define $\rho(k)$ recursively by the requirement that $\rho(k) = \rho(\psi(k))$. (Note that $\psi(k)$ is of an earlier generation and lies in $J(6)$, so we can assume that $\rho(\psi(k))$ has already been defined, by induction.)

3.6. THE STOPPING-TIME ARGUMENT

Let us estimate the size of $\rho^{-1}(l)$ for a given l.

LEMMA 3.127. *We have the estimates*

$$\sum_{k\in J(6)\,:\,\rho(k)=-1} |\mathscr{D}_k| \leq \tau'(\epsilon), \tag{3.128}$$

where the constant $\tau'(\epsilon)$ might depend on N and C_2 but tends to 0 with ϵ, and

$$\sum_{k\in J(6)\,:\,\rho(k)=l} |\mathscr{D}_k| \leq C|\mathscr{D}_l| \tag{3.129}$$

for every $l \geq 0$ such that $l \notin J(6)$. Here C is a constant that may depend on C_1, N, and C_2 but does not depend on η, η_0, η_1, η_2, θ, or ϵ.

In the statement of this lemma, the sums are (implicitly) restricted to those integers $k \in J(6)$ for which none of the stopping-time conditions (3.109)–(3.113) yet hold. Also, for this lemma it is necessary to require that ϵ be sufficiently small.

To prove the lemma, notice first that the inverse image of a given l by ρ consists of $\phi^{-1}(l)$ and then all inverse images of l and elements of $\phi^{-1}(l)$ by various iterates of ψ^{-1}. The estimate (3.129) therefore follows from (3.124) and multiple applications of (3.125) (with ϵ so small that $\tau(\epsilon) < \frac{1}{2}$, say). Similarly, (3.128) follows from (3.126) and (3.125).

We are now ready to combine our estimates to prove that the stopping-time conditions (3.112) and (3.113) never hold. Let k_0 be, as before, an integer such that none of the stopping-time conditions have obtained for any $k \leq k_0$. We want to estimate $\partial(E_{k_0+1}) = \delta \prod_{0 \leq k \leq k_0} \mu(k)$, which we write as

$$\partial(E_{k_0+1}) = \delta \prod_{l \in A} \hat{\mu}(l), \tag{3.130}$$

where $A = [-1, k_0] \cap (\{-1\} \cup J(1) \cup \cdots \cup J(5))$,

$$\hat{\mu}(-1) = \prod_{\substack{k \leq k_0 \\ \rho(k)=-1}} \mu(k), \tag{3.131}$$

and

$$\hat{\mu}(l) = \mu(l) \prod_{\substack{k \leq k_0 \\ \rho(k)=l}} \mu(k) \tag{3.132}$$

for $l \geq 0$. [Of course, only k's in $J(6)$ arise in these last two products, since $\rho(k)$ is defined only when $k \in J(6)$.]

It follows from (3.128) and many applications of (3.120) that

$$\hat{\mu}(-1) \geq 1 - C\eta_2 \tau'(\epsilon). \tag{3.133}$$

Similarly, (3.129) and many applications of (3.120) yield

$$\hat{\mu}(l) \geq \mu(l)\{1 - C\eta_2|\mathscr{D}_l|\} \tag{3.134}$$

for $l \geq 0$, where we now adopt the (new) convention that C denotes any constant which may depend on C_1, N, C_2, etc., but which does not depend on η, η_0, η_1, η_2, ϵ, or θ. Next we apply (3.116), (3.118), and (3.119) to obtain the estimate

$$(3.135) \qquad \hat{\mu}(l) \geq 1 + C^{-1}\eta_1 |\mathscr{D}_l|$$

for $l \in J(1) \cup J(4) \cup J(5)$. For this, we just have to make sure that $\eta_1 \leq \eta$ and that $C\eta_2$ is small enough with respect to η and η_1 (which is compatible with the order in which we said that we would choose the constants). For $l \in J(2) \cup J(3)$, we use (3.117) together with (3.134) to get

$$(3.136) \qquad \hat{\mu}(l) \geq 1 - C\delta^{-1}|\mathscr{D}_l|.$$

From our construction we have that each $\mathscr{D}_l$, $l \in J(2) \cup J(3)$, is associated to one of the cubes of $G(m) \setminus G(m-1)$ or $\mathscr{C}(m) \setminus \mathscr{C}(m-1)$, so that (3.103) and the fact that (3.109) and (3.110) have not occurred gives

$$\sum_{l \in J(2) \cup J(3)} |\mathscr{D}_l| \leq CN^{1/3}\theta.$$

We now require θ and ϵ to be so small that this, (3.136), and (3.133) imply that

$$(3.137) \qquad \prod_{\substack{l \in \{-1\} \cup J(2) \cup J(3) \\ l \leq k_0}} \hat{\mu}(l) \geq \left(1 - \frac{\eta_0}{10}\right).$$

On the other hand, (3.135) yields

$$(3.138) \qquad \prod_{l \in A'} \hat{\mu}(l) \geq 1 + C^{-1}\eta_1 \sum_{l \in A'} |\mathscr{D}_l|,$$

where $A' = \{l \leq k_0 : l \in J(1) \cup J(4) \cup J(5)\}$ is the rest of A.

Notice that (3.130), (3.137), and (3.138) already imply that $\partial\left(E_{k_0+1}\right) \geq (1 - \eta_0/10)\delta$, so the stopping-time condition (3.113) is still not satisfied for $k_0 + 1$. Also, it follows from the fact that (3.111) does not hold for k_0 that $\partial\left(E_{k_0+1}\right) \leq (1 + 3\eta_0)\delta$. [For this to work, we have to take ϵ so small that $|\mathscr{D}_{k_0}|$ will be much smaller than $|E_{k_0}|$ (which is $> \frac{1}{2}|E_0|$, since (3.112) fails for k_0) so that $\partial\left(E_{k_0+1}\right)$ is very close to $\partial(E_{k_0})$.] Using this, (3.137), (3.138), and (3.130) again, we now get that

$$(3.139) \qquad \sum_{l \in A'} |\mathscr{D}_l| \leq C\frac{\eta_0}{\eta_1}.$$

Using again the fact that $\sum_{l \in J(2) \cup J(3)} |\mathscr{D}_l| \leq CN^{1/3}\theta$, we now get that

$$(3.140) \qquad \sum_{\substack{l \leq k_0 \\ l \notin J(6)}} |\mathscr{D}_l| \leq C\left(\frac{\eta_0}{\eta_1} + \theta\right),$$

3.6. THE STOPPING-TIME ARGUMENT

with a constant C that may depend on N but not on the various η's, θ, or ϵ. Using now the two estimates of Lemma 3.127, we get

$$|E_0 \setminus E_{k_0+1}| = \sum_{l \leq k_0} |\mathscr{D}_l| \leq C\left(\frac{\eta_0}{\eta_1} + \theta + \tau'(\epsilon)\right),$$

which implies that $|E_{k_0+1}| > \frac{1}{2}|E_0|$ if we choose η_0 small enough compared to η_1 and θ and ϵ small enough. Thus neither of the events (3.112) and (3.113) ever takes place.

(c) *The end of the proof.* The reader will be pleased to know that the delicate part of the proof is now over. In particular, all the constants except a_1 are essentially chosen now, depending on our initial constants A, a, and a_2, and also innocuous geometric quantities. (Actually, we shall impose a couple of additional requirements on ϵ and θ, but these are probably weaker than the various constraints that we have already imposed on them.) We still have to find the set $Z_0 \subset Z \cap Q_0$ and the collection $\mathscr{C}$ of minimal cubes contained in Q_0 with the properties (3.56) and (3.57).

Let us assume for the moment that, whenever Q is a cube of S such that $\partial(Q) \geq (1+\eta_0)\delta$, we can find a closed set $Z_0(Q) \subset Z \cap Q$ and a collection $\mathscr{C}(Q)$ of minimal cubes of S that are contained in Q, with the properties that

$$(3.141) \qquad \left| Z_0(Q) \cup \left(\bigcup_{T \in \mathscr{C}(Q)} T \right) \right| \geq a_1' |Q|$$

and

$$(3.142) \qquad |\Pi(x) - \Pi(y)| \geq a_1' |x - y|$$

whenever $x, y \in Z_0(Q) \cup \{b_T : T \in \mathscr{C}(Q)\}$, where a_1' is some (very small) constant. We shall explain how to get rid of this assumption later, either by an iteration of the construction, or by an induction argument.

Let us now show how we can construct Z_0 and $\mathscr{C}$. We start with the case where we never have to stop because of (3.109), (3.110), or (3.111). In this case, we may as well throw away the sets $G(m)$ and $\mathscr{E}(m)$ (which are small anyway), and take $\mathscr{C} = \emptyset$ and $Z_0 = \bigcap_{m \geq 1} \{E(m) \setminus [G(m) \cup \mathscr{E}(m)]\}$.

The fact that $Z_0 \subset Z \cap Q_0$ is essentially obvious (every point of Z_0 is contained in arbitrarily small cubes in S). The inequality (3.56) follows easily from the fact that $|E(m)| \geq \frac{1}{2}|E_0| = \frac{1}{2}|Q_0|$ and $|G(m) \cup \mathscr{E}(m)| \leq 2\theta|Q_0|$ for each m. Of course, Z_0 is not necessarily closed, but this can easily be remedied. (For instance, we can take the closure of Z_0, after possibly removing a little piece of Z_0 which is too close to $E \setminus Q_0$.) Thus we have only to check (3.57).

Let $x, y \in Z_0$ be given, with $x \neq y$. Let m be the smallest integer such that $|x - y| > KC_3\epsilon^m$, where K is a geometric constant which will be specified in a moment. Let $Q(x)$ and $Q(y)$ be the m-cubes that contain

x and y, respectively. We choose K to be sufficiently large so that we must have $\text{dist}(Q(x), Q(y)) \geq C_3 \epsilon^m$. Since x and y lie in Z_0, $Q(x)$ and $Q(y)$ are still contained in $E(m) \setminus [G(m) \cup \mathscr{E}(m)]$. Hence they do not satisfy (3.108), and therefore $\text{dist}(p(Q(x)), p(Q(y))) \geq \epsilon^m$. This implies that $|\Pi(x) - \Pi(y)| \geq \epsilon^m$, because $\Pi(x) \in p(Q(x))$ since $x \in Z \cap Q(x)$ (see (3.95)), and similarly $\Pi(y) \in p(Q(y))$. This implies (3.57) with $a_1 = (KC_3)^{-1}\epsilon$, because of our choice of m.

Our next case is when $|\mathscr{E}(m_0)| \geq \theta |Q_0|$ for some m_0. In this case we simply take $Z_0 = \varnothing$ and $\mathscr{E}$ to be the collection of all the minimal cubes T that make up $\mathscr{E}(m_0)$. We have (3.56) with $a_1 = \theta$, so it is enough to check that $|\Pi(x) - \Pi(y)| \geq a_1 |x - y|$ whenever $x = b_T$ and $y = b_{T'}$ are the "centers" of two of the minimal cubes contained in $\mathscr{E}(m_0)$.

The proof is essentially the same as in the previous case. Let m be the smallest integer which satisfies $\epsilon^m \leq |x - y|$, so that

$$(3.143) \qquad \epsilon^{m-1} > |x - y|.$$

From (3.143) and the fact that the cubes T and T' satisfy (3.104) (with m replaced with the generation of T or T', as appropriate) we deduce that neither T nor T' can be of a generation $\leq m - 1$. If T or T' is a cube of generation m, then (3.105) gives that $\text{dist}(p(T), p(T')) \geq \epsilon^m$. Otherwise, T and T' are of generations $\geq m + 1$, so we may consider the cubes Q and Q' of generation $m + 1$ that contain T and T', respectively. These cubes are contained in $E(m+1)$, and they are at a distance $> C_3 \epsilon^{m+1}$ from each other because $|x - y| \geq \epsilon^m$. (Remember that ϵ is chosen after C_3, so we may require that ϵ be much smaller than C_3^{-1}.) Since they do not satisfy (3.108), we have $\text{dist}(p(Q), p(Q')) \geq \epsilon^{m+1}$. In both cases, we get $|\Pi(x) - \Pi(y)| \geq \text{dist}(p(T), p(T')) \geq \epsilon^{m+1} \geq \epsilon^2 |x - y|$ (by (3.143)). This gives (3.57) with $a_1 = \epsilon^2$.

Our third case is when $|G(m_0)| \geq \theta |Q_0|$ for some m_0. In this case, we use our additional assumption to obtain, for each of the cubes Q which make up $G(m_0)$, a set $Z_0(Q) \subset Z \cap Q$ and a set $\mathscr{E}(Q)$ of minimal cubes contained in Q with the properties (3.141) and (3.142). We let Z_0 be the union of the $Z_0(Q)$'s and $\mathscr{E}$ be the union of the $\mathscr{E}(Q)$'s. The inequality (3.56) is then satisfied with $a_1 = \theta a_1'$, so we only have to check (3.57). If the points x and $y \in Z_0 \cup \{b_T : T \in \mathscr{E}\}$ are contained in the same cube Q of $G(m_0)$, then (3.57) follows from (3.142). Otherwise, we can use the same argument as in the second case. For this it is helpful to notice that $x \in p(Q)$ when $x \in Z_0(Q) \cup \{b_T : T \in \mathscr{E}(Q)\}$, by the definition (3.95) of $p(Q)$.

To be honest, we need to modify Z_0 slightly to get it to be closed, as before.

Our last case is when (3.111) occurs first. Let us suppose that this happens sometime during the mth generation of the construction, and let k be the moment when (3.111) occurs for the first time. Let V be the union of all the m-cubes in $E_k \setminus [G(m-1) \cup \mathscr{C}(m-1)]$ that satisfy (3.98), let W be the union $G(m-1) \cup \mathscr{C}(m-1)$, and set $X = E_k \setminus (V \cup W)$. We have the estimates

$$\begin{aligned}(1 + 2\eta_0)\delta|E_k| &\leq \partial(E_k)|E_k| = |p(E_k)| \\ &\leq |p(V)| + |p(W)| + |p(X)| \\ &\leq |p(V)| + C|W| + (1 + \eta_0)\delta|X| \\ &\leq |p(V)| + 2C\theta + (1 + \eta_0)\delta|E_k|.\end{aligned}$$

(We can safely assume that $|G(m-1) \cup \mathscr{C}(m-1)| \leq 2\theta|Q_0|$, since otherwise we are in one of the previous cases.)

Thus $\eta_0 \delta |E_k| \leq |p(V)| + 2C\theta$, and hence

(3.144) $$|p(V)| \geq \frac{\eta_0 \delta |Q_0|}{4}$$

if we take θ small enough.

Note that our stopping-time condition (3.111) has occurred some time before the end of Stage 2, for otherwise (3.109) would have held first; if θ is small enough, depending on η_0, then (3.144) is much stronger than (3.109). We now proceed as in Stage 2 of our construction. We choose a maximal set R_i, $i \in I$, of m-cubes R_i in the set V such that $\mathrm{dist}(p(R_i), p(R_j)) \geq \epsilon^m$ for $i \neq j$. The maximality of this collection ensures that $p(V)$ is contained in a union of balls centered on the $p(R_i)$'s and with radius $C\epsilon^m$. Therefore, it follows from (3.144) that the set I has at least $C^{-1}\epsilon^{-md}\eta_0\delta|Q_0|$ elements. The rest of the argument is almost the same as in the third case. Each R_i comes with a set $Z_0(R_i) \subset Z \cap R_i$ and a collection $\mathscr{C}(R_i)$ of minimal cubes contained in R_i, and we take $Z_0 = \bigcup_{i \in I} Z_0(R_i)$ and $\mathscr{C} = \bigcup_{i \in I} \mathscr{C}(R_i)$ (modulo the usual modification to get Z_0 to be closed). The estimate (3.56) follows immediately from (3.141) and our lower bound for the number of elements of I. The estimate (3.57) can be obtained using arguments very similar to those that we have use already. [If x and y lie in the same R_i, we use (3.142). If $x \in R_i$ and $y \in R_j$ with $j \neq i$ and $|x - y|$ is not much larger than ϵ^{m-1}, then we use the fact that $\mathrm{dist}(p(R_i), p(R_j)) \geq \epsilon^m$. If $|x - y|$ is much larger than ϵ^{m-1}, we can use the properties of the first $m-1$ generations of the construction (in particular Stage 6), as we did before.]

To finish the proof of Lemma 3.55, and thereby Theorem 3.9, we still have to get rid of our additional assumption. What we have accomplished, up to now, is to reduce the problem of finding the sets $Z_0 = Z_0(Q_0)$ and $\mathscr{C} = \mathscr{C}(Q_0)$ when $\partial(Q_0) = \delta$ to the case where $\partial(Q_0) \geq (1 + \eta_0)\delta$. In fact, this argument also works, with the same value of the various constants (and η_0 in particular), as soon as we know that $\delta \geq a_2$ (see (3.97)). By applying the argument a second time we reduce to the case where $\partial(Q_0) \geq (1 + \eta_0)^2 \delta$. Using it many times, we can reduce to the case where $\partial(Q_0) \geq (1 + \eta_0)^M \delta$. But we can choose M so large that $\partial(Q_0) \geq (1 + \eta_0)^M \delta$ can never happen, so that the result is trivially true in that case. This proves the result we wanted, with a constant a_1 which is of course extremely small if we have to use a large M. (Each time we have to apply the argument, a_1 goes down by a factor of at least θ.)

Another way of doing the same thing would be to start the argument again on each of the good cubes that are left after the first stopping-time (3.109) or (3.111), but this time with δ replaced by $(1+\eta_0)\delta$ or some even larger number depending on the cube. After a certain number of iterations of this, we would get to the point where no good cube can possibly exist, and we would be able to conclude directly as in the case when none of the stopping-time events ever occurs or when we have enough minimal cubes left.

CHAPTER 4

Other Conditions in the Codimension 1 Case

4.1. Introduction.
In this chapter we consider some other criteria for uniform rectifiability which are more topological in nature and, in particular, less "rigid" than the BWGL or the WEC, for instance. Although our methods will work only when the codimension is 1, there probably are some analogous results in higher codimensions, but with more complicated statements and proofs.

Of course, when the codimension is maximal—i.e., when $d = 1$—we already have the WCC (see §1.1), which is nicely nonrigid. When $d = 1$ and $n = 2$, so that both the dimension and the codimension are 1, then the WCC is more general than the criterion for uniform rectifiability that we are going to present now.

The main condition considered in this chapter — the "weakly topologically nice condition" — is, roughly speaking, a condition of bilateral approximation by sets with exactly two complementary components, each of which is a uniform domain with estimates. The precise condition is as follows.

Let E be a d-dimensional regular set in $\mathbf{R}^{d+1}$. For each choice of constants $k_0 > 1$ and $\epsilon > 0$, let $\mathscr{G}(\epsilon, k_0)$ be the good set of all $(x, t) \in E \times \mathbf{R}_+$ for which there exist two disjoint open sets U_1 and U_2 in $B(x, 10t) \setminus E$ with the following properties:

(4.1)   if $z \in B(x, 10t)$ and $\mathrm{dist}(z, E) > \epsilon t$, then $z \in U_1 \cup U_2$;

(4.2)   if F is a connected compact set contained in $B(x, 5k_0 t)$ such that $\mathrm{dist}(F, E) > \epsilon t$, then it cannot meet both U_1 and U_2;

(4.3)   there exist points $z_1 \in U_1$ and $z_2 \in U_2$ such that $z_i \in B(x, t)$ and $\mathrm{dist}(z_i, E) \geq (1 - \epsilon) k_0^{-1} t$ for $i = 1, 2$;

(4.4)   if y and z are two points in the same U_i, $i = 1$ or 2, and if $\mathrm{dist}(y, E) \geq \epsilon t$ and $\mathrm{dist}(z, E) \geq \epsilon t$, then there is a curve γ in $\mathbf{R}^{d+1} \setminus E$ which joins y to z and which satisfies $\mathrm{diam}\, \gamma \leq k_0 |y - z|$ and $\mathrm{dist}(\gamma, E) \geq k_0^{-1} \min\{\mathrm{dist}(y, E), \mathrm{dist}(z, E)\}$.

[The purpose of the $(1-\epsilon)$ in (4.3) is to prevent a minor technical problem that would otherwise arise due to the tension between (4.2) on the one hand and (4.3) and (4.4) on the other. Notice that (4.2) becomes stronger as k_0 increases, while (4.3) and (4.4) become weaker.]

DEFINITION 4.5. A d-dimensional regular set E in $\mathbf{R}^{d+1}$ is said to satisfy the WTN (Weakly Topologically Nice condition) if there exists $k_0 > 1$ so that $E \times \mathbf{R}_+ \setminus \mathscr{G}(\epsilon, k_0)$ is a Carleson set for every $\epsilon > 0$.

The WTN is a little weaker a priori than the following condition of bilateral approximation. For each $k_0 > 1$, denote by $\mathscr{C}(k_0)$ the class of all closed sets $F \subset \mathbf{R}^{d+1}$ with exactly two complementary components V_1 and V_2 which have the following properties:

(i) each $V_j \cap B(0, (50k_0)^{-1})$ contains a ball of radius $(50k_0^2)^{-1}$;

(ii) if y and z lie in the same $V_j \cap B(0, (5k_0)^{-1})$, then there is a curve $\gamma \subset V_j$ such that

$$\operatorname{diam} \gamma \leq k_0 |y - z| \quad \text{and} \quad \operatorname{dist}(\gamma, F) \geq k_0^{-1} \min\{\operatorname{dist}(y, F), \operatorname{dist}(z, F)\}.$$

Let $\operatorname{Approx}(\mathscr{C}(k_0))$ be the class of regular sets that satisfy the condition of bilateral approximation by elements of $\mathscr{C}(k_0)$, as defined in §2.2 of Part I. It is easy to check that E satisfies the WTN as soon as it lies in some $\operatorname{Approx}(\mathscr{C}(k_0))$. Thus, by taking various classes of sets which are contained in the $\mathscr{C}(k_0)$'s, we obtain various conditions of bilateral approximation that imply the WTN. Let us describe two of these.

DEFINITION 4.6. . Let E be a d-dimensional regular set in $\mathbf{R}^n$. We say that E satisfies the BALG (bilateral approximation by Lipschitz graphs condition) if there exists a constant $M \geq 0$ such that, for each $\epsilon > 0$, $E \times \mathbf{R}_+ \setminus \mathscr{G}_{lg}(\epsilon, M)$ is a Carleson set. Here $\mathscr{G}_{lg}(\epsilon, M)$ denotes the set of all $(x, t) \in E \times \mathbf{R}_+$ for which there is a d-dimensional Lipschitz graph $\Gamma \subset \mathbf{R}^n$, with Lipschitz constant $\leq M$, such that

(4.7) $$\sup_{y \in E \cap B(x, t)} \operatorname{dist}(y, \Gamma) + \sup_{y \in \Gamma \cap B(x, t)} \operatorname{dist}(y, E) \leq \epsilon t.$$

(See the page of notation for the definition of a d-dimensional Lipschitz graph.)

Thus E satisfies the BALG if and only if E lies in $\operatorname{Approx}(LG(M))$ for some $M > 0$, where $LG(M)$ is the collection of all Lipschitz graphs with constant $\leq M$.

Notice that this definition makes good sense even when the codimension is larger than 1, while the WTN does not (although it is not difficult to envision suitable generalizations of the WTN to higher codimensions). Unfortunately we know how to prove that the BALG implies uniform rectifiability only when the codimension is 1 or when $d = 1$, the latter situation being handled by the WCC.

The BABI (condition of bilateral approximation by bilipschitz images of $\mathbf{R}^d$ in $\mathbf{R}^n$) is defined the same way as the BALG, except that Lipschitz graphs with constants $\leq M$ are replaced by sets of the form $F = \{\phi(z) : z \in \mathbf{R}^d\}$, where $\phi : \mathbf{R}^d \to \mathbf{R}^n$ satisfies

(4.8) $\quad M^{-1}|z' - z| \leq |\phi(z') - \phi(z)| \leq M|z' - z| \quad$ for all $z, z' \in \mathbf{R}^d$.

When $d = n - 1$, the BABI also implies the WTN. This follows from the fact that every bilipschitz image of $\mathbf{R}^d$ in $\mathbf{R}^{d+1}$ belongs to the set $\mathscr{C}(k_0)$ for some $k_0 = k_0(M, d)$. See [**V**].

More generally, we could formulate the condition of bilateral approximation by chord-arc surfaces. If we were to define this condition, then it would be immediately obvious to the reader that every regular set which satisfies it must be in the class Approx($\mathscr{C}(k_0)$) for k_0 sufficiently large. We shall, however, not give this definition, but the reader may consult [**S3**], [**S4**] to find out what a chord-arc surface is. (See also [**DS3**].)

As the reader may have surmised, the main result of this chapter is the following.

THEOREM 4.9. *Let E be a d-dimensional regular set in $\mathbf{R}^{d+1}$. If E satisfies the WTN, then it is uniformly rectifiable.*

COROLLARY 4.10. *If E is a d-dimensional regular set in $\mathbf{R}^{d+1}$ which satisfies the BALG or the BABI, then E is uniformly rectifiable.*

As usual, the converses of these results are true and easy. Also, the Carleson set conditions in the hypotheses are not really needed for all values of ϵ but only for some small value of ϵ which may be computed in terms of all the other constants.

The proof of Theorem 4.9 will use an argument from [**DS3**], combined with some of the techniques of Chapter 2. Thus we begin with a brief review of [**DS3**].

DEFINITION 4.11. Let E be a d-dimensional regular set in $\mathbf{R}^{d+1}$. We say that E satisfies Condition B if there is a constant $C > 0$ so that, for every $x \in E$ and $R > 0$, there exist points y and z which lie in different connected components of $\mathbf{R}^{d+1} \setminus E$ and which satisfy $|x - y| \leq R$, $|x - z| \leq R$, dist$(y, E) \geq C^{-1}R$, and dist$(z, E) \geq C^{-1}R$.

This condition (in a somewhat more restrictive form) first appeared in [**S1**] as a criterion for the L^2-boundedness of certain singular integral operators. In [**D3**] it was shown that E has BPLG if it satisfies Condition B, so that all "reasonable" singular integral operators must then be bounded on $L^2(E)$. This result was simplified, generalized, and improved in [**DJ**]. Related references include [**D4**] and [**S4**].

Another proof of the fact that Condition B implies BPLG was given in [**DS3**]. The starting point for the approach used in [**DS3**] was the observation that a regular set has BPLG if it has big projections and satisfies the WGL. (See Theorem 1.76 in Part I.) If E satisfies Condition B, then it certainly has big projections. This follows immediately from the definitions (see

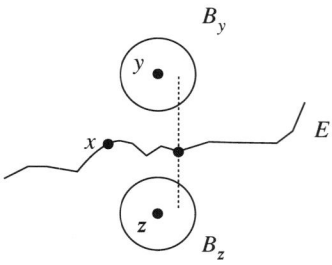

FIGURE 4.1. Every line segment from B_y to B_z meets E.

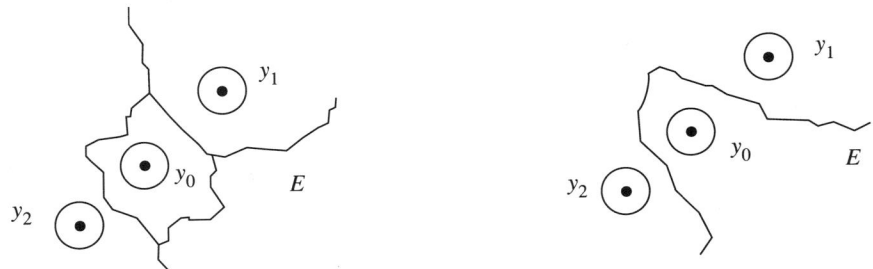

FIGURE 4.2. Two cases where $(x, t) \in \mathscr{B}(\epsilon)$.

Figure 4.1). To show that E also satisfies the WGL, it was convenient to employ the fact that the local symmetry condition LS implies the WGL (not to mention the BWGL). (See Proposition I.1.81.) The next result (Proposition 5.0 from [**DS3**]) was the main step in the proof of the fact that Condition B implies LS.

For each $\epsilon > 0$, let $\mathscr{B}(\epsilon)$ denote the set of $(x, t) \in E \times \mathbf{R}_+$ for which there exist three points y_0, y_1, y_2 in $B(x, t) \setminus E$ such that $\text{dist}(y_i, E) \geq \epsilon t$ for $i = 0, 1, 2$, y_0 lies on the line segment that joins y_1 to y_2, but y_0 does not lie in the same component of the complement of E as y_1 or y_2. (See Figure 4.2.)

PROPOSITION 4.12. *If E is a d-dimensional regular set in $\mathbf{R}^{d+1}$, then $\mathscr{B}(\epsilon)$ is a Carleson set for all $\epsilon > 0$.*

It is not difficult to show that Condition B implies LS once you have Proposition 4.12.

Our proof of Theorem 4.9 will rely on an extension of the method of [**DS3**] and, in particular, a generalization of Proposition 4.12. We shall make use of the fact that the proof of Proposition 4.12 did not use the assumption that various points lie in different connected components of the complement at full strength. Rather, it used the following two ingredients:

(a) if two points p and q lie in "different components" of $\mathbf{R}^d \setminus E$, then the line segment which joins them intersects E;

(b) the notion of "different components" has some sort of global meaning.

In our generalization of Proposition 4.12 we are going to work with a notion of labellings of points of $\mathbf{R}^{d+1} \setminus E$ which has the two features (a) and (b) but which is more flexible than simply taking complementary components. We shall then show how to produce a suitable labelling under the assumptions of Theorem 4.9.

4.2. Labellings.

DEFINITION 4.13. A labelling of a set $U \subset \mathbf{R}^{d+1}$ is a mapping l from U into a set L_* of labels. This set L_* has a distinguished element $*$, and elements of U which are assigned to this label are considered to be unlabelled. We denote by L the set $L_* \setminus \{*\}$. A labelling is said to be admissible if $x, y \in U$, $l(x), l(y) \in L$, and $l(x) \neq l(y)$ imply that the line segment which joins x to y intersects the complement of U.

For example, we could take L to be the set of components of U, and l to be the obvious mapping. This labelling is certainly admissible. Note that the trivial labelling, which sends every element of U to $*$, is admissible (but not very useful).

Given $E \subset \mathbf{R}^{d+1}$, a labelling $l : \mathbf{R}^{d+1} \setminus E \to L_*$, and $\epsilon > \delta > 0$, let $\mathscr{B}_l(\epsilon, \delta)$ be the set of $(x, t) \in E \times \mathbf{R}_+$ for which there exist $y_0, y_1, y_2 \in B(x, t) \setminus E$ with the following properties:

(4.14) $\text{dist}(y_i, E) \geq \epsilon t$;
(4.15) y_0 lies on the line segment which joins y_1 to y_2;
(4.16) $l(y_i) \in L$ for $i = 0, 1, 2$;
(4.17) $l(z) = l(y_i)$ whenever $|z - y_i| < \delta t$ and $i = 0, 1,$ or 2;
(4.18) $l(y_0) \neq l(y_1)$ and $l(y_0) \neq l(y_2)$.

PROPOSITION 4.19. *If E is a d-dimensional regular set in $\mathbf{R}^{d+1}$ and if $l : \mathbf{R}^{d+1} \setminus E \to L_*$ is an admissible labelling, then*

(4.20) $\mathscr{B}_l(\epsilon, \delta)$ *is a Carleson set for all choices of $\epsilon > \delta > 0$.*

The proof of Proposition 4.19 is the same as that of Proposition 5.0 in [**DS3**] except for cosmetic changes. [The reader might welcome the knowledge that most of the excitement takes place in the proof of Lemma 5.7 in [**DS3**], although (5.5) in [**DS3**] must also be changed to accomodate the labelling.] We omit the details.

Using Proposition 4.19, it is easy to give a generalization of the fact that Condition B implies BPLG. Before we state it we need to give a couple of definitions.

DEFINITION 4.21. Let E be a d-dimensional regular set in $\mathbf{R}^{d+1}$. We say that a labelling $l : \mathbf{R}^{d+1} \setminus E \to L_*$ satisfies Condition B if there is a $C > 0$ so that for each $x \in E$ and $R > 0$ there exist $y, z \in \mathbf{R}^{d+1} \setminus E$ with the following properties: $y, z \in B(x, R)$; $B(y, C^{-1}R)$ and $B(z, C^{-1}R)$ are both contained in $\mathbf{R}^{d+1} \setminus E$; $l(\cdot)$ is constant on each of these two balls; $l(y)$ and $l(z)$ belong to L; and finally $l(y) \neq l(z)$.

DEFINITION 4.22. Let E be a d-dimensional regular set in $\mathbf{R}^{d+1}$. We say that a labelling $l : \mathbf{R}^{d+1} \setminus E \to L_*$ is nearly ubiquitous if there is an $\eta > 0$ such that the complement in $\mathbf{R}^{d+1} \setminus E$ of

(4.23) $\quad A_\eta = \{z \in \mathbf{R}^{d+1} \setminus E : l(z) \neq * \text{ and } l|_{B(z, \eta \operatorname{dist}(z, E))} \text{ is constant}\}$

is a Carleson set.

Recall from (I.2.44) that this last means that there is a constant C such that

(4.24) $$\iint_{(\mathbf{R}^{d+1} \setminus A_\eta) \cap B(x, R)} \operatorname{dist}(z, E)^{-1} dz \leq C R^d$$

for all $x \in E$ and $R > 0$. In particular, $\{z \in \mathbf{R}^{d+1} \setminus E : l(z) = *\}$ should be a Carleson set.

THEOREM 4.25. *Suppose that E is a d-dimensional regular set in $\mathbf{R}^{d+1}$ and that $l : \mathbf{R}^{d+1} \setminus E \to L_*$ is a labelling which is admissible, nearly ubiquitous, and satisfies Condition B. Then E has BPLG.*

We shall derive this from Proposition 4.19 using practically the same argument as in [DS3]. Let us sketch the key points. First notice that if l is admissible and satisfies Condition B, then E has big projections. This is an easy consequence of the definitions. (See Figure 4.1 again.) Thus it suffices to show that E satisfies the local symmetry condition LS (which we know implies the WGL by Proposition I.1.81).

Let $x \in E$ and $t > 0$ be given. Let $\tau > 0$ be given, and suppose that there exist u, v in $E \cap B(x, t)$ such that $z = 2u - v$ is at distance $\geq \tau t$ from E (so that (x, t) lies in the bad set (I.1.80) for the local symmetry condition). We want to show that this implies that (x, t) belongs to some Carleson set.

Notice that, since l is nearly ubiquitous, the set

(4.26) $\quad \{(x, t) \in E \times \mathbf{R}_+ : \text{ there is a point } w \in B(x, t) \text{ such that}$
$\quad \operatorname{dist}(w, E) \geq \tau t \text{ and } z \notin A_{\eta/2}\}$

is a Carleson set, where η is as in Definition 4.22. Thus we may as well assume that $l(z) \neq *$ and that l is constant on $B\left(z, \frac{\eta \tau t}{2}\right)$. Once we have these conditions on z, we can continue as in [DS3]. We first use Condition B to find a point $y_0 \in B\left(u, \frac{\eta \tau t}{100}\right)$, at a distance $\geq \frac{\eta \tau t}{100C}$ from E, and such that $l(\cdot)$ is constant on a small ball centered at y_0 and takes a value $l(y_0)$ which is different from $*$ and $l(z)$. We then use Condition B again to find a point $y_1 \in B\left(v, \frac{\eta \tau t}{100}\right)$ at a distance $\geq \frac{\eta \tau t}{100C}$ from E such that $l(\cdot)$ is constant on a small ball centered at y_1 and takes a value $l(y_1)$ which is different from $*$ and $l(y_0)$. Finally, we choose $y_2 = 2y_0 - y_1$. (See Figure 4.3.)

4.2. LABELLINGS

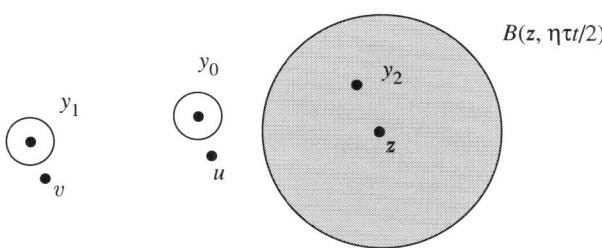

FIGURE 4.3

It is easy to check that the points y_i, $i = 0, 1, 2$, satisfy the conditions (4.14)–(4.18) for suitable choices of ϵ and δ, so $(x, 3t) \in \mathscr{B}_l\left(\frac{\epsilon}{3}, \frac{\delta}{3}\right)$. Theorem 4.25 now follows from Proposition 4.19.

Notice that we only used our hypothesis that l is admissible to get the big projections and the Carleson estimate (4.20). Thus we have also proved the following slightly different result:

(4.27) *if E is a d-dimensional regular set in $\mathbf{R}^{d+1}$ and if $l: \mathbf{R}^{d+1} \setminus E \to L_*$ is a labelling which is nearly ubiquitous and satisfies* (4.20) *and Condition* B, *then E satisfies the local symmetry condition.*

As a practical matter, Theorem 4.25 tends not to be so useful. It will be more useful to us to have a version of this theorem which is better suited to the corona-type arguments that we have been using in Chapters 2 and 3. In particular, our proof of Theorem 4.9 will require a version of Theorem 4.25 where we do not demand that the labelling be admissible but rather that it be "compatible" with a given coronization of E. Let us be more precise.

Let E be a regular set in $\mathbf{R}^{d+1}$ with codimension 1, and choose (once and for all) a collection Δ of dyadic cubes as in §3.1 of Part I. Let a coronization $(\mathscr{B}, \mathscr{G}, \mathscr{F})$ of E be given. (See Definition I.3.13.) For each stopping-time region $S \in \mathscr{F}$ define a function $d(x) = d_S(x)$ on $\mathbf{R}^n$ by

$$(4.28) \qquad d(x) = \inf_{Q \in S}\{\operatorname{dist}(x, Q) + \operatorname{diam} Q\}.$$

Also, for each $S \in \mathscr{F}$, set
(4.29)
$$\widehat{S} = \{x \in \mathbf{R}^{d+1} \setminus E : \operatorname{dist}(x, Q(S)) \leq \operatorname{diam} Q(S) \text{ and } \operatorname{dist}(x, E) \geq \tfrac{1}{2} d_S(x)\},$$

where $Q(S)$ denotes the top cube of S. Roughly speaking, $\widehat{S}$ is the subset of $\mathbf{R}^{d+1} \setminus E$ that corresponds naturally to S in the same way that Δ corresponds to $\mathbf{R}^{d+1} \setminus E$.

DEFINITION 4.30. Let E be a d-dimensional regular set in $\mathbf{R}^{d+1}$, let $(\mathscr{B}, \mathscr{G}, \mathscr{F})$ be a coronization of E, and let $l : \mathbf{R}^{d+1} \setminus E \to L_*$ be a labelling of $\mathbf{R}^{d+1} \setminus E$. We say that l is compatible with the coronization $(\mathscr{B}, \mathscr{G}, \mathscr{F})$ if, for each choice of $S \in \mathscr{F}$ and $x, y \in \widehat{S}$ such that $l(x)$,

$l(y) \in L$ and $l(x) \neq l(y)$, we have that the line segment that joins x to y contains a point z such that $\text{dist}(z, E) \leq \frac{1}{10} d_S(z)$.

THEOREM 4.31. *Let E be a d-dimensional regular set in $\mathbf{R}^{d+1}$, and let $(\mathscr{B}, \mathscr{G}, \mathscr{F})$ be a coronization of E. Suppose that there is a labelling $l: \mathbf{R}^{d+1} \setminus E \to L_*$ which is compatible with the coronization, nearly ubiquitous and which satisfies Condition B. Then E is uniformly rectifiable.*

The main step in the proof of this theorem is the following. Let $\mathscr{B}_l(\epsilon, \delta)$ be as in Proposition 4.19.

LEMMA 4.32. *If E is a d-dimensional regular set in $\mathbf{R}^{d+1}$, $(\mathscr{B}, \mathscr{G}, \mathscr{F})$ is a coronization of E, and l is a labelling of $\mathbf{R}^{d+1} \setminus E$ which is compatible with $(\mathscr{B}, \mathscr{G}, \mathscr{F})$, then l satisfies (4.20).*

Theorem 4.31 is an immediate consequence of Lemma 4.32, our observation (4.27), and the fact that LS implies the BWGL and hence uniform rectifiability (Proposition I.2.5 and Theorem I.2.4). The resulting proof of Theorem 4.31 is more complicated than necessary, because the proofs of Lemma 4.32 and Theorem I.2.4 overlap substantially; in particular, a simpler proof would involve only one coronization.

Let us now prove Lemma 4.32. Let E, $(\mathscr{B}, \mathscr{G}, \mathscr{F})$, and a labelling $l: \mathbf{R}^{d+1} \setminus E \to L_*$ be as in the lemma. Let $\epsilon > \delta > 0$ be given. We claim that it suffices to show that
(4.33)
$$\{(x, t) \in E \times \mathbf{R}_+ : (x, t) \in \mathscr{B}_l(\epsilon, \delta), x \in Q(S), \text{ and } d_S(x) \leq t \leq \text{diam } Q(S)\}$$

is a Carleson set for each $S \in \mathscr{F}$, with an estimate that does not depend on S. The claim is fairly easy to prove, using the fact that the cubes Q, $Q \in \mathscr{B}$, and $Q(S)$, $S \in \mathscr{F}$, satisfy Carleson packing conditions. The argument is practically the same as one given in the proof of Proposition I.3.32 (see (I.3.36) and its proof), so we omit it.

Fix $S \in \mathscr{F}$. We want to reduce the problem to Proposition 4.19 applied to some other regular set and labelling. Let us begin with the construction of an approximating set $E(S)$. We could quite probably take the same set $E(S)$ as in Chapters 2 and 3, but, for the sake of variety, let us use a slightly different construction.

Define $Z = Z(S) \subset E$ by

(4.34) $$Z = \{x \in E : d(x) = 0\}.$$

Let A be a maximal subset of $E \setminus Z$ with the property that $|x - y| > \frac{1}{100} d(x)$ for all $x, y \in A$ with $y \neq x$. Thus, if $x \in (E \setminus Z) \setminus A$, then there is a point $a \in A$ such that either $|x - a| \leq \frac{1}{100} d(x)$ or $|x - a| \leq \frac{1}{100} d(a)$. In the first case we have $|x - a| \leq \frac{1}{100}(d(a) + |x - a|)$, and hence $|x - a| \leq \frac{1}{99} d(a)$. This last is also true in the second case, and $d(x) \leq d(a) + |x - a| \leq \frac{100}{99} d(a)$

in both cases. Let us use this to prove that

$$(4.35) \quad \left\{ z \in \mathbf{R}^{d+1} : \mathrm{dist}(z, E) \leq \frac{1}{10} d(z) \right\} \subset Z \cup \left\{ \bigcup_{a \in A} B\left(a, \frac{d(a)}{8}\right) \right\}.$$

Suppose that z lies in the set on the left side. If $d(z) = 0$, then $z \in Z$ and there is nothing to prove. Otherwise, choose $x \in E$ such that $|x - z| = \mathrm{dist}(z, E)$. Then $|x - z| \leq \frac{d(z)}{10} \leq \frac{d(x)}{10} + \frac{|x-z|}{10}$, and hence $|x - z| \leq \frac{d(x)}{9}$. On the other hand, there is an $a \in A$ such that $|x - a| \leq \frac{1}{99} d(a)$ and $d(x) \leq \frac{100}{99} d(a)$. This yields

$$|z - a| \leq |z - x| + |x - a| \leq \frac{d(x)}{9} + \frac{d(a)}{99} \leq \frac{109}{9} \frac{d(a)}{99} < \frac{d(a)}{8},$$

and (4.35) follows.

Set

$$(4.36) \quad E(S) = E \cup \bigcup_{a \in A} \partial B\left(a, \frac{d(a)}{8}\right).$$

LEMMA 4.37. *$E(S)$ is a d-dimensional regular set in $\mathbf{R}^{d+1}$.*

Let us first check that $E(S)$ is closed. Let $\{x_j\}$ be a sequence of points of $E(S)$ which converges to a point x, and let us show that $x \in E(S)$. If infinitely many x_j belong to E or to some fixed $\partial B\left(a, \frac{d(a)}{8}\right)$, then the result is obvious. Thus, extracting a subsequence if necessary, we may assume that $x_j \in \partial B\left(a_j, d(a_j)/8\right)$, where the a_j's are distinct. This assumption implies that $|d(x_j) - d(a_j)| \leq |a_j - x_j| = \frac{1}{8} d(a_j)$ and $\frac{7}{8} d(a_j) \leq d(x_j) \leq \frac{9}{8} d(a_j)$.

If $\liminf_{j \to \infty} d(a_j) = 0$, then $d(x) = 0$, so $x \in Z \subset E(S)$. Otherwise, the a_j's eventually leave every compact set of $\mathbf{R}^{d+1}$. This follows from the fact that the a_j's are distinct (and therefore far away from each other, by definition of A). This implies that $\lim_{j \to \infty} d(a_j) = +\infty$, which contradicts the fact that $\lim_{j \to \infty} d(x_j) = d(x) < +\infty$. Thus $E(S)$ is closed.

Let us now prove that

$$(4.38) \quad C^{-1} t^d \leq H^d(E(S) \cap B(x, t)) \leq C t^d$$

for all $x \in E(S)$ and $t > 0$. Notice that the first inequality is true (even when $x \notin E$) because E is regular. Thus it is enough to check that

$$(4.39) \quad H^d\left(B(x, t) \cap \left\{\bigcup_{a \in A} \partial B\left(a, \frac{d(a)}{8}\right)\right\}\right) \leq C t^d$$

for all x and t.

Set

$$A(x, t) = \left\{ a \in A : B(x, t) \cap \partial B\left(a, \frac{d(a)}{8}\right) \neq \varnothing \right\}.$$

Let $A_0(x, t)$ and $A_1(x, t)$ be the subsets of $A(x, t)$ where $d(a) \geq 10t$ or $d(a) < 10t$, respectively. If $a, a' \in A_0(x, t)$, then

$$|d(a) - d(a')| \leq |a - a'| \leq \tfrac{1}{8}d(a) + 2t + \tfrac{1}{8}d(a') < \tfrac{1}{2}\max(d(a), d(a')),$$

so $\tfrac{1}{2}d(a) < d(a') < \tfrac{3}{2}d(a)$. A simple geometric argument (using the definition of A) implies that $A_0(x, t)$ can have at most a bounded number of elements, so its contribution to the left side of (4.39) is under control. On the other hand, the balls $\overline{B}(a, 10^{-4}d(a))$, $a \in A_1(x, t)$, are all contained in $B(x, 3t)$, and they are also pairwise disjoint, by the definition of A. Hence the contribution of $A_1(x, t)$ to the left side of (4.39) is less than

$$C \sum_{a \in A_1(x,t)} d(a)^d \leq C \sum_{a \in A_1(x,t)} H^d(B(a, 10^{-4}d(a)) \cap E)$$

$$\leq CH^d(B(x, 3t) \cap E) \leq Ct^d.$$

This proves (4.39) and Lemma 4.37.

Let us now define a labelling $\tilde{l}$ on $\mathbf{R}^{d+1} \setminus E(S)$. We use the same set of labels L_* as for l, and we define $\tilde{l} : \mathbf{R}^{d+1} \setminus E(S) \to L_*$ by

(4.40) $$\begin{cases} \tilde{l}(x) = l(x) & \text{when } x \in \widehat{S}, \\ \tilde{l}(x) = * & \text{otherwise.} \end{cases}$$

LEMMA 4.41. *$\tilde{l}$ is an admissible labelling of $\mathbf{R}^{d+1} \setminus E(S)$.*

Indeed, let x, y be two points of $\mathbf{R}^{d+1} \setminus E(S)$ such that $\tilde{l}(x), \tilde{l}(y) \neq *$ and $\tilde{l}(x) \neq \tilde{l}(y)$. By definition of $\tilde{l}$, x and y must lie in $\widehat{S}$, and $l(x) = \tilde{l}(x)$, $l(y) = \tilde{l}(y)$. Since l is compatible with the coronization, the line segment $\mathscr{L}$ from x to y contains a point z such that $\text{dist}(z, E) \leq \tfrac{1}{10}d(z)$. Because of (4.35), $z \in Z \cup \{\bigcup_{a \in A} B(a, \tfrac{d(a)}{8})\}$. If $z \in Z$, then $z \in E(S)$. Otherwise, z belongs to some $B(a, \tfrac{d(a)}{8})$. If we prove that neither x nor y lie in $\overline{B}(a, \tfrac{d(a)}{8})$, then we shall be able to deduce that $\mathscr{L}$ meets $\partial B(a, \tfrac{d(a)}{8})$, and hence $E(S)$, and Lemma 4.41 will follow at once. Thus it is enough to show that $\overline{B}(a, \tfrac{d(a)}{8})$ never meets $\widehat{S}$. If $u \in \overline{B}(a, \tfrac{d(a)}{8})$, then $|u - a| \leq \tfrac{d(a)}{8} \leq \tfrac{d(u)}{8} + \tfrac{|u-a|}{8}$, which implies that $|u - a| \leq \tfrac{d(u)}{7}$ and then $\text{dist}(u, E) \leq \tfrac{d(u)}{7}$. This implies that $u \notin \widehat{S}$, as desired, because of the definition (4.29) of $\widehat{S}$. Notice, incidentally, that we have also proved that

(4.42) $$E(S) \subset \mathbf{R}^{d+1} \setminus \widehat{S}.$$

In view of Lemmas 4.37 and 4.41, we can apply Proposition 4.19 to $E(S)$ and the labelling $\tilde{l}$. We obtain that $\widetilde{\mathscr{B}}(\epsilon, \delta)$ is a Carleson set for all choices of $\epsilon > \delta > 0$, where $\widetilde{\mathscr{B}}$ is the analogue of $\mathscr{B}_l$ for $E(S)$ and $\tilde{l}$. We want to use this to show that (4.33) is a Carleson set, with estimates that do not depend on S.

Let $H = H(S, \epsilon, \delta)$ denote the subset of $E \times \mathbf{R}_+$ given by (4.33). The desired Carleson estimate on H, and hence Lemma 4.32 and Theorem 4.31, will be an immediate consequence of the next lemma.

LEMMA 4.43. $H \setminus \widetilde{\mathscr{B}}\left(\frac{\epsilon}{2}, \frac{\delta}{2}\right)$ *is a Carleson set in* $E \times \mathbf{R}_+$, *with an estimate that does not depend on* S.

The statement of this lemma may seem a little strange, since H is a subset of $E \times \mathbf{R}_+$, while $\widetilde{\mathscr{B}}\left(\frac{\epsilon}{2}, \frac{\delta}{2}\right)$ is a subset of $E(S) \times \mathbf{R}_+$. However, $E(S)$ contains E, so $\widetilde{\mathscr{B}}\left(\frac{\epsilon}{2}, \frac{\delta}{2}\right)$ may well contain elements of $E \times \mathbf{R}_+$, and we are removing those elements of $E \times \mathbf{R}_+$ from H.

To prove the lemma, define an auxiliary set $W \subset E \times \mathbf{R}_+$ by

(4.44) $$W = \{(x, t) \in E \times \mathbf{R}_+ : \text{ there exists } u \in B(x, t) \setminus E \text{ such that}$$
$$\text{dist}(u, E) \geq \epsilon t \text{ but } \text{dist}(u, \mathbf{R}^{d+1} \setminus \widehat{S}) < \tfrac{1}{2} \text{dist}(u, E)\}.$$

The lemma will follow once we have verified the following two assertions:

(4.45) $\quad H \setminus \widetilde{\mathscr{B}}\left(\frac{\epsilon}{2}, \frac{\delta}{2}\right) \subset W$;

(4.46) $\quad H \cap W$ is a Carleson set, with an estimate that does not depend on S.

Let us check (4.45) first. Fix $(x, t) \in H \setminus \widetilde{\mathscr{B}}\left(\frac{\epsilon}{2}, \frac{\delta}{2}\right)$. Since $(x, t) \in \mathscr{B}_l(\epsilon, \delta)$ (remember that H was defined to be the set in (4.33)), there exist $y_0, y_1, y_2 \in B(x, t) \setminus E$ that satisfy the properties (4.14)–(4.18). Because $(x, t) \notin \widetilde{\mathscr{B}}\left(\frac{\epsilon}{2}, \frac{\delta}{2}\right)$, the points y_0, y_1, y_2 do not satisfy all the corresponding properties with E and l replaced by $E(S)$ and $\tilde{l}$, and ϵ, δ replaced by $\frac{\epsilon}{2}, \frac{\delta}{2}$. Let us distinguish cases depending on which of the properties (4.14)–(4.18) fails, and let us show that in each case one of the y_i must satisfy

(4.47) $$\text{dist}(y_i, \mathbf{R}^{d+1} \setminus \widehat{S}) < \tfrac{1}{2} \text{dist}(y_i, E).$$

This will give (4.45). (Take $u = y_i$ in the definition of W, and use (4.14).)

The first possibility is that $y_i \in E(S)$ for some i. Then (4.42) implies that $y_i \in \mathbf{R}^{d+1} \setminus \widehat{S}$, and (4.47) is obvious. Similarly, if the analogue of (4.16) or (4.18) fails, then we must have $\tilde{l}(y_i) \neq l(y_i)$ for some i. By the definition (4.40) of $\tilde{l}$, this implies that $y_i \in \mathbf{R}^{d+1} \setminus \widehat{S}$, and again (4.47) follows. Our next case is when the analogue of (4.17) fails, i.e., when there is a z such that $|z - y_i| < \frac{\delta t}{2}$ for some i, but $\tilde{l}(z) \neq \tilde{l}(y_i)$. Comparing this with the true (4.17), we conclude that $\tilde{l}(z) \neq l(z)$ or $\tilde{l}(y_i) \neq l(y_i)$. Since we already know what to do when $\tilde{l}(y_i) \neq l(y_i)$, we are left with the case where $\tilde{l}(z) \neq l(z)$, so that $z \in \mathbf{R}^{d+1} \setminus \widehat{S}$. Then

$$\operatorname{dist}\left(y_i, \mathbf{R}^{d+1} \setminus \widehat{S}\right) \leq |z - y_i| < \frac{\delta t}{2} \leq \frac{\epsilon t}{2} \leq \frac{1}{2}\operatorname{dist}(y_i, E)$$

(by (4.14)). In this case too we have (4.47).

The case where the analogue of (4.15) fails is impossible, so we are left with the case where the analogue of (4.14) fails, i.e., $\operatorname{dist}(y_i, E(S)) < \frac{\epsilon t}{2}$ for some i. When that happens we have $\operatorname{dist}(y_i, \mathbf{R}^{d+1} \setminus \widehat{S}) < \frac{\epsilon t}{2}$ because of (4.42), and (4.47) again follows from (4.14). This completes the proof of (4.45).

It remains to check (4.46). The idea is that elements of $H \cap W$ correspond to cubes on the "boundary" of S. Let $(x, t) \in H \cap W$ be given, and let u be as in the definition (4.44) of W. Let w be a point of $\mathbf{R}^{d+1} \setminus \widehat{S}$ such that $|w - u| < \frac{1}{2}\operatorname{dist}(u, E)$. Then $\operatorname{dist}(w, E) \geq \frac{t}{2}$, and also $|w - u| < \frac{1}{2}\operatorname{dist}(u, E) \leq \frac{t}{2}$. Thus, if $(x, t) \in H \cap W$, then

(4.48)   there exists $w \in B\left(x, \frac{3t}{2}\right) \setminus \widehat{S}$ such that $\operatorname{dist}(w, E) \geq \frac{\epsilon t}{2}$.

Also remember that H is the set defined by (4.33), so $x \in Q(S)$ and $d(x) \leq t \leq \operatorname{diam} Q(S)$ for all $(x, t) \in H \cap W$. Since the pairs (x, t) with $x \in Q(S)$ and $\frac{1}{2}\operatorname{diam} Q(S) \leq t \leq \operatorname{diam} Q(S)$ obviously form a Carleson set, it suffices to prove that

(4.49)
$$V = \{(x, t) \in E \times \mathbf{R}_+ : x \in Q(S),\ d(x) \leq t \leq \tfrac{1}{2}\operatorname{diam} Q(S) \text{ and (4.48) holds}\}$$

is a Carleson set, with an estimate that does not depend on S.

Let $(x, t) \in V$ be given. We want to prove that (x, t) is close to the boundary of the region $U = \{(x, t) : x \in Q(S) \text{ and } d(x) \leq t \leq \tfrac{1}{2}\operatorname{diam} Q(S)\}$, in the sense that there is a cube $Q \in \Delta \setminus S$ such that

(4.50) $\operatorname{dist}(x, Q) \leq 3t$ and $k^{-1}t \leq \operatorname{diam} Q \leq t$,

where k is some constant that does not depend on S. The Carleson estimate will then follow rather easily.

Let w be as in (4.48). Then

(4.51) $\operatorname{dist}(w, Q(S)) \leq |x - w| < \frac{3t}{2} < \operatorname{diam} Q(S)$,

and hence

(4.52) $\operatorname{dist}(w, E) < \tfrac{1}{2}d(w)$,

since $w \notin \widehat{S}$. (See (4.29).) Choose $u \in E$ such that $|u - w| = \operatorname{dist}(w, E)$. Let us first assume that $u \notin Q(S)$. Let Q be the largest cube of Δ such that $u \in Q$ and $\operatorname{diam} Q \leq t$. Since $u \notin Q(S)$, Q is not contained in $Q(S)$, so $Q \notin S$. Also,

$$\operatorname{dist}(x, Q) \leq |x - u| \leq |x - w| + |w - u|$$
$$= |x - w| + \operatorname{dist}(w, E) \leq 2|x - w| \leq 3t,$$

while the other half of (4.50) is true by definition of Q.

4.2. LABELLINGS

Assume now that $u \in Q(S)$. Using (4.52), we get

$$\operatorname{dist}(w, E) < \tfrac{1}{2}d(w) \leq \tfrac{1}{2}(d(u) + |u - w|) = \tfrac{1}{2}(d(u) + \operatorname{dist}(w, E)),$$

so $d(u) \geq \operatorname{dist}(w, E)$, which is $\geq \frac{\epsilon t}{2}$ because we chose w as in (4.48). Let Q be the largest element of Δ which contains u and which satisfies $\operatorname{diam} Q < \frac{\epsilon t}{2}$. Then $Q \in \Delta \setminus S$, because otherwise we would have $d(u) \leq \operatorname{diam} Q < \frac{\epsilon t}{2}$. Also, $\operatorname{diam} Q \geq C^{-1}\epsilon t$, and $\operatorname{dist}(x, Q) \leq |x - u| \leq 3t$ as before. Thus Q satisfies the condition (4.50), with $k = C\epsilon^{-1}$.

We are left with the task of verifying that the set

$$(4.53) \quad U_k = \{(x, t) \in V : \text{there is a cube } Q \in \Delta \setminus S \text{ which satisfies (4.50)}\}$$

is a Carleson set, with estimates that do not depend on S. This is pretty easy, so we only sketch the argument. Let $\mathscr{A}_k$ be the set of cubes $Q \in \Delta \setminus S$ which correspond to some $(x, t) \in U_k$ as in the definition of U_k. It suffices to show that the cubes in $\mathscr{A}_k$ satisfy a Carleson packing condition, with an estimate that does not depend on S.

Suppose that $Q \in \mathscr{A}_k$. A first possibility is that $Q \subset Q(S)$. Let R be the minimal cube of S that contains Q. We may assume that Q is much smaller than R, because otherwise Q is a neighbor of R, and we know that the set of neighbors of minimal cubes of S satisfies a Carleson packing condition, because the minimal cubes of S are disjoint. (See Lemma I.3.27.) Now let $(x, t) \in U_k$ be associated to Q as in (4.53), and let Q' be a cube of S such that $\operatorname{dist}(x, Q') + \operatorname{diam} Q' \leq 2d(x)$. Let Q'' be the largest ancestor of Q' such that $\operatorname{diam} Q'' \leq 2t$. [We know it exists because $d(x) \leq t$, by (4.49).] Then $\operatorname{dist}(Q, Q'') \leq Ck \operatorname{diam} Q$, and also $Q'' \in S$ because $t \leq \tfrac{1}{2} \operatorname{diam} Q(S)$. If $\operatorname{diam} Q$ is small enough compared to $\operatorname{diam} R$, then Q'' cannot meet R (since R is a minimal cube of S), so

$$(4.54) \quad \operatorname{dist}(Q, E \setminus R) \leq Ck \operatorname{diam} Q.$$

We now use our small boundary estimate (I.3.4) to conclude that the collection of cubes $Q \subseteq R$ for which (4.54) holds satisfies a Carleson packing condition and, in particular, that the sum of their measures is less than $C|R|$. The desired Carleson packing condition on the cubes of $\mathscr{A}_k$ which are contained in $Q(S)$ follows at once.

Thus we are left with the case when Q is not contained in $Q(S)$. The same argument as before gives a cube $Q'' \in S$ of roughly the same diameter as Q such that $\operatorname{dist}(Q'', Q) \leq Ck \operatorname{diam} Q$. Notice that Q does not meet $Q(S)$, because $\operatorname{diam} Q \leq t \leq \tfrac{1}{2} \operatorname{diam} Q(S)$. Thus Q is contained in $E \setminus Q(S)$ but lies at a distance $\leq Ck \left(\frac{\operatorname{diam} Q}{\operatorname{diam} Q(S)}\right) \operatorname{diam} Q(S)$ from $Q(S)$. The desired Carleson packing condition on the collection of such cubes Q can again be derived from (I.3.4).

This completes our proof of Lemma 4.43, and Theorem 4.31 follows as announced.

4.3. The derivation of Theorem 4.9 from Theorem 4.31.

Let E be a d-dimensional regular set in $\mathbf{R}^{d+1}$ which satisfies the WTN. Choose a family Δ of cubes on E, as in §3.1 of Part I. We want to find a coronization of E and a labelling which is compatible with this coronization and which is also nearly ubiquitous and satisfies Condition B. We begin by specifying a set of good cubes.

Let $k_0 > 1$ be such that $E \times \mathbf{R}_+ \setminus \mathscr{G}(\epsilon, k_0)$ is a Carleson set for every $\epsilon > 0$. (We are using here the notation of Definition 4.5.) Let $\epsilon > 0$ be small and $A > 1$ be large, to be chosen later. We shall choose ϵ after A, and in particular we shall require that $\epsilon A \ll 1$. Let $\mathscr{G}_0$ denote the set of cubes $Q \in \Delta$ such that

(4.55) $\quad (x, t) \in \mathscr{G}(\epsilon, k_0)$ whenever $x \in AQ$ and $A^{-1} \operatorname{diam} Q \leq t \leq A \operatorname{diam} Q$.

LEMMA 4.56. *$\Delta \setminus \mathscr{G}_0$ satisfies a Carleson packing condition.*

This is proved in the usual way. Notice first that if $(x, t) \in \mathscr{G}\left(\frac{\epsilon}{10}, k_0\right)$ and if $\eta = \eta(\epsilon, k_0)$ is small enough, then all pairs $(x', t') \in E \times \mathbf{R}_+$ with $B(x', t') \subseteq B(x, t)$, $|x' - x| \leq \eta t$, and $|t' - t| < \eta t$ lie in $\mathscr{G}(\epsilon, k_0)$. Lemma 4.56 now follows from a simple accounting argument.

We now apply Lemma I.3.22 to obtain a coronization $(\mathscr{B}, \mathscr{G}, \mathscr{F})$ of E such that $\mathscr{G} \subset \mathscr{G}_0$.

Before we define our labelling we want to dispense with some simple preliminary matters. Given a stopping-time region $S \in \mathscr{F}$, let $\widehat{S} \subset \mathbf{R}^{d+1} \setminus E$ be as in (4.29). Set

(4.57) $$V_1 = \mathbf{R}^{d+1} \setminus \left\{ \bigcup_{S \in \mathscr{F}} \widehat{S} \right\}$$

and

(4.58) $$V_2 = \bigcup_{\substack{S, S' \in \mathscr{F} \\ S \neq S'}} \{\widehat{S} \cap \widehat{S'}\}.$$

Define $\widetilde{V}_1$ by

(4.59) $\quad \widetilde{V}_1 = \{z \in \mathbf{R}^{d+1} \setminus E : \operatorname{dist}(z, V_1) < \tfrac{1}{2} \operatorname{dist}(z, E)\}$,

and similarly for $\widetilde{V}_2$.

LEMMA 4.60. *There is a constant C (which depends on A, ϵ, etc.) such that*

$$\int_{\widetilde{V}_j \cap B(x, R)} \operatorname{dist}(y, E)^{-1} dy \leq C R^d$$

for all $x \in E$ and $R > 0$ and for $j = 1, 2$.

These Carleson measure estimates will make it easier for us to prove that the labelling that we shall define is nearly ubiquitous.

To prove the lemma, the following notation will be convenient. Given $Q \in \Delta$ and $a > 1$, set
$$N_a(Q) = \{z \in \mathbf{R}^{d+1} \setminus E : \operatorname{dist}(z, Q) < a \operatorname{diam} Q \text{ and } \operatorname{dist}(z, E) > a^{-1} \operatorname{diam} Q\}. \tag{4.61}$$

LEMMA 4.62. *Let $\mathscr{A}$ be a subset of Δ which satisfies a Carleson packing condition. Then for each $a > 1$ there is a $C > 0$ so that*
$$\int_{\{\bigcup_{Q \in \mathscr{A}} N_a(Q)\} \cap B(x, R)} \operatorname{dist}(y, E)^{-1} dy \leq CR^d \tag{4.63}$$

for all $x \in E$ and $R > 0$.

This is very easy to check.

LEMMA 4.64. *Given $T > 1$, let $\mathscr{B}(T)$ denote the set of cubes Q in Δ which are T-close to an element of $\mathscr{B}$ or T-close to cubes Q_1 and Q_2 in $\mathscr{G}$ that lie in different elements of $\mathscr{F}$. (See Definition I.3.23 for an explanation of T-closeness.) Then $\mathscr{B}(T)$ satisfies a Carleson packing condition.*

This is also easy to verify, so we only mention some key points. (See also the proof of Lemma I.3.26.) The cubes that are T-close to an element of $\mathscr{B}$ are easily taken care of using the Carleson packing condition on $\mathscr{B}$. Thus it is enough to prove a Carleson packing estimate on the set of cubes Q_1 that belong to some stopping-time region S_1 but for which there is a cube Q_2 that is $2T^2$-close to Q_1 and belongs to a different region S_2. The case where Q_1 or Q_2 is close to the top cube or to a minimal cube of S_1 or S_2 is easily taken care of using the Carleson packing condition on the family of top cubes. If this does not happen and if $Q(S_1)$ is smaller than $Q(S_2)$ (i.e., of a "lower" generation), then Q_2 does not meet $Q(S_1)$ and so $\operatorname{dist}(Q_1, E \setminus Q(S_1)) \leq 2T^2 \operatorname{diam} Q_1$. This case can be taken care of using the "small boundary condition" (I.3.4) for $Q(S_1)$, and also the Carleson packing condition on $\{Q(S)\}_{S \in \mathscr{F}}$. The case where $Q(S_1)$ is larger than $Q(S_2)$ is treated in a similar fashion, and Lemma 4.64 follows.

Because of Lemmas 4.62 and 4.64, Lemma 4.60 will follow immediately once we prove that
$$\tilde{V}_1 \cup \tilde{V}_2 \subset \bigcup_{Q \in \mathscr{B}(T)} N_a(Q) \tag{4.65}$$

for some choice of a and T.

To prove (4.65), first let $z \in \tilde{V}_1$ be given. Let $w \in V_1$ be such that $|z - w| < \frac{1}{2} \operatorname{dist}(z, E)$. Let $x \in E$ minimize the distance to w, and let

$Q \in \Delta$ be the largest cube containing x such that $\operatorname{diam} Q \leq \frac{1}{10} \operatorname{dist}(w, E)$. If $Q \in \mathscr{B}(T)$, then we are done, because $z \in N_a(Q)$ (if a is large enough). Otherwise, Q belongs to some region $S \in \mathscr{F}$, and then

$$d_S(w) \leq \operatorname{dist}(w, Q) + \operatorname{diam} Q \leq |x - w| + \tfrac{1}{10} \operatorname{dist}(w, E) < 2 \operatorname{dist}(w, E).$$

We also have that $\operatorname{dist}(w, Q(S)) \leq \operatorname{dist}(w, E) \leq C \operatorname{diam} Q < \operatorname{diam} Q(S)$ (if T is large enough), because we are assuming that $Q \notin \mathscr{B}(T)$. But then $w \in \widehat{S}$ (see (4.29)), which contradicts the definition of w. This proves that $\widetilde{V}_1$ is contained in the right-hand side of (4.65). The proof for $\widetilde{V}_2$ is very easy and left to the reader. [The main point is that if $z \in \widehat{S}$, then there is a $Q \in S$ such that $\operatorname{dist}(z, E)$, $\operatorname{dist}(z, Q)$, and $\operatorname{diam} Q$ are all of comparable size.] This concludes our proof of Lemma 4.60.

Now let us proceed to the definition of our labelling $l : \mathbf{R}^{d+1} \setminus E \to L_*$. Take $L_* = \{1, 2, *\}$ and define l on $V_1 \cup V_2$ to be $*$, the unlabel. We still need to define l on the sets $\widehat{S} \setminus V_2$, $S \in \mathscr{F}$. Notice, however, that these sets are disjoint, and so we may define l on each $\widehat{S} \setminus V_2$ without paying attention to the others.

Fix a region $S \in \mathscr{F}$. For each cube $Q \in S$, choose some $q \in Q$, and let $U_1(Q)$, $U_2(Q)$ be the two disjoint open subsets of $\mathbf{R}^{d+1} \setminus E$ that we get from the fact that the pair $(q, \operatorname{diam} Q)$ lies in $\mathscr{G}(\epsilon, k_0)$. We wish to take $l(z) = i$ for $z \in U_i(Q)$, but first we should deal with the obvious coherence issue.

For each $Q \in S$ and $i = 1, 2$, choose a point $z_i(Q) \in U_i(Q)$ such that

(4.66) $\quad \operatorname{dist}(z_i(Q), Q) < \operatorname{diam} Q \quad \text{and} \quad \operatorname{dist}(z_i(Q), E) \geq (2k_0)^{-1} \operatorname{diam} Q.$

Such a point exists because of (4.3).

As things currently stand, the choice of names of $U_1(Q)$ and $U_2(Q)$—i.e., 1 versus 2—does not matter.

LEMMA 4.67. *It is possible to change the names of the $U_i(Q)$ in such a way that*

(4.68) $\quad\quad\quad z_i(Q) \in U_i(Q) \cap U_i(\widetilde{Q}) \quad \text{for } i = 1, 2$

whenever $Q \in S$ and its parent $\widetilde{Q}$ also lies in S.

In order to prove the lemma, we are going to start with the top cube $Q(S)$, leaving the names of the $U_i(Q(S))$ unchanged, and then proceed to rename the $U_i(Q)$ for each child Q of $Q(S)$ as needed, and then do the same for the grandchildren, etc.

Let $Q \in S$, $Q \neq Q(S)$, be given, and assume that we have already renamed the $U_i(\widetilde{Q})$ (if necessary), where $\widetilde{Q}$ is the parent of Q. Suppose first that $z_1(Q) \in U_1(\widetilde{Q})$. In this case we keep the old names for the $U_i(Q)$. Observe that $z_2(Q)$ belongs to $U_1(\widetilde{Q}) \cup U_2(\widetilde{Q})$, if we choose ϵ small enough, because of (4.1) (applied to $\widetilde{Q}$). Moreover, it cannot lie in $U_1(\widetilde{Q})$, because it would then be connected to $z_1(Q)$ by a path as in (4.4) (applied to $\widetilde{Q}$), and the existence of this path would contradict (4.2) (applied to Q), provided that we take ϵ small enough (depending on k_0). Thus our choice of new names satisfies (4.68).

If $z_1(Q) \notin U_1(\widetilde{Q})$, then it lies in $U_2(\widetilde{Q})$, again because of (4.1). The same argument as above yields $z_2(Q) \in U_1(\widetilde{Q})$. In this case we exchange the names of the $U_i(Q)$ (and consequently of the $z_i(Q)$), and we get (4.68) with the new choice of names.

By repeating this argument from level to level, we see that we can indeed rename the $U_i(Q)$, $Q \in S$, so that (4.68) holds. This proves Lemma 4.67.

Assume now that the $U_i(Q)$, $Q \in S$, have been renamed in accordance with the lemma. Let $a > 1$ be large, to be chosen soon. We intend to define our labelling l on each set $N_a(Q)$, $Q \in S$, separately, and then prove that all these definitions are compatible. Before we do this, let us check that

$$(4.69) \qquad \widehat{S} \subset \bigcup_{Q \in S} N_a(Q)$$

if a is large enough.

Actually, we are going to need a slightly stronger version of this fact. Fix $q(S) \in Q(S)$, and let us show that

$$(4.70) \quad \{u \in B(q(S), 2 \operatorname{diam} Q(S)) : \operatorname{dist}(u, E) > \tfrac{1}{10} d(u)\} \subset \bigcup_{Q \in S} N_a(Q)$$

if a is sufficiently large (with a not depending on S, A, or ϵ). Notice that this implies (4.69) because $\widehat{S}$ is contained in the left side of (4.70), by the definition (4.29) of $\widehat{S}$.

Given u in the left side of (4.70), choose $Q_0 \in S$ so that $\operatorname{dist}(u, Q_0) + \operatorname{diam} Q_0 < 2d(u)$. Let Q_1 be the maximal ancestor of Q_0 such that $\operatorname{diam} Q_1 \leq 20 \operatorname{dist}(u, E)$ and $Q_1 \subset Q(S)$; then $\operatorname{diam} Q_1 \geq C^{-1} \operatorname{dist}(u, E)$, because $\operatorname{dist}(u, E) \leq |u - q(S)| \leq 2 \operatorname{diam} Q(S)$. Also,

$$\operatorname{dist}(u, Q_1) \leq \operatorname{dist}(u, Q_0) \leq 2d(u) \leq 20 \operatorname{dist}(u, E) \leq C \operatorname{diam} Q_1.$$

Since $\operatorname{dist}(u, Q_1) \geq \operatorname{dist}(u, E) \geq \tfrac{1}{20} \operatorname{diam} Q_1$, we have that $u \in N_a(Q_1)$ if a is large enough. [In fact, a depends only on the regularity constant for E and our choice of Δ.]

Let such a choice of a be made and fixed from now on. We are now ready to define l on each $N_a(Q) \setminus V_2$.

For each $Q \in S$, we apply our hypothesis that $(q, A \operatorname{diam} Q) \in \mathcal{G}(\epsilon, k_0)$, where q is (as before) any element of Q, to get two open sets which we call $U_1'(Q)$ and $U_2'(Q)$. By the same sort of argument as in the proof of Lemma 4.67, we can exchange the names of the $U_i'(Q)$ if necessary and obtain that $z_i(Q) \in U_i'(Q)$ for each i. [This requires ϵ to be sufficiently small compared to A^{-1}.] Also, if A is large enough and if ϵ is small enough (still depending on A as well as on a), then we have that

(4.71) $$N_{2a}(Q) \subset U'_1(Q) \cup U'_2(Q).$$

We want to set

(4.72) $$l(z) = i \quad \text{when } z \in \widehat{S} \cap N_{2a}(Q) \cap U'_i(Q) \setminus V_2;$$

the consistency of this definition will come from the following.

LEMMA 4.73. *If A is large enough and ϵA is small enough, then*

(4.74) $$U'_1(Q) \cap N_{2a}(Q) \cap U'_2(Q') \cap N_{2a}(Q') = \varnothing$$

for all Q, $Q' \in S$ (and any choice of S).

Suppose not. Let Q, $Q' \in S$ be given, and let w be a point in the left side of (4.74). From the definition (4.61) of $N_a(\cdot)$ we get that
(4.75)
$$C^{-1} \operatorname{diam} Q \leq \operatorname{diam} Q' \leq C \operatorname{diam} Q \quad \text{and} \quad \operatorname{dist}(Q, Q') \leq C \operatorname{diam} Q$$

(where C may depend on a, but we do not care). The points w and $z_1(Q)$ both belong to $U'_1(Q)$ and both lie at distance $\geq \epsilon A \operatorname{diam} Q$ from E, so (4.4) applied with $x = q$ and $t = A \operatorname{diam} Q$ implies that there is a curve γ that joins them and satisfies

(4.76) $$\operatorname{diam} \gamma \leq C' k_0 \operatorname{diam} Q \quad \text{and} \quad \operatorname{dist}(\gamma, E) \geq (C' k_0^2)^{-1} \operatorname{diam} Q.$$

The same argument applied to $z_2(Q')$ and w gives us a path γ' from $z_2(Q')$ to w that also satisfies the appropriate version of (4.76).

The union $F = \gamma \cup \gamma'$ is a connected compact set which is contained in $B(q, A \operatorname{diam} Q)$ if A is large enough, and we have that $\operatorname{dist}(F, E) > \epsilon A \operatorname{diam} Q$ if ϵ is small enough. Thus (4.2) says that F cannot meet $U'_2(Q)$. [Actually, for this we can simply use (4.1) and the requirement that $U'_1(Q)$ and $U'_2(Q)$ be open and disjoint.] The desired contradiction (and hence Lemma 4.73) will follow at once from the next lemma.

LEMMA 4.77. *The point $z_2(Q')$ lies in $U'_2(Q)$ whenever Q, Q' are two cubes of S that satisfy (4.75) (if A is large enough and ϵA is small enough).*

The truly devoted reader will notice that we encountered a similar situation in Chapter 2 of this part, when we were dealing with compatibility issues for orientations of the approximating d-planes $P(Q)$. (See Lemma 2.22 in particular.) As in that case, the proof of Lemma 4.77 becomes easier once we notice that the property we want to prove is hereditary: we claim that $z_2(Q') \in U'_2(Q)$ if and only if $z_2(\widetilde{Q}') \in U'_2(\widetilde{Q})$, where $\widetilde{Q}$ and $\widetilde{Q}'$ denote the respective parents of Q and Q'.

To prove the claim, notice first that

$$z_2(Q') \in U'_1(Q) \cup U'_2(Q) \quad \text{and} \quad z_2(\widetilde{Q}') \in U'_1(\widetilde{Q}) \cup U'_2(\widetilde{Q})$$

if A is large enough and ϵA is small enough, by (4.1). Also, from (4.68) we get that

$$z_2(Q), z_2(\widetilde{Q}) \in U_2(\widetilde{Q}) \quad \text{and} \quad z_2(Q'), z_2(\widetilde{Q}') \in U_2(\widetilde{Q}').$$

This implies that we can join $z_2(Q)$ to $z_2(\widetilde{Q})$ by a path γ that satisfies (4.76) (perhaps with a different value of C') and that we can join $z_2(Q')$ to $z_2(\widetilde{Q}')$ by a path γ' that also satisfies the appropriate version of (4.76).

If $z_2(Q') \in U_2'(Q)$, then we can join $z_2(Q')$ to $z_2(Q)$ by a path α that satisfies (4.76), since $z_2(Q) \in U_2'(Q)$. Thus $z_2(\widetilde{Q})$ is connected to $z_2(\widetilde{Q}')$ by $\gamma \cup \alpha \cup \gamma'$, and we can use (4.2) to conclude that $z_2(\widetilde{Q}') \in U_2'(\widetilde{Q})$, since $z_2(\widetilde{Q}) \in U_2'(\widetilde{Q})$. [This uses again the assumption that A is sufficiently large and ϵA is sufficiently small.]

The proof that $z_2(Q') \in U_2'(Q)$ when $z_2(\widetilde{Q}') \in U_2'(\widetilde{Q})$ is similar. This proves the claim.

To prove Lemma 4.77, notice that if Q, Q' satisfy (4.75), if m is a positive integer, and if $Q^{(m)}$, $Q'^{(m)}$ are the mth-order parents of Q and Q', then $Q^{(m)}$ and $Q'^{(m)}$ satisfy (4.75) with C replaced by C'', where C'' may depend on C and the regularity constant for E but not on m. Applying our claim (or rather its proof, since we may have to increase the constant in (4.75)) successively to Q and Q', $Q^{(1)}$ and $Q'^{(1)}$, $Q^{(2)}$ and $Q'^{(2)}$, etc., we can reduce to the case when Q or Q' is the top cube $Q(S)$. This case is easy, because we can apply (4.68) and (4.4) repeatedly to find a path from $z_2(Q)$ to $z_2(Q')$ that satisfies an estimate similar to (4.76). [We are using here the trivial observation that one of the cubes Q, Q' is now an ancestor of the other.] The fact that $z_2(Q')$ belongs to $U_2'(Q)$ now follows from (4.2); this proves Lemma 4.77 and, hence, Lemma 4.73 also.

Let $A > 0$ and $\epsilon > 0$ be chosen and fixed in such a way that all the preceding arguments work.

From Lemma 4.73 we conclude that our definition (4.72) of $l(z)$ on the union of the sets $\widehat{S} \cap N_{2a}(Q) \cap U_i'(Q) \setminus V_2$, $Q \in S$, is consistent. Because $\widehat{S} \setminus V_2$ is equal to this union (by (4.69) and (4.71)), we get a consistent definition of $l(z)$ on $\widehat{S} \setminus V_2$ for each region S. All these sets are disjoint by the definition of V_2, and we already decided to take $l(z) = *$ when $z \in (V_1 \cup V_2) \setminus E$. Thus our labelling $l : \mathbf{R}^{d+1} \setminus E \to \{1, 2, *\}$ is now unambiguously defined.

It remains to check that the labelling l satisfies the hypothesis of Theorem 4.31.

LEMMA 4.78. *There is an $\eta \in (0, \frac{1}{10})$ so that if $S \in \mathscr{F}$ and $u, v \in \widehat{S} \setminus V_2$ are such that $|u - v| < \eta \operatorname{dist}(u, E)$, then $l(u) = l(v)$.*

Let S, u, and v be given. Choose $Q \in S$ so that $u \in N_a(Q)$. Such a cube exists because of (4.69). If η is small enough, then the closed line segment σ from u to v is contained in $N_{2a}(Q)$. [This is an easy consequence of the definition (4.61) of $N_a(\cdot)$.] Thus σ is contained in $U'_1(Q) \cup U'_2(Q)$ by (4.71). The sets $U'_1(Q)$ and $U'_2(Q)$ are open and disjoint, so u and v must be contained in the same $U'_i(Q)$. Hence $l(u) = l(v) = i$, and Lemma 4.78 follows.

It is now easy to check that l is nearly ubiquitous (Definition 4.22). If u lies in $\widehat{S} \setminus (\widetilde{V}_1 \cup \widetilde{V}_2)$ for some region S, then Lemma 4.78 and the definition (4.59) of $\widetilde{V}_1 \cup \widetilde{V}_2$ imply that u lies in the set A_η of (4.23). The desired Carleson measure estimate (4.24) on the complement of A_η follows immediately from Lemma 4.60 and (4.57).

Let us now show that l is compatible with the coronization $(\mathscr{B}, \mathscr{G}, \mathscr{F})$. [See Definition 4.30.] Fix a region $S \in \mathscr{F}$ and $x, y \in \widehat{S}$ such that $l(x), l(y) \in L$ and $l(x) \neq l(y)$. We have to prove that the line segment σ from x to y contains a point z such that $\mathrm{dist}(z, E) \leq \frac{1}{10} d(z)$.

Let $q(S)$ be the point of $Q(S)$ that was chosen just before (4.70). Because $x, y \in \widehat{S}$, we have that their distances to $Q(S)$ are $\leq \mathrm{diam}\, Q(S)$, and so $x, y \in B(q(S), 2 \mathrm{diam}\, Q(S))$. Hence $\sigma \subset B(q(S), 2 \mathrm{diam}\, Q(S))$.

Suppose that there is no $z \in \sigma$ such that $\mathrm{dist}(z, E) \leq \frac{1}{10} d(z)$. Then $\sigma \subset \bigcup_{Q \in S} N_a(Q)$ by (4.70). Set

$$(4.79) \qquad \sigma_i = \sigma \cap \left(\bigcup_{Q \in S} [N_a(Q) \cap U'_i(Q)] \right)$$

for $i = 1, 2$.

Lemma 4.73 implies that $\sigma_1 \cap \sigma_2 = \varnothing$, and since we know that $\sigma = \sigma_1 \cup \sigma_2$, we conclude that one of σ_1 and σ_2 must be empty. This contradicts our assumption that $l(x), l(y) \in L$ and $l(x) \neq l(y)$.

This proves that l is compatible with the coronization. We are left with the task of verifying that l satisfies Condition B (Definition 4.21).

Let us first make a preliminary observation. Consider the collection of cubes given by

(4.80)
$$\mathscr{B} \cup \{Q \in \Delta : Q \in \mathscr{G} \text{ but } \mathrm{dist}(z_i(Q), V_1 \cup V_2) < \tfrac{1}{10} \mathrm{dist}(z_i(Q), E)$$
$$\text{for either } i = 1 \text{ or } 2\}.$$

This collection satisfies a Carleson packing condition, because $\mathscr{B}$ does, and because of Lemma 4.60. [We are implicitly using (4.66) as well.]

Now let us check Condition B. Let $x \in E$ and $R > 0$ be given, and let $\tau > 0$ be small. Consider the collection
(4.81)
$$\{Q \in \Delta : Q \in \mathscr{G}, \; z_i(Q) \in B(x, R) \text{ for } i = 1, 2,$$
$$\mathrm{dist}(z_i(Q), V_1 \cup V_2) \geq \tfrac{1}{10} \mathrm{dist}(z_i(Q), E) \text{ for } i = 1, 2,$$
$$\text{and } \mathrm{diam}\, Q \geq \tau R\}.$$

If τ is sufficiently small, then this collection of cubes must be nonempty, for otherwise the Carleson constant for (4.80) would be too large. If Q lies in (4.81), then $z_1(Q)$ and $z_2(Q)$ have the properties required in Definition 4.21.

Thus our labelling also satisfies Condition B. This completes the proof of Theorem 4.9.

PART III

Applications

CHAPTER 1

Uniform Rectifiability and Singular Integral Operators

This chapter is devoted to the proofs of Theorems I.2.32 and I.2.33.

1.1. Preliminaries.
Let E be a d-dimensional regular set in $\mathbf{R}^n$. Define the sublinear operator T^* acting on functions on E by

$$(1.1) \qquad T^*(f)(x) = \sup_{r,R>0} \left| \int_{E \cap (B(x,R) \setminus B(x,r))} \frac{x-y}{|x-y|^{d+1}} f(y) dy \right|.$$

PROPOSITION 1.2. *If T^* is bounded on $L^2(E)$ and if E satisfies the WGL, then E satisfies the BWGL, and hence is uniformly rectifiable.*

Recall that the definitions of the WGL and the BWGL were given in Definitions I.1.71 and I.2.2, respectively, and that the fact that the BWGL implies uniform rectifiability was stated as Theorem I.2.4 and proved in Chapter II.2. Notice that Proposition 1.2 implies Theorems I.2.32 and I.2.33. [The discussion around (I.1.25) is relevant here.]

The special form of the kernel $(x-y)/|x-y|^{d+1}$ is not too significant for the purposes of the proof of Proposition 1.2. One could formulate general nondegeneracy conditions on a kernel under which the obvious analogue of Proposition 1.2 would still be true, but we shall not bother. Of course, we are especially interested in this kernel because it is the obvious generalization of the Cauchy kernel and because it is so simple and symmetric.

In this section we shall begin the proof of Proposition 1.2, confining ourselves to notational and organizational matters. The main computations will be carried out in the next two sections.

Let Δ denote the set of cubes on E, as in §3.1 of Part I. Given $\delta > 0$ set

$$\mathscr{G}(\delta) = \{Q \in \Delta : \text{ there is a } d\text{-plane } P \text{ such that } \operatorname{dist}(x, P) \leq \delta \operatorname{diam} Q$$
$$\text{when } x \in 2Q, \text{ and } \operatorname{dist}(y, E) \leq \delta \operatorname{diam} Q$$
$$\text{for all } y \in P \text{ such that } \operatorname{dist}(y, Q) \leq \operatorname{diam} Q\}.$$

In order to prove that E satisfies the BWGL, we must show that $\Delta \setminus \mathscr{G}(\delta)$ satisfies a Carleson packing condition for each $\delta > 0$. Let $\delta > 0$ be given.

For each $\epsilon > 0$ set

$$\mathscr{H}(\epsilon) = \{Q \in \Delta : \text{ there is a } d\text{-plane } P = P_Q \text{ such that }$$
$$\operatorname{dist}(x, P) \leq \epsilon \operatorname{diam} Q \text{ whenever } x \in 100Q\}.$$

Since E satisfies the WGL, $\Delta \setminus \mathscr{H}(\epsilon)$ satisfies a Carleson packing condition. Thus it suffices to show that we can choose ϵ so that $\mathscr{H}(\epsilon) \setminus \mathscr{G}(\delta)$ satisfies a Carleson packing condition. Here ϵ is allowed to depend on δ, as well as pretty much everything else. We shall choose ϵ later, but we shall certainly require that it be small compared to δ.

LEMMA 1.3. $\mathscr{H}(\epsilon) \setminus \mathscr{G}(\delta) \subseteq \mathscr{J}(\epsilon, \delta)$, *where* $\mathscr{J}(\epsilon, \delta)$ *is the set of* $Q \in \Delta$ *for which there exists a d-plane P and an open ball B centered on P with radius R such that*:

(1.4) $\qquad B \cap E = \varnothing$;

(1.5) $\qquad \partial B \cap E \neq \varnothing$;

(1.6) $\qquad \delta \operatorname{diam} Q \leq R \leq \operatorname{diam} Q$;

(1.7) $\qquad \operatorname{dist}(u, Q) \leq 2 \operatorname{diam} Q$ *for all* $u \in B$;

(1.8) $\qquad 10B \cap E \subseteq \{u \in \mathbf{R}^n : \operatorname{dist}(u, P) \leq \epsilon \operatorname{diam} Q\}$.

This lemma is basically trivial. If $Q \in \mathscr{H}(\epsilon)$ then we have a candidate P for the d-plane needed in order to have $Q \in \mathscr{G}(\delta)$. If $Q \notin \mathscr{G}(\delta)$, then there is a $y \in P$ such that $\operatorname{dist}(y, Q) \leq \operatorname{diam} Q$ but $\operatorname{dist}(y, E) \geq \delta \operatorname{diam} Q$. Let B be the largest open ball centered at y which is disjoint from E. Then (1.4)–(1.7) are automatic, and (1.8) follows from the assumption that $Q \in \mathscr{H}(\epsilon)$.

Thus it suffices to show that $\mathscr{J}(\epsilon, \delta)$ satisfies a Carleson packing condition if ϵ is small enough. We shall prove this in two steps. We first show that $T^*(\chi_{10Q})$ has to be large near at least one point when $Q \in \mathscr{J}(\epsilon, \delta)$, and then we show that that cannot happen too frequently.

1.2. Step one.

We continue with the same notation and assumptions as in the preceding section.

The constants C that appear in the inequalities in this section are allowed to depend on n, d, and the regularity constant for E without further mention. Any dependence on ϵ or δ will be made explicit.

Let $Q \in \mathscr{J}(\epsilon, \delta)$ be given, and let B, P, and R be as in Lemma 1.3. Let y denote the center of B, and let z be an element of $\partial B \cap E$. We are going to estimate quantities like $T^*(\chi_{10Q})(z)$ from below. In order to do that we need first to have some control on how the mass of E is distributed near z.

Let v be the unit vector given by $v = \frac{z-y}{|z-y|}$. Notice that

(1.9) there is a unit vector $\tilde{v}$ such that
$$|v - \tilde{v}| \leq C\epsilon\delta^{-1} \text{ and } \tilde{v} \text{ is parallel to } P,$$

because $y \in P$ and $\text{dist}(z, P) \leq \epsilon \operatorname{diam} Q \leq \epsilon \delta^{-1} R$.

Let H_+ and H_- denote the two half-spaces defined by
$$H_+ = \{x \in \mathbf{R}^n : (x - z) \cdot v \geq 0\},$$
$$H_- = \{x \in \mathbf{R}^n : (x - z) \cdot v < 0\}.$$

Given $\theta > 0$, set
$$H_+(\theta) = \{x \in \mathbf{R}^n : (x - z) \cdot v \geq \theta |x - z|\}.$$

Let us show that there is not too much of E in H_- near z. More precisely,

(1.10) $$|E \cap H_- \cap B(z, r)| \leq Cr^{d-1}(\epsilon R + r^2 R^{-1})$$

when $\epsilon R \leq r \leq R$. To prove this, it is enough to show that

(1.11) $$\text{dist}(x, \partial H_- \cap P) \leq C(\epsilon R + r^2 R^{-1})$$

when $x \in E \cap H_- \cap B(z, r)$ and $r \leq R$. Indeed, (1.10) easily follows from (1.11) and the regularity of E, because $\partial H_- \cap P$ is a $(d-1)$-plane. [More precisely, we can cover the set $\{x \in B(z, r) : (1.11) \text{ holds}\}$ by a family of $\leq Cr^{d-1}(\epsilon R + r^2 R^{d-1})^{-(d-1)}$ balls of radius $(\epsilon R + r^2 R^{-1})$, so the measure of the intersection of this set with E is $\leq Cr^{d-1}(\epsilon R + r^2 R^{-1})$, since E is regular.]

To verify (1.11) we observe that

(1.12) $$\text{dist}(x, \partial H_-) \leq Cr^2 R^{-1},$$
(1.13) $$\text{dist}(x, P) \leq C\epsilon R$$

when $x \in E \cap H_- \cap B(z, r)$ and $r \leq R$. We get (1.13) from (1.8), while (1.12) is an immediate consequence of the fact that $E \cap H_- \subseteq H_- \setminus B$. (See Figure 1.1, next page.) Because the hyperplane ∂H_- is almost perpendicular to P if ϵ is small enough, by (1.9), it is easy to derive (1.11) from (1.12) and (1.13). This completes the proof of (1.10).

We also need to control the amount of E inside $H_+ \setminus H_+(\theta)$ when θ is small. Let us show that

(1.14) $$|E \cap (H_+ \setminus H_+(\theta)) \cap B(z, r)| \leq Cr^{d-1}(\theta r + \epsilon R)$$

when $\epsilon R \leq r \leq R$. This is proved in almost the same way as (1.10). As before it is easy to reduce to showing that

$$\text{dist}(x, P \cap \partial H_+) \leq C(\theta r + \epsilon R)$$

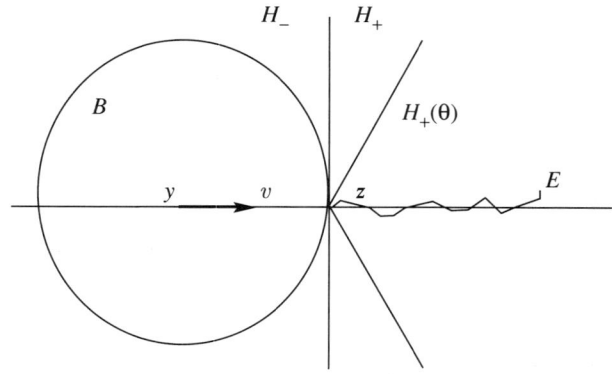

FIGURE 1.1

when $x \in E \cap (H_+ \setminus H_+(\theta)) \cap B(z, r)$, and this is easily reduced further to checking that

$$\operatorname{dist}(x, \partial H_+) \leq C\theta r \quad \text{and} \quad \operatorname{dist}(x, P) \leq C\epsilon R$$

when $x \in E \cap (H_+ \setminus H_+(\theta)) \cap B(z, r)$. The first of these last two inequalities follows from the definitions, while the second again comes from (1.8).

An important consequence of (1.10) and (1.14) is that E has plenty of mass in $H_+(\theta)$ near z if θ and ϵ are small enough. The precise statement is as follows.

LEMMA 1.15. *There is a constant C such that if ϵ and θ are small enough, then*

(1.16) $$C^{-1} r^d \leq |E \cap H_+(\theta) \cap B(z, r)| \leq C r^d$$

when $C\epsilon R \leq r \leq C^{-1} R$.

This is an immediate consequence of (1.10), (1.14), and the regularity of E.

We want to use these mass bounds to estimate the effect of T^* on certain functions. Set

(1.17) $$f(r) = \int_{E \cap (B(z, R) \setminus B(z, r))} \frac{(x - z) \cdot v}{|x - z|^{d+1}} \, dx.$$

LEMMA 1.18.

(1.19) $$f(r) \geq C^{-1} \log\left(\frac{R}{r + \epsilon R}\right) - C$$

when $0 < r \leq R$.

1.2. STEP ONE

Let $f_+(r)$ and $f_-(r)$ be defined as in (1.17) but with the domain of integration intersected with H_+ and H_-, respectively. Thus $f = f_+ + f_-$, $f_+ \geq 0$, and $f_- \leq 0$.

Notice first that

(1.20) $\qquad |f_-(r)| \leq C \quad \text{when} \quad 0 < r \leq R.$

Indeed,

$$|(x-z) \cdot v| \leq C|x-z|^2 R^{-1}$$

when $x \in E \cap H_- \cap B(z, R)$, since $E \cap H_- \subseteq H_- \setminus B$, so

$$|f_-(r)| \leq \int_{E \cap B(z,R) \cap H_-} C|x-z|^{-d+1} R^{-1} dx \leq C.$$

This last inequality uses the regularity of E.

Next we want a suitable lower bound on $f_+(r)$. We may as well assume that ϵ is sufficiently small so that there is a θ such that the conclusion of Lemma 1.15 is true. [If ϵ were larger than some fixed constant, then (1.19) would follow from (1.20) and the fact that $f_+ \geq 0$. Of course we are ultimately interested only in small ϵ's anyway.] Fix such a θ, once and for all, and let $f_1(r)$ be defined as in (1.17), but with the domain of integration intersected with $H_+(\theta)$. Then

(1.21) $\qquad f_+(r) \geq f_1(r).$

By definitions we have that

$$f_1(r) \geq \theta \int_{E \cap (B(z,R) \setminus B(z,r)) \cap H_+(\theta)} |x-z|^{-d} dx.$$

Using (1.16) it is not hard to show that

$$f_1(r) \geq C^{-1} \log\left(\frac{R}{(r + \epsilon R)}\right) - C.$$

Combining this with (1.21) and (1.20) gives (1.19). This proves Lemma 1.18.

The next lemma is a minor variation of Lemma 1.18 whose conclusion is easier to work with.

LEMMA 1.22. *Let $g : E \to \mathbf{R}$ be any locally integrable function that satisfies $g \equiv 1$ on $10Q$. Then*

(1.23) $\qquad T^*(g)(x) \geq C(\delta)^{-1} \log\left(\frac{\operatorname{diam} Q}{|x-z| + \epsilon \operatorname{diam} Q}\right) - C(\delta)$

for all $x \in E$.

We should perhaps say that $C(\delta)$ is allowed, as usual, to depend on innocuous quantities like the regularity constant of E. We do not care about the specific nature of its dependence on δ, but it is very important that it does not depend on ϵ.

Let us prove the lemma. We may as well assume that $|x - z| \leq R/4$; if not, then (1.23) is automatic if $C(\delta)$ is large enough, since $T^*(g) \geq 0$ and R satisfies (1.6).

Set $r = |x - z|$ so that $r \leq R/4$. Then

$$(1.24) \qquad T^*(g)(x) \geq \left| \int_{E \cap (B(x, R) \setminus B(x, r))} \frac{x - u}{|x - u|^{d+1}} \, du \right|,$$

since $g \equiv 1$ on $10Q$. The difference between the right side of (1.24) and

$$(1.25) \qquad \left| \int_{E \cap (B(z, R) \setminus B(z, r))} \frac{z - u}{|z - u|^{d+1}} \, du \right|$$

is $\leq C$, by straightforward estimates. From here we get (1.23) using the lower bound for (1.25) that comes from Lemma 1.18 and also (1.6). This proves the lemma.

Thus we have shown that if $Q \in \mathscr{F}(\epsilon, \delta)$ and ϵ is small enough, then $T^*(g)$ has to be large somewhere, for suitable functions g. In the next section we use this to prove a Carleson packing condition on $\mathscr{F}(\epsilon, \delta)$ when ϵ is small enough.

1.3. Step two.

Again we continue with the same notation and assumptions as before. We begin by stating a general criterion for a family of cubes to satisfy a Carleson packing condition.

LEMMA 1.26. *Let $\mathscr{A}$ be a subset of Δ. Suppose that there is a constant $\alpha > 0$ so that for each $Q_0 \in \Delta$ we have*

$$(1.27) \qquad |\cup \{Q \in \mathscr{A} : Q \subseteq Q_0, \, \operatorname{diam} Q \leq \alpha \operatorname{diam} Q_0\}| \leq \tfrac{1}{2} |Q_0|.$$

Then $\mathscr{A}$ satisfies a Carleson packing condition.

Let us assume the lemma for the moment and use it to prove that $\mathscr{F}(\epsilon, \delta)$ satisfies a Carleson packing condition if ϵ is small enough. We are not going to apply Lemma 1.26 to $\mathscr{F}(\epsilon, \delta)$ itself but to an auxiliary class of cubes defined below.

Let $Q \in \mathscr{F}(\epsilon, \delta)$ be given. Let B, R be as in Lemma 1.3, and let z be as in the preceding section. From (1.7) we have in particular that $z \in 3Q$. Choose $Q' \in \Delta$ so that $z \in Q'$ and

$$(1.28) \qquad \epsilon \operatorname{diam} Q \leq \operatorname{diam} Q' \leq C\epsilon \operatorname{diam} Q.$$

Let $\mathscr{A}$ be the collection of cubes Q' that arise from elements of $\mathscr{F}(\epsilon, \delta)$ in this way.

It is not hard to check that $\mathscr{F}(\epsilon, \delta)$ satisfies a Carleson packing condition if $\mathscr{A}$ does (with estimates which depend on ϵ). To do this it is helpful to notice that a given cube in Δ can be the Q' associated to only a bounded number of Q's in $\mathscr{F}(\epsilon, \delta)$. This bounded number depends on ϵ, but that does not matter.

Thus it suffices to show that $\mathscr{A}$ satisfies the hypotheses of Lemma 1.26 if ϵ is small enough. For this the key fact is the following:

(1.29)   If $S \in \mathscr{A}$, then $T^*(g) \geq C(\delta)^{-1} \log \frac{1}{\epsilon} - C(\delta)$ on S
whenever $g: E \to \mathbf{R}$ satisfies $g \equiv 1$ on $(100\epsilon^{-1})S$.

Indeed, suppose that S is the cube Q' associated to some $Q \in \mathscr{F}(\epsilon, \delta)$. By definition of Q', $\operatorname{dist}(Q', Q) \leq 2 \operatorname{diam} Q$, so $100\epsilon^{-1}S = 100\epsilon^{-1}Q' \supseteq 10Q$. This uses also (1.28). From here (1.29) follows from Lemma 1.22.

Let us now verify that $\mathscr{A}$ does satisfy the hypothesis of Lemma 1.26 with $\alpha = 10^{-2}\epsilon$. Let Q_0 be given, and set $g = \chi_{2Q_0}$. If $S \in \mathscr{A}$, $S \subseteq Q_0$, and $\operatorname{diam} S \leq 10^{-2}\epsilon \operatorname{diam} Q_0$, then (1.29) implies that

(1.30) $T^*(g) \geq C(\delta)^{-1} \log \frac{1}{\epsilon} - C(\delta)$ on S.

Thus

$$\bigcup \{S \in \mathscr{A} : S \subseteq Q_0, \ \operatorname{diam} S \leq 10^{-2}\epsilon \operatorname{diam} Q_0\}$$
$$\subseteq \{x \in Q_0 : T^*(g)(x) \geq C(\delta)^{-1} \log \frac{1}{\epsilon} - C(\delta)\},$$

and the measure of this last set is $\leq \frac{1}{2}|Q_0|$ if ϵ is small enough, because

$$\int_{Q_0} T^*(g)^2 \leq C|Q_0|,$$

by the L^2-boundedness of T^*.

Thus $\mathscr{A}$, and hence $\mathscr{F}(\epsilon, \delta)$, satisfies a Carleson packing condition if ϵ is small enough. This proves Proposition 1.2, modulo Lemma 1.26.

Lemma 1.26 is a cousin of well-known results of John and Nirenberg and of Strömberg for BMO. We shall reduce it to the following simpler fact.

LEMMA 1.31. *Let $\mathscr{A}$ be a subset of Δ. Suppose that for each $Q_0 \in \mathscr{A}$ we have that*

$$\left|\bigcup \{Q \in \mathscr{A} : Q \subseteq Q_0, Q \neq Q_0\}\right| \leq \frac{1}{2}|Q_0|.$$

Then $\mathscr{A}$ satisfies a Carleson packing condition.

Let us first check that Lemma 1.31 implies Lemma 1.26. Given a large integer N, define $\mathscr{A}_j$, $j = 1, 2, \ldots, N$, by

$$\mathscr{A}_j = \mathscr{A} \cap \left(\bigcup_{l \equiv j \bmod N} \Delta_l\right).$$

If $\mathscr{A}$ satisfies the hypothesis of Lemma 1.26 and if N is large enough (depending on α), then each $\mathscr{A}_j$ satisfies the hypothesis of Lemma 1.31. Thus Lemma 1.31 does imply Lemma 1.26.

Now let us prove Lemma 1.31. Fix $Q_0 \in \mathscr{A}$. Let $\{Q_{1,j}\}$ be an enumeration of the maximal proper subcubes of Q_0 that lie in $\mathscr{A}$. Let $\{Q_{2,j}\}$ denote the maximal elements of $\mathscr{A}$ that are proper subcubes of one of the $Q_{1,l}$'s. Continue this process indefinitely. Thus every element of $\{Q \in \mathscr{A} : Q \subseteq Q_0, Q \neq Q_0\}$ appears in the doubly indexed sequence $\{Q_{i,j}\}$ exactly once. The hypothesis of Lemma 1.31 implies that

$$\sum_j |Q_{i+1,j}| \leq \frac{1}{2} \sum_j |Q_{i,j}|$$

for each i, and hence

(1.32) $$\sum_{\substack{Q \in \mathscr{A} \\ Q \subseteq Q_0}} |Q| = |Q_0| + \sum_{i=1}^{\infty} \sum_j |Q_{i,j}| \leq |Q_0|\left(1 + \sum_{i=1}^{\infty} 2^{-i}\right) = 2|Q_0|.$$

To prove that $\mathscr{A}$ satisfies a Carleson packing condition, we must show that

$$\sum_{\substack{Q \in \mathscr{A} \\ Q \subseteq Q_0}} |Q| \leq C|Q_0|$$

holds even when $Q_0 \notin \mathscr{A}$. This is easy. Let $Q_0 \notin \mathscr{A}$ be given, and let $\{Q_k\}$ denote the maximal subcubes of Q_0 which lie in $\mathscr{A}$. Then

$$\sum_{\substack{Q \in \mathscr{A} \\ Q \subseteq Q_0}} |Q| = \sum_k \left(\sum_{\substack{Q \in \mathscr{A} \\ Q \subseteq Q_k}} |Q|\right) \leq \sum_k 2|Q_k| \leq 2|Q_0|,$$

by (1.32). This completes the proof of Lemma 1.31.

1.4. An abstraction of §3.

The portion of the proof of Proposition 1.2 given in the preceding section has an abstract formulation which we record in this section.

Let E be a d-dimensional regular set in $\mathbf{R}^n$. Suppose that there is a mapping $T^* : L^2(E) \to L^2(E)$ which is bounded in the sense that there is a constant K such that

(1.33) $$\int_E |T^*(f)|^2 \, dx \leq K \int_E |f|^2 \, dx$$

for all $f \in L^2(E)$. We do not require that T^* be sublinear or continuous or anything like that.

Given η, $N > 0$ let $\mathscr{N}(\eta, N)$ denote the set of cubes Q in Δ such that there exists a cube $Q' \in \Delta$ that satisfies

(1.34) $\qquad Q' \subseteq 4Q \quad \text{and} \quad \operatorname{diam} Q' \geq \eta \operatorname{diam} Q,$

and

(1.35) $\qquad T^*(g) \geq N \quad \text{on} \quad Q' \quad \text{whenever} \quad g \equiv 1 \quad \text{on} \quad 10Q.$

[There is, of course, nothing special about the numbers 4 and 10 in (1.34) and (1.35).]

PROPOSITION 1.36. *There is a constant N_0 that depends on K, d, n and the regularity constant of E but not on η so that $\mathscr{N}(\eta, N)$ satisfies a Carleson packing condition when $N \geq N_0$.*

The price for shrinking η is that the packing constant gets larger.

Proposition 1.36 can be proved using the same arguments as in the preceding section.

CHAPTER 2

Uniform Rectifiability and Square Function Estimates for the Cauchy Kernel

This chapter is devoted to the proof of Theorems I.2.41 and I.2.45 when $d = 1$.

2.1. Some general comments about square function estimates.

Let E be a one-dimensional regular set in $\mathbf{C}$. Recall from Definition I.2.35 that E is said to satisfy the usual square function estimates for the Cauchy kernel (USFE) if there exists a $C > 0$ such that

$$(2.1) \qquad \iint_{\mathbf{C}} |F'(z)|^2 \operatorname{dist}(z, E) \, dz \leq C \int_E |f(w)|^2 \, dw$$

for all $f \in L^2(E)$, where $F(z)$ is given by

$$(2.2) \qquad F(z) = \int_E \frac{1}{z - w} f(w) \, dw \quad \text{for } z \in \mathbf{C} \setminus E.$$

We are abusing our notation slightly here by letting the dz in (2.1) denote 2-dimensional Lebesgue measure while the dw's in (2.1) and (2.2) represent 1-dimensional Hausdorff measure. We shall employ the convention of using a double-integral sign when we are integrating against Lebesgue measure on $\mathbf{C}$ and a single-integral sign when we are integrating against H^1 on a one-dimensional set.

Square function estimates of this type have a long history in analysis, particularly when E is a line (or a circle). Unfortunately, many of the special properties of these estimates do not persist when you consider more general classes of sets, as we are doing. Still, there are a couple of interesting reformulations of this condition, which we shall now review.

Given $z \in \mathbf{C} \setminus E$, define $h_z : E \to \mathbf{C}$ by

$$h_z(w) = \frac{\operatorname{dist}(z, E)^{3/2}}{(z - w)^2}.$$

This lies in $L^2(E)$, and its norm is bounded above and below by positive constants that do not depend on z, because of the regularity of E.

LEMMA 2.3. *E satisfies the USFE if and only if there is a constant $C > 0$ so that if $g(z)$ is any continuous function on $\mathbf{C}$ with compact support contained in $\mathbf{C} \setminus E$, then*

(2.4)
$$\int_E \left| \iint_{\mathbf{C}} h_z(w) g(z) \operatorname{dist}(z, E)^{-2} dz \right|^2 dw \leq C \iint_{\mathbf{C}} |g(z)|^2 \operatorname{dist}(z, E)^{-2} dz.$$

This is easily proved using the fact that the operator that sends $f \in L^2(E)$ to
$$\phi(z) = \int_E h_z(w) f(w) \, dw, \qquad z \in \mathbf{C} \setminus E,$$
is a bounded linear mapping from $L^2(E)$ into $L^2(\mathbf{C}, \operatorname{dist}(z, E)^{-2} dz)$ if and only if E satisfies the USFE, while (2.4) simply means that the adjoint of this operator is bounded.

The point of (2.4) is that it is basically a statement about the extent to which the functions h_z on E are approximately orthogonal, at least for z's which are not too close together. When E is a line this sort of thing is easily calculated explicitly.

There is a discrete version of this orthogonality condition which perhaps makes the ideas clearer. Let us say that a sequence $\{z_j\}$ in $\mathbf{C} \setminus E$ is separated if
$$|z_j - z_k| \geq \tfrac{1}{10} \max(\operatorname{dist}(z_j, E), \operatorname{dist}(z_k, E))$$
whenever $j \neq k$.

LEMMA 2.5. *E satisfies the USFE if and only if there is a constant $C > 0$ so that if $\{z_j\}$ is any separated sequence in $\mathbf{C} \setminus E$ and $\{a_j\}$ is any sequence of complex numbers, with all but finitely many of the a_j's equal to zero, then*

$$\left\| \sum a_j h_{z_j} \right\|_{L^2(E)} \leq C \left(\sum |a_j|^2 \right)^{1/2}.$$

This is a simple variant of Lemma 2.3. The "if" part can be derived from Lemma 2.3, while the "only if" part can be proved using a duality argument as before. To be honest, for this last part we are using also the fact that (2.1) automatically implies a slightly stronger version of itself, because

(2.6)
$$\iint_{\mathbf{C}} \left(\sup_{\zeta \in B(z, \operatorname{dist}(z, E)/2)} |F'(\zeta)| \right)^2 \operatorname{dist}(z, E) \, dz \leq C \iint_{\mathbf{C}} |F'(z)|^2 \operatorname{dist}(z, E) \, dz$$

is true for any holomorphic function F on $\mathbf{C} \setminus E$.

Notice, incidentally, that the condition in Lemma 2.5 is also equivalent to the following: there is a $C > 0$ so that if $\{z_j\}$ is any separated sequence in $\mathbf{C} \setminus E$, then $\{h_{z_j}\}$ is the image of an orthonormal sequence under some bounded linear mapping $T: H \to L^2(E)$ with norm $\leq C$, where H is some Hilbert space, which may as well be taken to be l^2.

The next lemma provides another formulation of the square function estimates for the Cauchy kernel.

LEMMA 2.7. *E satisfies the USFE if and only if there is a $C > 0$ so that*

(2.8) $$\int_E |G'(w)|^2 \, dw \leq C \iint_{\mathbf{C}} |\overline{\partial} G(z)|^2 \operatorname{dist}(z, E)^{-1} \, dz$$

whenever $G : \mathbf{C} \to \mathbf{C}$ is a C^1 function which tends to 0 at infinity and for which $\operatorname{supp} \overline{\partial} G$ is a compact subset of $\mathbf{C} \setminus E$.

Notice that these requirements on G ensure that it is holomorphic on a neighborhood of E so that G' is well defined on E.

The lemma is not hard to verify. For each $w \in E$, $G'(w)$ is given by

$$\iint_{\mathbf{C}} \frac{1}{(w-z)^2} \overline{\partial} G(z) \, dz$$

except for an inessential multiplicative constant. Using this it is easy to reduce to Lemma 2.3.

2.2. Uniform rectifiability implies the USFE when $d = 1$.

The fact that

(2.9) E satisfies the USFE if E is uniformly rectifiable

is basically known, but unfortunately it does not seem to be written down anywhere. In this section we provide an outline of a proof.

One way to deal with (2.9) would be to use the techniques based on good-λ inequalities in [**D1**]. (See also Part III of [**D4**], especially §§1–4.) This is probably not optimally pleasant from the reader's perspective. A simpler approach would be to use the methods of [**J1**] to reduce to the case of Lipschitz graphs. For this you must first reduce to the case of connected 1-dimensional regular sets in $\mathbf{C}$, which is not hard, because any uniformly rectifiable set E in $\mathbf{C}$ is contained in one. This approach is also not optimal for our purposes, because we are going to need a higher-dimensional version of (2.9) in the next chapter.

Instead we shall use a method — which is really a variant of the method in [**J1**] — based on the fact that a uniformly rectifiable set admits a corona decomposition. The argument is very similar to those given in §4 of [**S4**] and §15 of [**DS2**], and we shall not give all the details.

Although we shall confine ourselves to the $d = 1$ case here, the same argument works when $d > 1$ as well, with only minor changes. We shall assert this truth more fervently in §2 of the next chapter.

So let E be a 1-dimensional regular set in $\mathbf{C}$ which admits a corona decomposition. We want to get the square function estimates for E from the corresponding estimates for Lipschitz graphs. For the record let us state the case of Lipschitz graphs as a lemma.

LEMMA 2.10. *Lipschitz graphs in* $\mathbf{C}$ *satisfy the USFE, with estimates which depend only on the Lipschitz constants.*

By now there are many proofs of this. A simple one is given in [S2]. [Actually, for the argument that we are going to give, we really only need this for Lipschitz graphs with small constant, which is even easier.]

We should perhaps mention that the USFE condition has not appeared explicitly in the literature so often. For instance, the USFE is clearly "around" in [CMM], although it is never stated in this way. At any rate, Lemma 2.10 can certainly be derived from the results of [CMM]. (See also [K].)

It turns out that there is a minor technical difficulty with deriving the usual square function estimates on E from Lemma 2.10 and the existence of a corona decomposition for E. In order for the standard real-variable methods to work, we cannot deal with L^2-estimates only, we have to work with other kinds of estimates as well. We will explain how this difficulty arises later. For now we want to give a criterion for the square function estimates to hold that circumvents this problem.

Recall that a (positive) measure μ on $\mathbf{C} \setminus E$ is said to be a Carleson measure if there is a $K > 0$ so that

$$\mu(B(\zeta, R)) \leq KR$$

for all $\zeta \in E$ and $R > 0$. Also, let us adopt the notation that if f is a function on E, then $T(f)$ is the function on $\mathbf{C} \setminus E$ given by

$$Tf(z) = \int_E \frac{1}{(z-w)^2} f(w) \, dw.$$

LEMMA 2.11. *A necessary and sufficient condition for the usual square function estimates to be satisfied on* E *is that*

(2.12) $$|T(1)(z)|^2 \operatorname{dist}(z, E) dz$$

be a Carleson measure on $\mathbf{C} \setminus E$ *(where* dz *denotes Lebesgue measure).*

This is a well-known variant of the $T(1)$ theorem. The necessity part is very easy and will also be needed in the next section. [To estimate the integral of (2.12) over a ball $B = B(w, R)$, you use the L^2 estimate for the near part $T(\chi_{2B})$, and you control the far-away part $T(1 - \chi_{2B})$ with crude estimates on the kernel.] Let us sketch a proof of the sufficiency part for the sake of completeness.

Suppose that (2.12) is a Carleson measure. Then so is

(2.13) $$|T_*(1)(z)|^2 \operatorname{dist}(z, E) dz,$$

where $T_*(f)(z) = \sup\{|T(f)(w)| : |w - z| < \frac{1}{2} \operatorname{dist}(z, E)\}$. This follows easily from the observation that $|T(f)|$ is subharmonic on $\mathbf{C} \setminus E$.

It will be convenient for us to deal with a discretized version of the usual square function estimates. Let Δ be a family of cubes on E, as in §3.1 of

2.2. UNIFORM RECTIFIABILITY IMPLIES THE USFE

Part I. For each $Q \in \Delta$ and $a > 0$ define $\widehat{Q}(a) \subseteq \mathbf{C} \setminus E$ by

$$\widehat{Q}(a) = \{z \in \mathbf{C} : \operatorname{dist}(z, Q) \leq \operatorname{diam} Q, \ \operatorname{dist}(z, E) \geq a \operatorname{diam} Q\}.$$

Choose, once and for all, an $a_0 > 0$ so that $\bigcup_{Q \in \Delta} \widehat{Q}(a_0) = \mathbf{C} \setminus E$, and set $\widehat{Q} = \widehat{Q}(a_0)$.

As in the preceding section, define $h_z : E \to \mathbf{C}$ by

$$h_z(w) = \operatorname{dist}(z, E)^{3/2}(z-w)^{-2}.$$

LEMMA 2.14. *In order to show that E satisfies the USFE, it is enough to show that there is a $C_0 > 0$ so that if $\{z_Q : Q \in \Delta\}$ is any family of points with $z_Q \in \widehat{Q}$ for all Q, then*

$$(2.15) \qquad \sum_{Q \in \Delta} \left| \int_E h_{z_Q} f \right|^2 \leq C \int_E |f|^2$$

for all $f \in L^2(E)$.

This is very easy to derive, just from the definitions. [Notice that

$$\int_E h_{z_Q} f = \operatorname{dist}(z_Q, E)^{3/2} F'(z_Q),$$

where F is the Cauchy integral of f.] The converse is also true and not difficult to prove, but we will not need it. Note that Lemma 2.14 is very similar to Lemma 2.5.

To prove the sufficiency part of Lemma 2.11 it is enough to show that the criterion of Lemma 2.14 is satisfied if (2.13) is a Carleson measure. Let a family $\{z_Q : Q \in \Delta\}$ be given, as above, and define h_Q on E by

$$h_Q(w) = \frac{|Q|^{3/2}}{(z_Q - w)^2}.$$

Let us prove the analogue of (2.15) for h_Q.

Set $\eta_Q = \left(\int_E h_Q(w) \, dw \right)$. The fact that (2.13) is a Carleson measure implies that there is a $C > 0$ so that

$$\sum_{Q \subseteq Q_0} |\eta_Q|^2 = \sum_{Q \subseteq Q_0} |T(1)(z_Q)|^2 |Q|^3$$

$$\leq C \sum_{Q \subseteq Q_0} \int_{\widehat{Q}(a)} |T_*(1)(u)|^2 \operatorname{dist}(u, E) \, du \quad \text{(for some } a > 0\text{)}$$

$$\leq C |Q_0|$$

for all $Q_0 \in \Delta$. Define $g_Q \in L^2(E)$ by

$$g_Q = h_Q - \eta_Q |Q|^{-1} \chi_Q.$$

It is enough to show that

$$\sum_{Q \in \Delta} \left| \int_E g_Q f \right|^2 \leq C \int_E |f|^2$$

and

$$\sum_{Q \in \Delta} |\eta_Q|^2 \left(|Q|^{-1} \int_Q |f| \right)^2 \leq C \int_E |f|^2$$

for all $f \in L^2(E)$.

The second inequality follows from (a discrete version of) Carleson's inequality and our Carleson measure estimate on $|\eta_Q|^2$. Using duality the first inequality can be reduced to showing that

$$\left| \sum \alpha_Q \int_E g_Q f \right| \leq C \left(\sum_{Q \in \Delta} |\alpha_Q|^2 \right)^{1/2} \|f\|_2$$

for all families $\{\alpha_Q : Q \in \Delta\}$ of complex numbers and all $f \in L^2(E)$. This is equivalent to proving that

$$\int_E \left| \sum_{Q \in \Delta} \alpha_Q g_Q \right|^2 \leq C \sum_{Q \in \Delta} |\alpha_Q|^2$$

for all families $\{\alpha_Q\}$. This is in turn equivalent to saying that the inner product matrix $\langle g_Q, g_{Q'} \rangle$ defines a bounded operator on $l^2(\Delta)$, i.e.,

$$\sum_{Q \in \Delta} \left| \sum_{Q' \in \Delta} \alpha_{Q'} \langle g_Q, g_{Q'} \rangle \right|^2 \leq C \sum_{Q \in \Delta} |\alpha_Q|^2,$$

where $\langle \, , \, \rangle$ denotes the standard inner product on $L^2(E)$. The proof of this last estimate is based on well-known techniques, and we shall not give it here. The main point is to estimate $|\langle g_Q, g_{Q'} \rangle|$ using the facts that the g_Q's are fairly well localized, have integral zero, and are reasonably smooth. This "smoothness" requires a bit of interpretation; χ_Q has some smoothness, in effect, because our cubes have small boundary, in the sense of (3.4) in §I.3.1.

This completes our sketch of the proof of Lemma 2.11.

Before proceeding with the main part of the proof of (2.9), let us record a technical fact.

LEMMA 2.16. *Let A be a 1-dimensional regular set in $\mathbf{C}$. Set*

$$\Omega = \{z \in \mathbf{C} : \operatorname{dist}(z, E) > 10 \operatorname{dist}(z, A)\},$$

and let μ be the measure obtained by multiplying 2-dimensional Lebesgue measure by $\operatorname{dist}(z, E)^{-1} \chi_\Omega(z)$. Then μ is a Carleson measure, with a norm estimate that depends only on the regularity constant of A.

To prove this, we use Fubini's theorem. Given $z \in \Omega$ set

$$B(z) = B\left(z, \frac{1}{2} \operatorname{dist}(z, E)\right).$$

2.2. UNIFORM RECTIFIABILITY IMPLIES THE USFE

By definition of Ω we have that
$$|B(z) \cap A| \geq C^{-1} \operatorname{dist}(z, E).$$

Let $w \in E$ and $R > 0$ be given. Then
$$\mu(B(w, R)) = \iint_{B(w,R) \cap \Omega} \operatorname{dist}(z, E)^{-1} dz$$
$$\leq C \iint_{B(w,R) \cap \Omega} \left(\int_{B(z) \cap A} da \right) \operatorname{dist}(z, E)^{-2} dz,$$

where da denotes one-dimensional Hausdorff measure restricted to A. Because $w \in E$, we have that
$$\operatorname{dist}(z, E) \leq |z - w| \leq R$$

when $z \in B(w, R)$, so Fubini's theorem yields
$$\mu(B(w, R)) \leq C \int_{B(w,2R) \cap A} \left(\iint_{\{z \in \Omega : a \in B(z)\}} \operatorname{dist}(z, E)^{-2} dz \right) da.$$

Fix a for the moment. If $z \in \Omega$ and $a \in B(z)$, then
$$\tfrac{1}{2} \operatorname{dist}(z, E) \leq \operatorname{dist}(a, E) \leq 2 \operatorname{dist}(z, E)$$

and $z \in B(a, \operatorname{dist}(a, E))$. This implies that
$$\iint_{\{z \in \Omega : a \in B(z)\}} \operatorname{dist}(z, E)^{-2} dz \leq C,$$

and therefore $\mu(B(w, R)) \leq CR$. This proves Lemma 2.16.

Let us now bring in the corona decomposition. According to Definition I.3.19, there are two parameters, θ and η, which we get to choose. We may as well take $\eta = 1$, but it is convenient to require that θ be small, how small depending on simple geometric considerations that arise in the arguments below (mostly for Lemma 2.21). Once these two parameters are chosen, the assumption that E admits a corona decomposition implies that there is a coronization $(\mathscr{B}, \mathscr{G}, \mathscr{F})$ of E (see Definition I.3.13) and a family of 1-dimensional Lipschitz graphs $\{\Gamma(S) : S \in \mathscr{F}\}$, each having constant $\leq \eta = 1$, such that
$$\operatorname{dist}(x, \Gamma(S)) \leq \theta \operatorname{diam} Q \quad \text{whenever} \quad Q \in S \text{ and } x \in 2Q.$$

We want to use Lemma 2.10 and these approximating Lipschitz graphs to show that (2.12) is a Carleson measure. We first want to dispense with some garbage terms.

Given a subset $\mathscr{A}$ of Δ (like $\mathscr{B}$, $\mathscr{G}$, or $S \in \mathscr{F}$), set $\widehat{\mathscr{A}} = \bigcup_{Q \in \mathscr{A}} \widehat{Q}$. Thus $\widehat{\Delta} = \mathbf{C} \setminus E$.

LEMMA 2.17. $\chi_{\widehat{\mathscr{B}}}(z)|T(1)(z)|^2 \operatorname{dist}(z, E) dz$ is a Carleson measure on $\mathbf{C} \setminus E$.

This is easy to check, using the trivial estimate

(2.18) $\qquad |T(1)(z)| \leq C \operatorname{dist}(z, E)^{-1}$

and the fact that $\mathscr{B}$ satisfies a Carleson packing condition.

For each $S \in \mathscr{F}$ set

$$\Omega(S) = \{z \in \mathbf{C} : \operatorname{dist}(z, E) > 10 \operatorname{dist}(z, \Gamma(S))\}.$$

These regions would be annoying soon if we did not get rid of them now.

LEMMA 2.19. Set $U = \bigcup_{S \in \mathscr{F}} (\Omega(S) \cap \widehat{S})$. Then

$$\chi_U(z)|T(1)(z)|^2 \operatorname{dist}(z, E) dz$$

is a Carleson measure on $\mathbf{C} \setminus E$.

For each $S \in \mathscr{F}$ the measure

$$\chi_{\Omega(S) \cap \widehat{S}}(z)|T(1)(z)|^2 \operatorname{dist}(z, E) dz$$

is a Carleson measure with bounded norm, because of Lemma 2.16 and (2.18). Hence Lemma 2.19 is a consequence of the next observation.

LEMMA 2.20. Let $\{\mu_S : S \in \mathscr{F}\}$ be a family of positive measures on $\mathbf{C} \setminus E$ such that $\mu_S = 0$ on the complement of $\widehat{S}$ and each μ_S is a Carleson measure, with uniformly bounded norm. Then $\sum_{S \in \mathscr{F}} \mu_S$ is also a Carleson measure on $\mathbf{C} \setminus E$.

The proof of this is rather straightforward, using the properties of a coronization, and we omit the details. (Similar issues arose in the proof of Proposition I.3.32.)

The main step in this proof of (2.9) is the following.

LEMMA 2.21. Let $S \in \mathscr{F}$ be given, and let $Q(S)$ be its maximal element. There is a $C > 0$ so that if $g \in L^2(E)$ and $g = 0$ on $E \setminus (\tfrac{3}{2} Q(S))$, then

(2.22) $\qquad \displaystyle\int\!\!\int_{\widehat{S} \setminus \Omega(S)} |T(g)(z)|^2 \operatorname{dist}(z, E) dz \leq C \int_E |g|^2.$

Before explaining why this is true let us explain why it is what we want. Using Lemma 2.21 it is not hard to check that

$$\chi_{\widehat{S} \setminus \Omega(S)}(z)|T(1)(z)|^2 \operatorname{dist}(z, E) dz$$

is a Carleson measure for each $S \in \mathscr{F}$, with a bounded norm. [To estimate the integral of this measure over a ball B centered on E, write

$$1 = \chi_{2B \cap (3Q(S)/2)} + \{1 - \chi_{2B \cap (3Q(S)/2)}\},$$

use the lemma to control the first piece, and estimate the contribution of the second brutally, using the definition of T.] From Lemma 2.20 we get that

$$\chi_{\bigcup_{S \in \mathcal{F}(\widehat{S} \setminus \Omega(S))}}(z) |T(1)(z)|^2 \operatorname{dist}(z, E) dz$$

is a Carleson measure on $\mathbf{C} \setminus E$. This implies that (2.12) is a Carleson measure, since we have already dealt with the other pieces.

It might appear as though we should be able to use Lemma 2.21 to prove that E satisfies the usual square function estimates directly, without using Lemma 2.11. There is a problem with this, though; the direct argument gives (2.1) but with the right side replaced by the integral of the Hardy-Littlewood maximal function of $|f|^2$ on E, which is no good. This is not a serious problem, in the sense that Calderón-Zygmund theory provides many ways to fix it, and we have simply chosen one.

Lemma 2.21 can be proved using arguments which are very similar to those given on pp. 1028–1031 in §4 of [S4]. Roughly speaking you build a model of g on $\Gamma(S)$, and then you control the left side of (2.22) in terms of the corresponding quantity for the model of g on $\Gamma(S)$, plus some errors which can be estimated directly. There are some modifications in the argument in [S4] which are needed, due to the differences between square functions and singular integral operators, but these are rather straightforward, so we omit the details.

There is a slightly different approach to this that one could take, which exploits the complex-variable nature of the problem. The region $\widehat{S} \setminus \Omega(S)$ really has two pieces, on opposite sides of $\Gamma(S)$. Let us call these two pieces D_+ and D_-. It is not hard to build a Lipschitz graph $\Gamma_+(S)$ such that $\frac{3}{2}Q(S)$ and D_+ lie on opposite sides of $\Gamma_+(S)$. [To be honest, it might be a good idea first to throw away a little of D_+, as we did with $\Omega(S)$.] Let $U_+(S)$ denote the component of $\mathbf{C} \setminus \Gamma_+(S)$ that contains D_+. Then G is holomorphic on U_+, and you can apply the usual square function estimates on Lipschitz graphs directly. For this you need to be able to control the norm of $G|_{\Gamma_+(S)}$ in $L^2(\Gamma_+(S))$, and you can do that using the boundedness of the Cauchy integral operator on $\Gamma_+(S)$. More precisely, the Cauchy integral defines a bounded operator from $L^2(\Gamma_+(S))$ to $L^2(E)$, because $\Gamma_+(S)$ is a Lipschitz graph and E is regular (see [D1], [D4]), and we are using the boundedness of the Cauchy integral operator from $L^2(E)$ to $L^2(\Gamma_+(S))$, which follows from duality.

This second method is a little more elegant but less general, and it requires more machinery.

This concludes our sketch of the proof of (2.9).

2.3. From square function estimates to uniform rectifiability: Preliminary reductions and the plan of the proof.

Throughout this section E is a one-dimensional regular set in $\mathbf{C}$.

As in (I.2.42), we define $e(z)$ on $\mathbf{C} \setminus E$ by

$$e(z) = \operatorname{dist}(z, E) \left| \int_E \frac{1}{(z-\zeta)^2} \, d\zeta \right|.$$

From Lemma 2.11 in the preceding section we know that E satisfies the USFE if and only if we have the Carleson measure estimate

$$(2.23) \qquad \iint_{B(w,R)} e(z)^2 \operatorname{dist}(z, E)^{-1} dz \leq CR$$

for some $C > 0$ and all $w \in E$ and $R > 0$. It is more convenient to work with this Carleson measure estimate instead of (2.1).

Consider the following seemingly weaker condition, which we christened the WUSFE in Definition I.2.43:

$$(2.24) \qquad \{z \in \mathbf{C} \setminus E : e(z) > \epsilon\} \text{ is a Carleson set for all } \epsilon > 0.$$

[A set $A \subseteq \mathbf{C} \setminus E$ is said to be a Carleson set if $\operatorname{dist}(z, E)^{-1} \chi_A(z) dz$ is a Carleson measure on $\mathbf{C} \setminus E$, where "dz" denotes 2-dimensional Lebesgue measure.] From our remarks about (2.23) we see that (2.24) holds if E satisfies the USFE. The "if" part of Theorem I.2.41 is thus reduced to the following theorem, which is simply a restatement of Theorem I.2.45.

THEOREM 2.25. *If E satisfies (2.24), then E satisfies the Weak Exterior Convexity condition (WEC), and hence is uniformly rectifiable.*

A consequence of Theorem 2.25 (and (2.9)) is that (2.24) implies (2.23). This is quite amusing, because there is no general analytic result which implies this, as there was for the equivalence of (2.23) with the usual square function estimates. You have to use the geometry.

Before we explain how Theorem 2.25 is to be proved we want to reformulate (2.24) in more convenient ways.

Given $k > 0$ and $z \in \mathbf{C} \setminus E$ set

$$W_k(z) = \{p \in \mathbf{C} \setminus E : |p - z| \leq k \operatorname{dist}(z, E) \text{ and}$$
$$\operatorname{dist}(p, E) \geq k^{-1} \operatorname{dist}(z, E)\}.$$

Here "W" stands for Whitney; $W_k(z)$ is, roughly, the union of the Whitney cubes of $\mathbf{C} \setminus E$ that are not too far from z and not too close to E.

LEMMA 2.26. *If E satisfies (2.24), then for every $\delta > 0$ and every $k > 0$ we have that*

$$(2.27) \qquad \left\{ z \in \mathbf{C} \setminus E : \sup_{\zeta \in W_k(z)} e(\zeta) > \delta \right\}$$

is a Carleson set in $\mathbf{C} \setminus E$.

We leave the proof as an exercise, except for saying that the main point is

$$|e(w) - e(w')| \leq C|w - w'|\operatorname{dist}(w, E)^{-1}$$

whenever $w, w' \in \mathbf{C}\setminus E$ satisfy $|w-w'| \leq \frac{1}{2}\operatorname{dist}(w, E)$. [Thus if $e(w) > \delta$, then $e(w') > \delta/2$ if w' is close to w.]

Because we defined the WEC in terms of $E \times \mathbf{R}_+$ rather than $\mathbf{C}\setminus E$, it is helpful to adjust (2.27) accordingly.

LEMMA 2.28. *Suppose that E satisfies (2.24). Then there is a $C_0 > 0$, which depends only on the regularity constant for E, so that for every $\delta > 0$ and every $k > 0$ the set*

(2.29)
$$\{(p, t) \in E \times \mathbf{R}_+ : \sup_{\zeta \in W_k(z)} e(\zeta) > \delta \text{ for all } z \in \mathbf{C}\setminus E \text{ that satisfy}$$
$$|p - z| \leq C_0 t \text{ and } \operatorname{dist}(z, E) \geq t\}$$

is a Carleson set in $E \times \mathbf{R}_+$.

Again we leave this as an exercise, modulo a hint or two. We can choose $C_0 > 0$ so that for each $(p, t) \in E \times \mathbf{R}_+$ there is a $w \in \mathbf{C}\setminus E$ that satisfies $|p - w| \leq C_0 t/2$ and $\operatorname{dist}(w, E) \geq 2t$. Thus to each (p, t) in (2.29) you can associate a ball of z's in $\mathbf{C}\setminus E$ which lie in (2.27). This allows you to reduce the Carleson estimate for (2.29) to that of (2.27).

To prove Theorem 2.25, we want to show that if $(p, t) \in E \times \mathbf{R}_+$ does not lie in (2.29), then it lies in the good set for the WEC, for a suitable choice of parameters. To do this, we need to understand what happens when $e(z)$ is small a lot of the time. We shall use a compactness argument to reduce the problem to understanding what happens when $e(z) = 0$ on all of $\mathbf{C}\setminus E$.

DEFINITION 2.30. We denote by CF ("Cauchy flat") the collection of closed subsets A of $\mathbf{C}$ such that A is a 1-dimensional regular set and such that there is a measure α on $\mathbf{C}$ which satisfies the following three conditions:

(2.31) $\quad \operatorname{supp} \alpha = A$;

(2.32) $\quad \alpha$ is (Ahlfors) regular, i.e., there exists $C > 0$ so that
$$C^{-1}R \leq \alpha(B(a, R)) \leq CR \text{ for all } a \in A \text{ and } R > 0;$$

(2.33) $\quad \int_A \frac{1}{(z-\zeta)^2} d\alpha(\zeta) = 0$ for all $z \in \mathbf{C}\setminus A$.

Examples of Cauchy flat subsets of $\mathbf{C}$ include any finite union of lines, as well as any union of 3 half-lines that emanate from the same point and whose union is not contained in any closed half-space.

Here are a couple of amusing reformulations of (2.33):

(2.34) $\quad \int_A \frac{1}{z-\zeta} d\alpha(\zeta)$ is constant on each connected component of $\mathbf{C}\setminus A$;

(2.35) $\quad \int_A \log|z - \zeta| d\alpha(\zeta)$ is affine on each component of $\mathbf{C}\setminus A$.

These integrals require some interpretation. We view

$$\int_A \frac{1}{z-\zeta} d\alpha(\zeta)$$

as a holomorphic function on $\mathbf{C} \setminus A$ which is only defined up to an additive constant. For instance, if you fix $z_0 \in \mathbf{C} \setminus A$, then

$$\int_A \left(\frac{1}{z-\zeta} - \frac{1}{z_0-\zeta} \right) d\alpha(\zeta)$$

is well defined. Similarly $\int_A \log|z-\zeta| d\alpha(\zeta)$ is viewed as a harmonic function on $\mathbf{C} \setminus A$ which is only defined modulo affine functions. Actually, this one extends to be continuous and subharmonic on all of $\mathbf{C}$.

LEMMA 2.36. *For every $\epsilon > 0$ there exists $\delta > 0$ (small) and $k > 0$ (large) so that if $z \in \mathbf{C} \setminus E$ and if $\sup\{e(\zeta) : \zeta \in W_k(z)\} \leq \delta$, then there is an $A \in CF$ such that*
(2.37)
$$\sup\{\mathrm{dist}(p, A) : p \in E \cap \overline{B}(z, 10\,\mathrm{dist}(z, E))\}$$
$$+ \sup\{\mathrm{dist}(q, E) : q \in A \cap \overline{B}(z, 10\,\mathrm{dist}(z, E))\} \leq \epsilon\,\mathrm{dist}(z, E).$$

Notice that (2.37) is almost a statement about Hausdorff distances, but you have to be a little careful; you could have a piece of E on the boundary of the ball which is approximated by a piece of A which lies just outside the ball. We could have used Hausdorff distances here, but that would have added an extra complication with no real benefit.

In Lemma 2.36 we are permitting k and δ to depend on E. This is fine for the purposes of proving Theorem 2.25, but it is not aesthetically pleasing, and it is not good enough for some more refined results. However, one can modify the proof of Lemma 2.36 without much difficulty to show that k and δ can be chosen so that they depend on E only to the extent that they depend on the regularity constant for E. (Basically you simply treat E as another variable in the compactness argument.)

We shall prove Lemma 2.36 in the next section.

The combination of Lemmas 2.28 and 2.36 implies that if E satisfies (2.24), then, for most $(p, t) \in E \times \mathbf{R}_+$, we can find an $A \in CF$ so that A approximates E well inside $B(p, t)$. In the language of §I.2.2 (see especially Definition I.2.21), we have that $E \in \mathrm{Approx}(CF)$ as soon as (2.24) holds. Thus we need to understand CF better.

PROPOSITION 2.38. *If $A \in CF$, then every component of $\mathbf{C} \setminus A$ is convex.*

The proof of this is given in §§5–7.

Let $\mathscr{E}_0$ denote the collection of closed sets F in $\mathbf{C}$ such that every component of $\mathbf{C} \setminus F$ is convex. Proposition 2.38 says that $CF \subseteq \mathscr{E}_0$, so (in the language of §I.2.2) $\mathrm{Approx}(CF) \subseteq \mathrm{Approx}(\mathscr{E}_0)$. On the other hand, it is easy to see that every element in $\mathrm{Approx}(\mathscr{E}_0)$ satisfies the WEC. In fact, we know from Lemma I.2.23 that $\mathrm{Approx}(\mathscr{E}_0)$ is precisely the collection of 1-dimensional regular sets in $\mathbf{C}$ which satisfy the WEC. Thus Theorem 2.25 will follow once we have proved Lemma 2.36 and Proposition 2.38, because of Theorem I.2.18.

We should perhaps mention that part of the difficulty with this story was to realize that Proposition 2.38 was what we wanted. Cauchy flatness is a strong condition, but much of the information that comes from it is hard with which to work. The exterior convexity from Proposition 2.38 is still not so easy to use, but at least it is pretty clean.

2.4. The proof of Lemma 2.36.

We are going to use a compactness argument. This has the attractive features of being simple and brief, but of course it has the usual flaw of being nonconstructive to the point of almost being mystical. In particular, we have no idea how δ and k depend on ϵ.

Given a positive number C_0 let $\text{RM}(C_0)$ denote the space of 1-dimensional regular measures in $\mathbf{C}$ with constant at most C_0. That is, μ lies in $\text{RM}(C_0)$ if μ is a nonnegative locally finite Borel measure which is not identically zero and which satisfies

$$(2.39) \qquad C_0^{-1} R \leq \mu(B(p, R)) \leq C_0 R$$

for all $p \in \text{supp}\,\mu$ and $R > 0$.

Notice that our use of the term "regular" here, in the sense of Ahlfors regularity, is unfortunately somewhat at odds with the standard terminology from measure theory. We are using it to refer to the condition (2.39), as in Definition I.1.13. All of our measures are regular in the other, traditional, sense, because they are locally finite Borel measures on $\mathbf{R}^n$.

We should also point out that we are restricting our attention here to the case where $d = 1$ and $n = 2$ because that is all we need at the moment. It is easy to extend much of what we are doing now to any choices of d and n, and we shall in fact use these extensions in Chapters 3 and 5.

Let σ denote the restriction of 1-dimensional Hausdorff measure to E. Thus σ lies in $RM(C_0)$ if C_0 is large enough. Choose a C_0 with this property, and let it be fixed for the rest of this section.

We say that a sequence $\{\mu_j\}$ in $RM(C_0)$ converges weakly to a locally finite Borel measure μ on $\mathbf{C}$ if

$$\lim_{j \to \infty} \int_\mathbf{C} \phi \, d\mu_j = \int_\mathbf{C} \phi \, d\mu$$

for all continuous functions ϕ on $\mathbf{C}$ that have compact support.

Our compactness argument is going to be based on this notion of weak convergence. Let us record some simple facts about it.

LEMMA 2.40. *Given any sequence in* $\text{RM}(C_0)$, *there is a subsequence that converges weakly to a Borel measure on* $\mathbf{C}$.

This follows from standard results in functional analysis.

LEMMA 2.41. *Suppose that* $\{\mu_j\} \subseteq \text{RM}(C_0)$ *and that* $\mu_j \to \mu$ *weakly. Then for every ball B in $\mathbf{C}$ we have that*

$$\lim_{j \to \infty} \left(\sup_{p \in B \cap \mathrm{supp}\, \mu} \mathrm{dist}(p, \mathrm{supp}\, \mu_j) \right) = 0.$$

This is fairly easy to check, just using the definitions. The reader may find it helpful to consider first the case where the ball B is replaced by a finite set.

LEMMA 2.42. *Suppose that $\{\mu_j\} \subseteq \mathrm{RM}(C_0)$ and that $\mu_j \to \mu$ weakly. Then either $\mu \in \mathrm{RM}(C_0)$ or $\mu \equiv 0$.*

It is not difficult to derive this from the definitions and the previous lemma. This is even easier if you are willing to settle for $\mu \in RM(2C_0)$, which would be fine for our applications.

LEMMA 2.43. *Suppose that $\{\mu_j\} \subseteq \mathrm{RM}(C_0)$ and that $\mu_j \to \mu$ weakly. Then for every ball B in $\mathbf{C}$ we have*

$$\lim_{j \to \infty} \left(\sup_{B \cap \mathrm{supp}\, \mu_j} \mathrm{dist}(p, \mathrm{supp}\, \mu) \right) = 0.$$

For the purposes of this lemma we adopt that the convention that the supremum above is zero if $B \cap \mathrm{supp}\, \mu_j = \emptyset$.

Let B be given, and let $\epsilon > 0$ be given also. Let $B_1, \ldots, B_l$ be a finite collection of balls with radius ϵ that cover B. Let $\phi_1, \ldots, \phi_l$ be nonnegative continuous functions on $\mathbf{C}$ such that $\mathrm{supp}\, \phi_i \subseteq 3B_i$ and $\phi_i \equiv 1$ on $2B_i$ for $i = 1, \ldots, l$. Choose N so large that

$$\left| \int_\mathbf{C} \phi_i \, d\mu_j - \int_\mathbf{C} \phi_i \, d\mu \right| \leq (2C_0)^{-1} \epsilon$$

for all $j \geq N$ and $i = 1, \ldots, l$. For each such i and j we have that if B_i intersects $\mathrm{supp}\, \mu_j$, then

$$\int_\mathbf{C} \phi_i \, d\mu_j \geq C_0^{-1} \epsilon,$$

and hence

$$\int_\mathbf{C} \phi_i \, d\mu \geq (2C_0)^{-1} \epsilon,$$

which of course implies that $\mathrm{supp}\, \mu$ intersects $3B_i$. From here Lemma 2.43 follows easily.

Let us now prove Lemma 2.36. Let $\epsilon > 0$ be given, and assume that the conclusion of Lemma 2.36 is false. Thus for each $k = 1, 2, \ldots$, there is a point $z_k \in \mathbf{C} \setminus E$ such that

(2.44) $$\sup\{e(\zeta) : \zeta \in W_k(z_k)\} \leq k^{-1}$$

but for which there does not exist a suitable $A \in CF$. We want to obtain a contradiction.

[If instead of working with a fixed E we also allowed E to depend on k here, but keeping the regularity constants bounded, then the argument below would show that Lemma 2.36 still holds if we add the requirement that δ

and k depend on E only to the extent that they depend on its regularity constant.]

Set $\rho_k = \text{dist}(z_k, E)$. Let $a_k : \mathbf{C} \to \mathbf{C}$ denote the affine map given by
$$a_k(z) = \rho_k^{-1}(z - z_k).$$
Let σ_k be the measure on $\mathbf{C}$ defined by
$$\sigma_k(F) = \rho_k^{-1}\sigma(a_k^{-1}(F))$$
for all Borel sets F, where σ is, as before, the restriction of 1-dimensional Hausdorff measure to E. Since we chose C_0 so that $\sigma \in \text{RM}(C_0)$, we also have $\sigma_k \in \text{RM}(C_0)$ for all k. Set $E_k = a_k(E)$, so that $E_k = \text{supp}\,\sigma_k$. Notice that $\text{dist}(0, E_k) = 1$ for all k.

Let $\mu_j = \sigma_{k_j}$ be a subsequence of $\{\sigma_k\}$ that converges weakly to a measure μ. Clearly $\mu \not\equiv 0$ (since $\mu_k(B(0, 2)) \geq C_0^{-1}$ for all k), so $\mu \in \text{RM}(C_0)$. Set $M = \text{supp}\,\mu$. Then

(2.45) $$\limsup_{j \to \infty}\{\text{dist}(p, M) : p \in E_{k_j} \cap \overline{B}(0, 10)\} = 0$$

and

(2.46) $$\limsup_{j \to \infty}\{\text{dist}(p, E_{k_j}) : p \in M \cap \overline{B}(0, 10)\} = 0$$

by Lemmas 2.41 and 2.43.

Next we want to prove that

(2.47) $$\int_M \frac{1}{(z - \zeta)^2}\,d\mu(\zeta) = 0 \quad \text{for all } z \in \mathbf{C} \setminus M.$$

Let us begin by rewriting (2.44) as

(2.48) $$\left| \int_{E_k} \frac{\text{dist}(z, E_k)}{(z - \zeta)^2}\,d\sigma_k(\zeta) \right| \leq k^{-1}$$

whenever $z \in \mathbf{C} \setminus E_k$ satisfies $|z| \leq k$ and $\text{dist}(z, E_k) \geq k^{-1}$. In order to derive (2.47) from (2.48) it is enough to show that

(2.49) $$\lim_{j \to \infty} \int_{E_{k_j}} \frac{1}{(z - \zeta)^2}\,d\sigma_{k_j}(\zeta) = \int_M \frac{1}{(z - \zeta)^2}\,d\mu(\zeta)$$

for all $z \in \mathbf{C} \setminus M$. To see this we begin by observing that if $z \in M \setminus \mathbf{C}$, then there is an $\eta > 0$ so that $\text{dist}(z, E_{k_j}) \geq \eta$ for all sufficiently large j. This follows from Lemma 2.43. Using this observation it is easy to derive (2.49) from the weak convergence of σ_{k_j} to μ and the fact that we have enough control at infinity to prevent anything strange from happening out there.

Thus $M \in CF$. From here it is easy to get the contradiction we wanted, because M approximates E_{k_j} well in $\overline{B}(0, 10)$ if j is large (by (2.45) and (2.46)), and so $a_k^{-1}(M)$ approximates E well on $\overline{B}(z_k, 10\,\text{dist}(z_k, E))$. We omit the details. This proves Lemma 2.36.

2.5. A topological lemma.

Set $\widehat{C} = \mathbf{C} \cup \{\infty\}$, which we identify with the Riemann sphere.

LEMMA 2.50. *If $A \in CF$, then $\widehat{A} = A \cup \{\infty\}$ is a connected subset of $\widehat{\mathbf{C}}$.*

This lemma will be important for us because it implies that A is rectifiable if $A \in CF$. We shall discuss this further in the next section. The rest of this section will be devoted to the proof of Lemma 2.50.

Let $A \in CF$ be given, and let α be the associated measure (as in Definition 2.30). Suppose that $\widehat{A}$ is not connected, and let us show how that leads to a contradiction. Because A is unbounded, the disconnectedness of $\widehat{A}$ means precisely that we can decompose A as $F \cup K$, where F and K are disjoint nonempty subsets of $\mathbf{C}$, with F closed and K compact. By standard reasoning we can find a bounded open set in $\mathbf{C}$ which contains K and whose closure is disjoint from F. It is convenient for us to state this as a lemma.

LEMMA 2.51. *If $\widehat{A}$ is not connected, then we can find a bounded open set Ω in $\mathbf{C}$ such that A intersects Ω and $\mathbf{C} \setminus \overline{\Omega}$ but not $\partial \Omega$.*

Using some basic topology we can improve this as follows.

LEMMA 2.52. *Lemma 2.51 remains true if we require also that Ω be connected and simply connected.*

It is trivial to get a connected Ω; just replace Ω by any of its connected components which intersect A. It is not much more difficult to get Ω to be simply connected, by "filling in the holes". Let us be more precise. For this we are going to use the fact that a connected bounded open subset of $\mathbf{C}$ is simply connected if and only if its complement in $\mathbf{C}$ is connected.

Let Ω be as in Lemma 2.51, but with the additional feature of being connected. Let X be the unbounded component of $\mathbf{C} \setminus \Omega$, and set $\widetilde{\Omega} = \mathbf{C} \setminus X$. We want to show that $\widetilde{\Omega}$ satisfies the properties described in Lemma 2.51 and also that $\widetilde{\Omega}$ is connected and simply connected.

Clearly $\widetilde{\Omega}$ is a bounded open subset of $\mathbf{C}$ that contains Ω. Thus $\widetilde{\Omega}$ intersects A, and we also have that A intersects the complement of the closure of $\widetilde{\Omega}$, because A is unbounded by definition. We also have that A is disjoint from $\partial \widetilde{\Omega}$, because $\partial \widetilde{\Omega} \subseteq \partial \Omega$. (This last is easy to verify from the definitions.) Thus $\widetilde{\Omega}$ satisfies the conditions in Lemma 2.51.

The connectedness of $\widetilde{\Omega}$ is not hard to verify. By definition $\widetilde{\Omega}$ is the union of Ω and the connected components of $\mathbf{C} \setminus \Omega$ other than X. Because each of these components touches $\partial \Omega$, and because Ω is connected, it follows that $\widetilde{\Omega}$ is connected.

Since $\widetilde{\Omega}$ is connected and $\mathbf{C} \setminus \widetilde{\Omega} = X$ is connected we get that $\widetilde{\Omega}$ is simply connected. This proves Lemma 2.52.

LEMMA 2.53. *Lemma 2.51 remains true if we add the requirement that Ω and $\partial \Omega$ be connected.*

2.5. A TOPOLOGICAL LEMMA

This follows from Lemma 2.52 by invoking a theorem from plane topology. Perhaps a more pleasant approach would be to use the fact that a connected and simply connected planar domain is homeomorphic to a disk, which permits us to approximate Ω from within by Jordan domains.

Let us come back now to the proof of Lemma 2.50. Let A and α be as before, with $\widehat{A}$ disconnected, so that we can find Ω as in Lemma 2.51, with Ω and $\partial\Omega$ also being connected. Let z_0 be any fixed point in $\partial\Omega$. Define $H(z)$ on $\mathbf{C}\setminus A$ by

$$H(z) = \int_A \left(\frac{1}{(z-\zeta)} - \frac{1}{(z_0-\zeta)} \right) d\alpha(\zeta).$$

This integral converges when $z \in \mathbf{C}\setminus A$ because α satisfies (2.32), and $H(z)$ is clearly holomorphic on $\mathbf{C}\setminus A$. From (2.33) we conclude that

$$H'(z) \equiv 0 \quad \text{on } \mathbf{C}\setminus A,$$

so H must be constant on every connected component of $\mathbf{C}\setminus A$. Thus

$$H \equiv 0 \quad \text{on a neighborhood of } \partial\Omega,$$

since $\partial\Omega$ is connected, disjoint from A, and contains z_0.

The reader should not be disturbed if this seems strange. It is, and that is the point.

Define $\widetilde{H}(z)$ on $\mathbf{C}\setminus(A\cap\Omega)$ by $\widetilde{H} = H$ on $\Omega\setminus A$ and $\widetilde{H} = 0$ on $\mathbf{C}\setminus\Omega$. Notice that $\widetilde{H}$ is continuous and holomorphic on $\mathbf{C}\setminus(A\cap\Omega)$.

Define $G(z)$ on $\mathbf{C}\setminus(A\cap\Omega)$ by

$$G(z) = \int_{A\cap\Omega} \frac{1}{(z-\zeta)} \, d\alpha(\zeta).$$

Let us prove that $\widetilde{H}$ equals G everywhere on $\mathbf{C}\setminus(\Omega\cap A)$, and also the negation of this statement.

When $z \in \Omega\setminus A$ we have that

$$\widetilde{H}(z) - G(z) = H(z) - G(z)$$
$$= \int_{A\setminus\Omega} \left(\frac{1}{(z-\zeta)} - \frac{1}{(z_0-\zeta)} \right) d\alpha(\zeta) - \int_{A\cap\Omega} \frac{1}{(z_0-\zeta)} \, d\alpha(\zeta),$$

by definitions. Thus $\widetilde{H}-G$ extends to be holomorphic across $A\cap\Omega$, so $\widetilde{H}-G$ extends to be holomorphic on $\mathbf{C}$. Since $\widetilde{H}(z) - G(z) \to 0$ as $z \to \infty$, we get that $\widetilde{H} - G \equiv 0$.

On the other hand

$$\lim_{z\to\infty} z(\widetilde{H}(z) - G(z)) = \lim_{z\to\infty} -zG(z) = -\int_{A\cap\Omega} d\alpha \neq 0.$$

This is the contradiction that we wanted, and it completes the proof of Lemma 2.50.

2.6. The main step in the proof of Proposition 2.38.

Before stating the main result of this section we need to cover some preliminaries.

LEMMA 2.54. *If $A \in CF$, then A has a tangent line at almost all points in A. That is to say, there is a set $A' \subseteq A$ with $\alpha(A \setminus A') = 0$ such that for each $a \in A'$ there is a line L_a in $\mathbf{C}$ that passes through a and satisfies*

$$(2.55) \qquad \lim_{r \to 0} \sup_{p \in A \cap B(a,r)} \left(\frac{\text{dist}(p, L_a)}{r} \right) = 0.$$

Here α is, as always, the measure associated to A as in Definition 2.30. In particular, α is equivalent in size to the restriction of one-dimensional Hausdorff measure to A.

To prove the lemma, we begin by observing that if $A \in CF$ then the set $(A \cap \overline{B}) \cup \partial B$ is connected for any ball B in $\mathbf{C}$, by Lemma 2.50. This connected set also has finite 1-dimensional Hausdorff measure, and it is well known that that implies the existence almost everywhere of tangent lines. See, for instance, Chapter 3 of [Fl1]. We should point out that the results in [Fl1] are stated in such a way that they only allow us to conclude that A has approximate tangents almost everywhere, which means (roughly speaking) that instead of saying that all points on A near a are very close to the tangent line, you only demand that this be true except for a bad set which is rather thin at a. However, it is easy to check that an approximate tangent to a regular set must be a true tangent. (If the set is regular, then the bad set cannot be as thin as it is supposed to be, unless it is empty.)

We should also perhaps remind the reader that our terminology is somewhat different from that of [Fl1], particularly with regard to our use of the term "regular".

This rectifiability result that we are using here is of course closely related to Theorem 1.8 in Part I.

Using the existence of tangents we can define a certain differential operator associated to A as follows. Given $a \in A'$, let $\tau(a)$ be a complex number such that $|\tau(a)| = 1$ and

$$(2.56) \qquad L_a = \{a + t\tau(a) : t \in \mathbf{R}\}.$$

This determines $\tau(a)$ up to multiplication by -1. For each complex-valued C^1 function g on $\mathbf{C}$ and every $a \in A'$ define $Dg(a)$ by

$$(2.57) \quad Dg(a) = \tau(a)^{-1} \left\{ (\text{Re}\,\tau(a)) \left(\frac{\partial}{\partial x} g \right)(a) + (\text{Im}\,\tau(a)) \left(\frac{\partial}{\partial y} g \right)(a) \right\}.$$

Notice that this is not changed if we replace $\tau(a)$ by $-\tau(a)$.

The quantity in braces in (2.57) is nothing but the derivative of g at a in the direction $\tau(a)$. Thus $Dg(a)$ is a slightly fancier version of this directional derivative that has the additional feature of depending on $\tau(a)$

only to the extent that it depends on the tangent line L_a and not on a choice of orientation of L_a.

For us one of the crucial properties of D is that

(2.58) $$Dg(a) = g'(a) \quad \text{if } g \text{ is holomorphic.}$$

This is, in fact, true as soon as $\overline{\partial} g(a) = 0$.

We should point out, once and for all, that Dg is a measurable function on A. Indeed, we can choose $\tau(\cdot)$ to be measurable. This is not hard to derive using standard facts from geometric measure theory, such as those in Chapter 3 of [Fl1], to the effect that A is contained in a countable union of C^1 curves, except for a set of H^1-measure zero.

The following is the principal result of this section.

PROPOSITION 2.59. *If $A \in CF$, then*

(2.60) $$\int_A (Dg) \, d\alpha = 0$$

for all C^1 functions $g : \mathbf{C} \to \mathbf{C}$ with compact support, where α is as in the definition of the class CF.

Let us make some remarks before proving this.

Proposition 2.59 has an easy converse. Suppose that A is a 1-dimensional regular set in $\mathbf{C}$ that is also rectifiable, and let α be a measure on A that satisfies the regularity condition (2.32). (These assumptions can be weakened.) If (2.60) holds for all C^1 functions g on $\mathbf{C}$ with compact support, then it holds for $g(\zeta) = (\zeta - z)^{-1}$ for all $z \in \mathbf{C} \setminus A$, i.e., (2.33) holds. This follows easily from (2.58) and a simple approximation argument.

There is an important geometrical interpretation of Proposition 2.59 which we shall explain now. Suppose that we are given a rectifiable set A in $\mathbf{C}$ and a locally finite measure α on it which is absolutely continuous with respect to $H^1 \big|_A$. We can associate to A and α an object which is called (in the jargon of minimal surface professionals) a rectifiable varifold. This last is basically a notion of a generalized submanifold. The requirement that

(2.61) $$\int_A \operatorname{Re}(Dg) \, d\alpha = 0$$

for all compactly supported C^1 functions $g : \mathbf{C} \to \mathbf{C}$ is equivalent to the condition that this varifold be stationary. This means, roughly, that the appropriate version of the length functional is unchanged to first order by small deformations of the varifold. Thus Proposition 2.59 provides us with a characterization of Cauchy-flatness in terms of this geometric notion of a stationary varifold, since (2.60) holds for all $g : \mathbf{C} \to \mathbf{C}$ which are C^1 and have compact support iff the same is true for (2.61).

This characterization is very useful because it implies Proposition 2.38. For the convenience of the reader we provide a reasonably self-contained proof of this in the next section, rather than simply citing the relevant results from the literature on stationary varifolds. In our case the argument is pretty simple.

The reader should consult [AA] for more information about 1-dimensional stationary varifolds. In particular, one can find in [AA] (partial) regularity results for 1-dimensional stationary varifolds as well as an example that shows that these things can still be rather complicated. A nice reference for the subject of stationary varifolds of general dimension (and related topics) is [Si].

In order to understand better this business about stationary varifolds it is helpful to think about the special case where A is a smooth Jordan curve that looks like a line at infinity. Then stationary means that the mean curvature of A vanishes, which simply means that the curvature vanishes and A is a line, since A is a curve. On the other hand,

$$\int_A \frac{1}{(z-\zeta)^2} \, d\zeta = 0 \quad \text{for all} \quad z \in \mathbf{C} \setminus A$$

is automatically true if "$d\zeta$" denotes the usual complex measure that appears in contour integrals from complex analysis. This measure incorporates the turning of the tangent of A, and the requirement that it be a complex constant times a positive measure is exactly the requirement that the curvature of A vanishes. In view of this it is not so surprising that (2.33) should be a statement about vanishing curvature, since $d\alpha$ is a positive measure and thus cannot take into account the turning of the tangent.

The rest of this section will be devoted to the proof of Proposition 2.59. We shall, incidentally, need a similar result in higher dimensions in the next chapter, the proof of which is so similar to the following that we are going to omit it.

Let $A \in CF$ be given, and let α be the associated measure which satisfies (2.31)–(2.33). Fix a C^1 function $g: \mathbf{C} \to \mathbf{C}$ with compact support. We want to reduce (2.60) to (2.33). The idea is that the rectifiability of A implies that $g|_A$ can be approximated rather well by holomorphic functions and, in fact, by superpositions of Cauchy kernels. In order to produce this approximation, we first construct a function G on $\mathbf{C}$ which equals g on A and which has $\bar{\partial} G$ being pretty small on $\mathbf{C} \setminus A$.

Let $\{Q_i\}$ be a Whitney decomposition of $\mathbf{C} \setminus A$, as in §1 of Chapter 6 of [St]. Thus the Q_i's are dyadic cubes with disjoint interiors which satisfy $\bigcup Q_i = \mathbf{C} \setminus A$ and

(2.62) $\qquad \operatorname{diam} Q_i \leq \operatorname{dist}(Q_i, A) \leq 4 \operatorname{diam} Q_i$

for each i. Let q_i denote the center of Q_i, and for each i take L_i to be a line in $\mathbf{C}$ which intersects $(10 Q_i) \cap A$ and for which

(2.63) $\quad \sup\{\operatorname{dist}(z, L_i)(\operatorname{diam} Q_i)^{-1} : z \in A, \ |z - q_i| \leq 10^{10} \operatorname{diam} Q_i\}$

is as small as possible. Let β_i denote the quantity given in (2.63).

2.6. THE MAIN STEP IN THE PROOF OF PROPOSITION 2.38

Let A' be as in Lemma 2.54. Then we have the following.

LEMMA 2.64. *For every $\eta > 0$ and every $a \in A'$ we have that*
$$\lim_{r \to 0} \max\{\beta_i : Q_i \cap B(a, r) \neq \varnothing \text{ and } \operatorname{diam} Q_i \geq \eta r\} = 0.$$

Notice that this is weaker than the existence of a tangent to A at a, because it does not prevent the L_i's from spinning around as the Q_i's approach a.

The proof of the lemma is fairly simple and we omit it.

To build G, we are going to use the standard techniques for building good extensions of functions which are described in §2 of Chapter 6 in [St]. We first need to have the appropriate partition of unity.

Set $Q_i^* = \frac{9}{8} Q_i$. It is not hard to show that the Q_i^*'s have bounded overlap. (See Proposition 3 on p. 169 of [St].) We shall use this fact repeatedly in this section.

Let $\tilde{\phi}_i$ be a smooth bump function on $\mathbf{C}$ such that $0 \leq \tilde{\phi}_i \leq 1$ on $\mathbf{C}$, $\operatorname{supp} \tilde{\phi}_i \subseteq Q_i^*$, $\tilde{\phi}_i = 1$ on Q_i, and $|\nabla \tilde{\phi}_i| \leq C(\operatorname{diam} Q_i)^{-1}$. Set

$$\phi_i = \tilde{\phi}_i \left(\sum \tilde{\phi}_i \right)^{-1}$$

so that each ϕ_i is a smooth function which is supported in Q_i^* and which satisfies $0 \leq \phi_i \leq 1$, $|\nabla \phi_i| \leq C(\operatorname{diam} Q)^{-1}$, and

$$\sum \phi_i = 1 \quad \text{on } \mathbf{C} \setminus A.$$

For each i pick a point p_i in $L_i \cap 10 Q_i \cap A$. Let λ_i be a complex number such that $|\lambda_i| = 1$ and

$$L_i = \{p_i + t\lambda_i : t \in \mathbf{R}\}.$$

Of course, λ_i does not depend on p_i, and it is only determined up to multiplication by -1.

Given $\lambda \in \mathbf{C} \setminus \{0\}$, define the (sort of) directional differentiation operator D_λ by

$$D_\lambda f(w) = \lambda^{-1} \lim_{\substack{t \to 0 \\ t \in \mathbf{R}}} \frac{f(w + t\lambda) - f(w)}{t},$$

assuming the limit exists. Thus if f is holomorphic at w then $D_\lambda f(w) = f'(w)$ for all λ. Also, if $Dg(a)$ is as in (2.57), then $Dg(a) = D_{\tau(a)} g(a)$ in this new notation.

For each i let h_i be the holomorphic affine function on $\mathbf{C}$ that satisfies

(2.65) $\quad h_i(p_i) = g(p_i) \quad \text{and} \quad h_i'(p_i) = D_{\lambda_i} h_i(p_i) = D_{\lambda_i} g(p_i).$

Because g is C^1, h_i approximates g well on $L_i \cap 100 Q_i$ when Q_i is small. Define G on $\mathbf{C}$ by

$$G(z) = \begin{cases} g(z) & \text{when } z \in A \\ \sum_{\text{diam } Q_i \leq 1} h_i(z)\phi_i(z) & \text{when } z \in \mathbf{C} \setminus A. \end{cases}$$

Clearly G is smooth on $\mathbf{C} \setminus A$.

LEMMA 2.66. *G has compact support, G is Lipschitz of order 1, and for every $\eta > 0$ and each $a \in A'$ we have that*

$$\lim_{r \to 0} \left(\sup\{|\overline{\partial} G(z)| : z \in B(a, r) \text{ and } \operatorname{dist}(z, A) \geq \eta r \} \right) = 0.$$

The fact that G has compact support follows directly from the observation that $h_i \equiv 0$ unless $5Q_i$ intersects the support of g.

To prove the lipschitzness of G we split it into $G_1 + G_2$, where

$$G_1(z) = \begin{cases} g(z) & \text{on } A \\ \sum_{\text{diam } Q_i \leq 1} h_i(p_i)\phi_i(z) & \text{on } \mathbf{C} \setminus A \end{cases}$$

and

$$G_2(z) = \begin{cases} 0 & \text{on } A \\ \sum_{\text{diam } Q_i \leq 1} h_i'(p_i)(z - p_i)\phi_i(z) & \text{on } \mathbf{C} \setminus A. \end{cases}$$

The lipschitzness of G_1 can be derived from the corresponding property of g using a standard argument which can be found on pp. 174–175 in [St]. For G_2 we observe that $|h_i'(p_i)| \leq C$ and hence

$$|G_2(z)| \leq C \operatorname{dist}(z, A) \quad \text{for all } z \in \mathbf{C},$$

and

$$|\nabla G_2(z)| \leq C \quad \text{when } z \in \mathbf{C} \setminus A.$$

Using these two facts it is not hard to show that G_2 is also Lipschitz.

It remains to control $\overline{\partial} G$ on $\mathbf{C} \setminus A$. By definitions we have

(2.67) $$\overline{\partial} G(z) = \sum_{\text{diam } Q_i \leq 1} h_i(z) \overline{\partial} \phi_i(z) \quad \text{when } z \in \mathbf{C} \setminus A.$$

Fix $z \in \mathbf{C} \setminus A$, and choose j so that $z \in Q_j$. We may as well assume that $\operatorname{dist}(z, A)$ is small, so that $z \in Q_i^*$ implies that $\operatorname{diam} Q_i \leq 1$ and so that this is also true for all points in some neighborhood of z. With this assumption we have that

$$\sum_{\text{diam } Q_i \leq 1} \overline{\partial} \phi_i(z) = \overline{\partial} \left(\sum_{\text{diam } Q_i \leq 1} \phi_i \right) \bigg|_z = 0,$$

and hence

(2.68) $$\overline{\partial} G(z) = \sum_{\text{diam } Q_i \leq 1} (h_i(z) - h_j(z)) \overline{\partial} \phi_i(z).$$

Let us, for the time being, restrict our attention to i's such that z lies in Q_i^*, which are the only ones that matter for (2.68). We want to control $|h_i(z) - h_j(z)|$. To do this we first look at the relative position of L_i and L_j.

2.6. THE MAIN STEP IN THE PROOF OF PROPOSITION 2.38

Because $Q_j \cap Q_i^* \neq \varnothing$ we have that

$$\text{Angle}(L_i, L_j) \leq C(\beta_i + \beta_j),$$

or, equivalently,

(2.69) $$\min\left(|\lambda_i - \lambda_j|, |\lambda_i + \lambda_j|\right) \leq C(\beta_i + \beta_j).$$

The reason for this is the following. If either of β_i and β_j is not small, then these inequalities are trivial. Otherwise, L_i is a good approximation to A near Q_i, and L_j is a good approximation to A near Q_j. This is made precise by the definition of β_i (see (2.63)). Since Q_i^* intersects Q_j, there is a nontrivial piece of A which lies near both L_i and L_j. If (2.69) were false, then you could show that

$$\{a \in A : \text{dist}(a, L_i \cap L_j) \leq \text{diam}\, Q_j\}$$

would have to have too much mass in too small a space. (This uses the assumption that A is a regular set.) We omit the details. [Lemma 5.13 in [DS2] is a more general version of this fact for which we also did not provide a complete proof.]

Let us use (2.69) to estimate $|h_i'(p_i) - h_j'(p_j)|$. Since $h_i'(p_i) = D_{\lambda_i} g(p_i)$, by definition, we get from (2.69) that

$$|h_i'(p_i) - D_{\lambda_j} g(p_i)| \leq C(\beta_i + \beta_j).$$

Here C is allowed to depend on $\|\nabla g\|_\infty$. Set $\omega(r) = \sup\{|\nabla g(p) - \nabla g(q)| : p, q \in \mathbf{C}, |p-q| \leq 10^{10} r\}$, so that $\lim_{r \to 0} \omega(r) = 0$, since g is C^1 and has compact support. Then

$$|D_{\lambda_j} g(p_i) - D_{\lambda_j} g(p_j)| \leq \omega(\text{dist}(z, A)),$$

so

(2.70) $$|h_i'(p_i) - h_j'(p_j)| \leq C(\beta_i + \beta_j) + \omega(\text{dist}(z, A)).$$

Next we want to estimate $|h_i(p_i) - h_j(p_j)|$. Let $\hat{p}_i$ denote the projection of p_i onto L_j. Then

$$|p_i - \hat{p}_i| \leq \beta_j \,\text{diam}\, Q_j.$$

This uses the definition (2.63) of β_j, the fact that we chose p_i to lie in $A \cap 10 Q_i$, and the fact that Q_i^* intersects Q_j (so that p_i is not too far from Q_j). Because $h_i(p_i) = g(p_i)$, we have that

$$|h_i(p_i) - h_j(p_i)| \leq |g(p_i) - g(\hat{p}_i)| + |g(\hat{p}_i) - h_j(\hat{p}_i)| + |h_j(\hat{p}_i) - h_j(p_i)|.$$

Our bounds on ∇g and h_j' imply that

(2.71) $$|h_i(p_i) - h_j(p_i)| \leq C\beta_j \,\text{diam}\, Q_j + |g(\hat{p}_i) - h_j(\hat{p}_i)|.$$

On the other hand, $\hat{p}_i$ lies in L_j, and $h_j|_{L_j}$ is really just the degree 1 Taylor approximation to $g|_{L_j}$ at p_j, so we have (from the mean value theorem, really) that

$$|g(\hat{p}_i) - h_j(\hat{p}_i)| \leq C[\sup\{|\nabla g(q) - \nabla g(p_j)| : q \in L_j,$$
$$q \text{ lies between } p_j \text{ and } \hat{p}_i\}]|\hat{p}_i - p_j|$$
$$\leq C\omega(10^{-10}|\hat{p}_i - p_j|)|\hat{p}_i - p_j|$$
$$\leq C\omega(\text{dist}(z, A))\text{dist}(z, A).$$

Combining this with (2.71) we get

(2.72) $\quad |h_i(p_i) - h_j(p_i)| \leq C(\beta_j + \omega(\text{dist}(z, A)))\text{dist}(z, A),$

since $\text{diam } Q_j \leq C \text{dist}(z, A)$.

Now we are ready to deal with $|h_i(z) - h_j(z)|$. We have

$$|h_i(z) - h_j(z)| = |h_i(p_i) + h'_i(p_i)(z - p_i) - h_j(p_j) - h'_j(p_j)(z - p_j)|$$
$$\leq |h_i(p_i) - h_j(p_j) - h'_j(p_j)(p_i - p_j)|$$
$$+ |(h'_i(p_i) - h'_j(p_j))(z - p_i)|$$
$$= |h_i(p_i) - h_j(p_i)| + |h'_i(p_i) - h'_j(p_j)||z - p_i|.$$

Using (2.70) and (2.72) we obtain

$$|h_i(z) - h_j(z)| \leq C(\beta_i + \beta_j + \omega(\text{dist}(z, A)))\text{dist}(z, A).$$

This, together with (2.68) and

$$|\nabla \phi_i| \leq C(\text{diam } Q_i)^{-1} \leq C(\text{dist}(z, A))^{-1}$$

gives

(2.73) $\quad |\overline{\partial} G(z)| \leq C \left[\sum_{\{i \,:\, z \in Q_i^*\}} \beta_i + \omega(\text{dist}(z, A)) \right].$

The last part of Lemma 2.66 now follows easily from Lemma 2.64 and the fact that $\omega(r) \to 0$ as $r \to 0$. This completes the proof of Lemma 2.66.

Now we want to represent g in terms of Cauchy kernels, in order to reduce (2.60) to (2.33). Standard reasoning gives us that

(2.74) $\quad G(z) = \dfrac{1}{\pi} \iint_{\mathbf{C}} \dfrac{1}{z - w} \overline{\partial} G(w) \, dw \quad \text{for all } z \in \mathbf{C},$

where dw denotes Lebesgue measure on $\mathbf{C}$. This formula really does hold everywhere on $\mathbf{C}$, as opposed to almost everywhere, because G is Lipschitz and compactly supported. In particular, we have

(2.75) $\quad g(z) = \dfrac{1}{\pi} \iint_{\mathbf{C}} \dfrac{1}{z - w} \overline{\partial} G(w) \, dw \quad \text{for all } z \in A.$

This is the representation for g that we wanted.

Now we want to get our hands on Dg in terms of (2.75). Roughly speaking the idea is that $Dg = DG$ on A' and that $\overline{\partial} G = 0$ a.e. on A so that $DG = \partial G$ a.e. on A. Although we shall follow this line of reasoning at a philosophical level, as a practical matter it is better to do things a little differently (i.e., to make some regularizations).

LEMMA 2.76. *For each $a \in A'$ we have*

$$(2.77) \qquad \lim_{r \to 0} \left\{ \frac{1}{r^2} \iint_{B(a,r)} |\partial G(z) - Dg(a)| \, dw \right\} = 0.$$

Here the dw denotes Lebesgue measure on $\mathbf{C}$.

Fix $a \in A'$. Let $\eta > 0$ (small) be given. For each $t > 0$ set

$$A_t = \{ z \in \mathbf{C} : \text{dist}(z, A) \leq t \}.$$

Then
(2.78)
$$\frac{1}{r^2} \iint_{B(a,r) \cap A_{\eta r}} |\partial G(z) - Dg(a)| \, dw \leq (\|\partial G\|_\infty + \|\nabla g\|_\infty) r^{-2} |B(a,r) \cap A_{\eta r}|$$
$$\leq C\eta.$$

This last inequality relies heavily on the fact that A is a 1-dimensional regular set.

Consider now

$$\frac{1}{r^2} \iint_{B(a,r) \setminus A_{\eta r}} |\partial G(z) - Dg(a)| \, dw.$$

We want to split the integrand into two pieces, using the identity

$$\partial G(z) - Dg(a) = \sum_{\text{diam } Q_i \leq 1} \left(h'_i(z) - Dg(a) \right) \phi_i(z) + \sum_{\text{diam } Q_i \leq 1} h_i(z) \partial \phi_i(z)$$
$$= T_1(z, a) + T_2(z, a),$$

which is true when $z \in \mathbf{C} \setminus A$ is close to A. (The second equality is a definition.) Let us first look at the T_2 part.

We claim that

$$(2.79) \qquad \limsup_{r \to 0} \{ T_2(z, a) : z \in B(a, r) \setminus A_{\eta r} \} = 0.$$

Indeed, this can be proved using exactly the same computations as was used to prove the last part of Lemma 2.66. You simply replace (2.67) with the definition of $T_2(z, a)$, and hardly any other changes are needed.

For $T_1(z, a)$ we also claim that

$$(2.80) \qquad \limsup_{r \to 0} \{ T_1(z, a) : z \in B(a, r) \setminus A_{\eta r} \} = 0.$$

Once we have this, (2.77) follows from (2.78), (2.79), and the fact that η is arbitrary.

The proof of (2.80) is similar to the proof of Lemma 2.66, but it is simpler. We continue to use the same notation as before. In particular, L_a is the tangent line to A through a, and $\tau(a)$ is as defined at (2.56).

We begin with the observation that

$$\lim_{r \to 0} \max\{\text{angle}(L_i, L_a) : Q_i^* \cap (B(a, r) \setminus A_{\eta r}) \neq \emptyset\} = 0,$$

or, equivalently,

(2.81) $\lim_{r \to 0} \max\{\min(|\lambda_i - \tau(a)|, |\lambda_i + \tau(a)|) : Q_i^* \cap (B(a, r) \setminus A_{\eta r}) \neq \emptyset\} = 0.$

This is true for the same reason as (2.69). Namely, for r small all of the set $A \cap B(a, r)$ stays very close to L_a, while L_i is supposed to be an approximately optimal linear approximation of that part of A which is not too far from Q_i. If Q_i intersects $B(a, r) \setminus A_{\eta r}$ and r is very small, then L_a and L_i will have to be providing good linear approximations to the same piece of A, and this forces L_a and L_i to be almost the same line.

To control T_1 using (2.81) we use also the fact that

$$h_i'(z) = h_i'(p_i) = D_{\lambda_i} g(p_i),$$

which is true by definition. (See (2.65).) Since $Dg(a) = D_{\tau(a)} g(a)$ by definition ((2.57)), we have that

$$|T_1(z, a)| \leq \max\{|D_{\lambda_i} g(p_i) - D_{\tau(a)} g(a)| : i \text{ satisfies } z \in Q_i^*\}.$$

From here (2.80) follows from (2.81), the easy observation that $|p_i - a| \leq Cr$, and the fact that g is C^1. This proves Lemma 2.76.

We are now rather close to the end of the proof of Proposition 2.59. Let θ be a smooth radial bump function on $\mathbf{C}$ such that $\operatorname{supp} \theta \subseteq B(0, 1)$ and $\iint_{\mathbf{C}} \theta(z) \, dz = 1$. Define θ_t by $\theta_t(z) = t^{-2} \theta(t^{-1} z)$. Then Lemma 2.76 implies that

$$\lim_{t \to 0} \theta_t * (\partial G) = Dg$$

at every point in A', so

(2.82) $$\lim_{t \to 0} \int_A |\theta_t * (\partial G) - Dg| \, d\alpha = 0$$

by the dominated convergence theorem and the first half of Lemma 2.66. In order to prove (2.60) it therefore suffices to show that

(2.83) $$\lim_{t \to 0} \int_A \theta_t * (\partial G) \, d\alpha = 0.$$

Let T denote the singular integral operator which is defined by

$$Tf(z) = \frac{1}{\pi} \text{ p.v.} \iint_{\mathbf{C}} \frac{1}{(z - w)^2} f(w) \, dw.$$

2.6. THE MAIN STEP IN THE PROOF OF PROPOSITION 2.38

This operator converts $\bar{\partial}$ derivatives into ∂ derivatives (under suitable hypotheses), and in particular we have that

(2.84) $$\partial G = T(\bar{\partial} G) \quad \text{a.e. on } \mathbf{C}.$$

Set $k_t = T(\theta_t)$, so that k_t is a smooth function on $\mathbf{C}$ that satisfies

(2.85) $$|k_t(z)| \leq C(t + |z|)^{-2}.$$

From (2.84) and standard manipulations for convolution operators we obtain that

$$\theta_t * (\partial G) = k_t * (\bar{\partial} G) \quad \text{on } \mathbf{C}.$$

Notice that both sides of this equation are smooth functions. Integrating over A we get that

(2.86) $$\int_A \theta_t * (\partial G)\, d\alpha = \int_A k_t * (\bar{\partial} G)\, d\alpha = \iint_{\mathbf{C}} (\bar{\partial} G)(z)(k_t * \alpha)(z)\, dz.$$

Fubini's theorem was of course used in the last equality.

LEMMA 2.87. *The support of $k_t * \alpha$ is contained in A_t and*

$$|k_t * \alpha(w)| \leq Ct^{-1} \quad \text{for all } w \in \mathbf{C}.$$

The second part follows easily from (2.85) and (2.32). Let us prove the first part. Fix $z \in \mathbf{C} \setminus A_t$ so that $\text{dist}(z, A) > t$. Fubini's theorem gives us

$$k_t * \alpha(z) = \int_A k_t(z - \zeta)\, d\alpha(\zeta) = \iint_{B(z,t)} \theta_t(z - w) \left(\int_A \frac{1}{(w - \zeta)^2}\, d\alpha(\zeta) \right) dw.$$

This last vanishes because of (2.33).

Combining Lemma 2.87 with (2.86) we get that

(2.88) $$\left| \int_A \theta_t * (\partial G)\, d\alpha \right| \leq C \iint_{A_t} |\bar{\partial} G(z)| t^{-1}\, dz.$$

We want to dominate the right side by a more convenient expression. From (2.32) and Fubini's theorem we have that

$$\iint_{A_t} |\bar{\partial} G(z)| t^{-1}\, dz \leq C \int_A \left(t^{-2} \iint_{B(a, 2t)} |\bar{\partial} G(z)|\, dz \right) d\alpha(a).$$

Thus, in order to show that (2.83) is true and thereby finish the proof of Proposition 2.56, we are reduced to the following.

LEMMA 2.89.

$$\lim_{t \to 0} \int_A \left(t^{-2} \iint_{B(a, 2t)} |\bar{\partial} G(z)|\, dz \right) d\alpha(a) = 0.$$

Because $\bar{\partial}G$ lies in L^∞ and has compact support we need only show that

$$\lim_{t \to 0} t^{-2} \iint_{B(a,t)} |\bar{\partial}G(z)|\,dz = 0 \tag{2.90}$$

for all $a \in A'$. This is easy. Let $a \in A'$ be given, and let $\eta > 0$ be small. Then

$$\sup_{t>0} t^{-2} \iint_{B(a,t) \cap A_{\eta t}} |\bar{\partial}G(z)|\,dz \le \sup_{t>0} Ct^{-2}|B(a,t) \cap A_{\eta t}| \le C\eta.$$

For the last inequality we have used the fact that A is regular. On the other hand, Lemma 2.66 implies that

$$\lim_{t \to 0} t^{-2} \iint_{B(a,t) \setminus A_{\eta t}} |\bar{\partial}G(z)|\,dz = 0, \tag{2.91}$$

no matter how small η is. From here (2.90) follows easily. This completes the proof of Proposition 2.59.

2.7. The end of the proof of Proposition 2.38.

Let $A \in CF$ be given, and let α be the associated measure, as in Definition 2.30. We want to use the fact that A and α satisfy the conclusion of Proposition 2.59 to show that every component of $\mathbf{C} \setminus A$ is convex. As we said in the preceding section, we could simply invoke known theorems from the literature for this last step, but we believe that the inclusion of the argument given below might be helpful to some readers.

We begin with a simple well-known criterion for convexity. Given p, $q \in \mathbf{C}$ let $S(p,q)$ denote the closed line segment that joins p to q.

LEMMA 2.92. *A connected open set Ω in $\mathbf{C}$ is convex if for every $\epsilon > 0$ it is true that*

$$\begin{aligned}&p, q \in \Omega, \ \mathrm{dist}(p, \mathbf{C} \setminus \Omega) > \epsilon, \ \mathrm{dist}(q, \mathbf{C} \setminus \Omega) > \epsilon,\\ &S(p,q) \subseteq \Omega, \ \mathrm{and} \ \mathrm{dist}(S(p,q), \mathbf{C} \setminus \Omega) > \epsilon/4\end{aligned} \tag{2.93}$$

imply that

$$\mathrm{dist}(S(p,q), \mathbf{C} \setminus \Omega) \ge \epsilon/2. \tag{2.94}$$

Let us briefly indicate the proof of this. Let $\Omega \subseteq \mathbf{C}$ be given, and suppose that Ω has the properties listed above. Let $a, b \in \Omega$ be given. We want to show that $S(a,b) \subseteq \Omega$.

Because Ω is connected, there is a path $\gamma(t)$, $0 \le t \le 1$, in Ω such that $\gamma(0) = a$ and $\gamma(1) = b$. Choose $\epsilon > 0$ so that $\mathrm{dist}(\gamma(t), \mathbf{C} \setminus \Omega) > \epsilon$ for all $t \in [0, 1]$. Consider the set

$$U = \{t \in [0,1] : S(a, \gamma(t)) \subseteq \Omega \ \text{and} \ \mathrm{dist}(S(a, \gamma(t)), \mathbf{C} \setminus \Omega) > \tfrac{\epsilon}{4}\}.$$

Of course $0 \in U$, and U is open. Our assumptions on Ω imply that

$$U = \{t \in [0,1] : S(a, \gamma(t)) \subseteq \Omega \ \text{and} \ \mathrm{dist}(S(a, \gamma(t)), \mathbf{C} \setminus \Omega) \ge \tfrac{\epsilon}{2}\},$$

2.7. THE END OF THE PROOF OF PROPOSITION 2.38

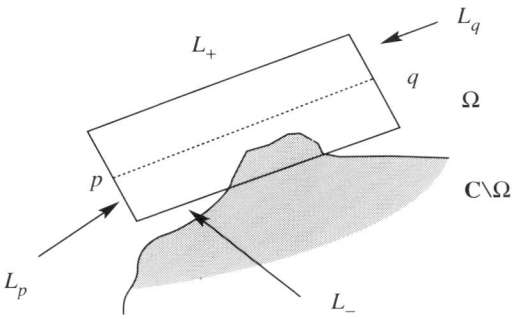

FIGURE 2.1

so U is also closed. Hence $U = [0, 1]$, so $S(a, b) \subseteq \Omega$, as desired. This proves Lemma 2.92.

Now let A be as before, and let Ω be a component of $\mathbf{C} \setminus A$. Assume that there exist $\epsilon > 0$ and $p, q \in \Omega$ such that (2.93) holds but (2.94) does not, and let us show that this leads to a contradiction.

Before proceeding there is a notational convention that we should establish. Given $F \subseteq \mathbf{C}$ and $z \in \mathbf{C}$ define $F + z$ by

$$F + z = \{w + z : w \in F\}.$$

Set $v = i\frac{q-p}{|q-p|}$, so that v is a unit vector which is normal to $S(p, q)$. Since (2.93) holds but (2.94) fails, we must have that

$$(S(p, q) + tv) \cap (\mathbf{C} \setminus \Omega) \neq \varnothing$$

for some $t \in \mathbf{R}$ such that $\epsilon/4 < |t| < \epsilon/2$. We may as well assume that this happens for a t which is negative.

Let R be the closed rectangle which is bounded by the following four line segments:

$$L_p = \{p + sv : s \in \mathbf{R}, \; -\epsilon/2 \leq s \leq \epsilon/4\},$$
$$L_q = \{q + sv : s \in \mathbf{R}, \; -\epsilon/2 \leq s \leq \epsilon/4\},$$
$$L_+ = S(p, q) + (\epsilon/4)v,$$
$$L_- = S(p, q) - (\epsilon/2)v.$$

(See Figure 2.1.) Then $L_p \cup L_q \cup L_+$ are contained in Ω, but L_- and the interior of R intersect $\mathbf{C} \setminus \Omega$.

We want to show that this cannot happen by constructing a suitable function g such that (2.60) cannot be true. To understand what is happening, it is helpful to think geometrically in terms of the stationary varifold business, which basically says that if you make a first-order deformation of the varifold, then there is no change in the "mass" to first order. However, if Ω behaves as above, then we can choose our deformation in such a way that it does not move A outside R, while it pushes A "down", in the direction of $-v$, inside R. This does indeed cause a nontrivial change in the mass, which gives the desired contradiction.

Although that is the idea of what we are doing, we shall carry out the details without mentioning varifolds. Fix $t_0 \in (\epsilon/4, \epsilon/2)$ such that

(2.95) $$(S(p,q) - t_0 v) \cap (\mathbf{C} \setminus \Omega) \neq \varnothing,$$

and fix $t_1 \in \mathbf{R}$ such that $t_0 < t_1 < \epsilon/2$. Let $f(r)$ be a smooth real-valued function on $\mathbf{R}$ that satisfies

$$f(r) = 0 \quad \text{when } r \leq -\epsilon/2 + 2\eta \text{ for some } \eta > 0,$$
$$f'(r) \geq 0 \quad \text{for all } r,$$
$$f'(r) \geq 1 \quad \text{when } r \geq -t_1.$$

Let $\widetilde{R}$ be the subrectangle of R obtained by translating L_- by ηv, i.e.,

$$\widetilde{R} = \{z \in R : \text{Re}[(z-p)\bar{v}] \geq -\epsilon/2 + \eta\}.$$

(Keep in mind that $\text{Re}[(z-p)\bar{v}]$ is just the usual inner product of $z-p$ and v as elements of $\mathbf{R}^2$.) Let ϕ be a C^∞ function on $\mathbf{C}$ that vanishes on $\mathbf{C} \setminus R$ and which satisfies $\phi \equiv 1$ on a neighborhood of $A \cap \widetilde{R}$. This is possible because L_p, L_q, and L_+ are contained in Ω and hence are disjoint from A. Define $g : \mathbf{C} \to \mathbf{C}$ by

(2.96) $$g(z) = v\phi(z)f(\text{Re}[(z-p)\bar{v}]).$$

Thus g is smooth and compactly supported. Notice that $g = 0$ on $\mathbf{C} \setminus \widetilde{R}$ and not just $\mathbf{C} \setminus R$, because of the conditions that we have imposed on f.

Since A satisfies (2.60), we have that

(2.97) $$\int_A Dg \, d\alpha = 0.$$

Let us compute what this means. Let $a \in A'$ be given, and let $\tau(a)$ be as defined around (2.56). Then we have (from the definition (2.57) of $Dg(a)$) that

$$\text{Re}(Dg(a)) = \begin{cases} 0 & \text{when } a \in A' \setminus \widetilde{R}, \\ [\text{Re}(\tau(a)\bar{v})]^2 f'(\text{Re}[(z-p)\bar{v}]) & \text{when } a \in A' \cap \widetilde{R}. \end{cases}$$

Since $f'(r) \geq 0$ for all r and $f'(r) \geq 1$ when $r \geq -t_1$, we obtain from (2.97) that

(2.98) $$\text{Re}(\tau(a)\bar{v}) = 0 \quad \text{for almost all } a \in A' \cap R_1,$$

where R_1 is the subrectangle of R given by

$$R_1 = \{z \in R : \text{Re}[(z-p)\bar{v}] \geq -t_1\}.$$

Here "almost all" refers to the measure α or, equivalently, 1-dimensional Hausdorff measure.

2.7. THE END OF THE PROOF OF PROPOSITION 2.38

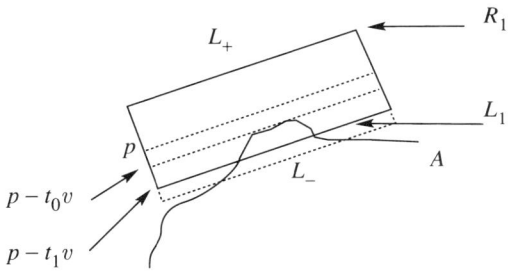

FIGURE 2.2

We are nearing our long-sought contradiction. By (2.98) the tangents to A at almost all points in $A \cap R_1$ are perpindicular to v (and hence parallel to $S(p,q)$). This clearly makes it hard for the situation pictured in Figure 2.1 to occur. Let us make this precise and rigorous.

Set $L_1 = S(p,q) - t_1 v$, so that L_1 is one of the pieces of the boundary of R_1. (See Figure 2.2.) Observe that $\widetilde{A} = (A \cap R_1) \cup L_1$ is a closed and connected set. This follows from Lemma 2.50 and the fact that

$$L_p \cup L_q \cup L_+ \subseteq \Omega \subseteq \mathbf{C} \setminus A.$$

From (2.95) we know that A intersects the interior of R_1. Because $\widetilde{A}$ is compact and connected and $H^1(\widetilde{A}) < \infty$, there is a Lipschitz mapping $\gamma : [0,1] \to \widetilde{A}$ such that $\gamma(0) \in L_1$ and $\gamma(1)$ lies in the intersection of A and the interior of R_1. (See Lemma 3.12 on p. 34 of [Fl1] or Theorem 1.8 in §1.1 of Part I.)

Consider the function $\hat{\gamma} : [0,1] \to \mathbf{R}$ which is defined by

(2.99) $$\hat{\gamma}(t) = \text{Re}[(\gamma(t) - p)\bar{v}].$$

Thus $\hat{\gamma}$ is Lipschitz since γ is, and we have that $\hat{\gamma}(0) = -t_1$ and $\hat{\gamma}(1) > -t_1$. In particular, the image of $\hat{\gamma}$ contains a nontrivial interval. This is incompatible with (2.98), as we now show.

Define $h : \mathbf{C} \to \mathbf{R}$ by

$$h(z) = \text{Re}[(z-p)\bar{v}]$$

so that $\hat{\gamma} = h \circ \gamma$. Set

$$X = \{t \in (0,1) : \gamma'(t) \text{ exists}\},$$
$$Y = \{t \in X : \hat{\gamma}'(t) \neq 0\},$$
$$Z = h(((A \cap R_1) \setminus A'') \cup L_1) \cup \hat{\gamma}([0,1] \setminus Y),$$

where A'' is the set of points a in $R_1 \cap A'$ such that (2.98) holds. Then Z is a subset of $\mathbf{R}$ with Lebesgue measure zero. Indeed, we already know that $H^1((A \cap R_1) \setminus A'') = 0$ and that $h(L_1) = \{-t_1\}$ by definitions. Also, the lipschitzness of γ implies that $[0,1] \setminus X$ has measure zero, and hence $\hat{\gamma}([0,1] \setminus Y)$ does too, because $\hat{\gamma}(X \setminus Y)$ has measure zero (since $\hat{\gamma}' = 0$ on $X \setminus Y$).

Let us now show that

(2.100) $$\hat{\gamma}([0, 1]) \subseteq Z,$$

which provides the contradiction that we seek. We prove (2.100) by contradiction. (Please forgive us.) Suppose that $t \in [0, 1]$ and that $\hat{\gamma}(t) \in \mathbf{R} \setminus Z$. Then $t \in Y$ and $\gamma(t) \in A'' \setminus L_1$. Hence there is a tangent $L_{\gamma(t)}$ to A at $\gamma(t)$, and this tangent line must be perpendicular to v. Because $t \in Y$ we have that $\gamma'(t)$ exists and is nonzero, so $\gamma'(t)$ must point in the same direction as $L_{\gamma(t)}$. Thus $\operatorname{Re}(\gamma'(t)\bar{v}) = 0$, which implies that $\hat{\gamma}'(t) = 0$, which is impossible since $t \in Y$. This proves (2.100).

On the other hand, (2.100) contradicts the facts that $|Z| = 0$ and $\hat{\gamma}([0, 1])$ contains a nontrivial interval. This proves that our original assumption, that there exist $\epsilon > 0$ and $p, q \in \Omega$ for which (2.93) is satisfied but (2.94) is not, is false. Thus Lemma 2.92 implies that Ω is convex, as desired.

This completes the proof of Proposition 2.38 (and hence Theorem 2.25 also).

CHAPTER 3

Square Function Estimates and Uniform Rectifiability in Higher Dimensions

The main goal of this chapter is the proof of Theorems I.2.41 and I.2.45 when $d > 1$. Many of the arguments are practically the same as in the $d = 1$ case, which was treated in the preceding chapter, but there are also some substantial differences. Clifford analysis will play an important role as a substitute for complex function theory.

3.1. A brief review of Clifford analysis.
A general reference for this section is [BDS].
Let $\mathscr{C}(k)$ denote the Clifford algebra over $\mathbf{R}$ with k generators $e_1, \ldots, e_k$. This algebra has an identity element, which will be denoted by 1, and the generators satisfy the relations

$$e_j^2 = -1, \qquad e_i e_j = -e_j e_i \quad \text{when } i \neq j,$$

and no others.

It is easy to see that $\mathscr{C}(1)$ is isomorphic to $\mathbf{C}$ while $\mathscr{C}(2)$ is isomorphic to the algebra of quarternions. In general $\mathscr{C}(k)$ is an (associative) algebra of dimension 2^k.

In this chapter we are going to be concerned with d-dimensional regular sets in $\mathbf{R}^{d+1}$, $d > 1$, and with functions living on $\mathbf{R}^{d+1}$, and some of the objects that we are going to work with will be defined in terms of $\mathscr{C}(d+1)$. Before getting precise let us establish the following.

CONVENTION 3.1. The set of real numbers will be identified with the set of real multiples of the identity in $\mathscr{C}(d+1)$, and $\mathbf{R}^{d+1}$ will be identified with the linear span of the generators $e_1, \ldots, e_{d+1}$ of $\mathscr{C}(d+1)$, via the map $x \mapsto \sum_{j=1}^{d+1} x_j e_j$.

Thus the reader should not be disturbed to see xy, where x and y lie in $\mathbf{R}^{d+1}$.

Notice, incidentally, that $x^2 = -|x|^2$ when $x \in \mathbf{R}^{d+1}$. In particular, $x \in \mathbf{R}^{d+1}$ is invertible in $\mathscr{C}(d+1)$ as long as $x \neq 0$.

We define a first-order differential operator δ by

(3.2) $$\delta f = \sum_{j=1}^{d+1} e_j \frac{\partial}{\partial x_j} f.$$

[Normally we would use D instead of δ, but that interacts poorly with other notation from the past and future.] We say that a $\mathscr{C}(d+1)$-valued function f is Clifford holomorphic on an open set if it satisfies $\delta f = 0$ there.

When $d = 1$ this reduces, after a bit of algebra, to something which is essentially equivalent to the usual notion of holomorphicity. More precisely, f splits into two pieces, one of which is holomorphic, while the other is antiholomorphic.

Many standard facts from complex analysis have reasonable counterparts in Clifford analysis. For instance,

$$(3.3) \qquad \delta^2 = -\Delta,$$

and in particular Clifford holomorphic functions are harmonic. We also get from (3.3) the simple fact that if u is harmonic, then δu is Clifford holomorphic.

Define $\mathscr{E} : \mathbf{R}^{d+1} \setminus \{0\} \to \mathscr{C}(d+1)$ by

$$(3.4) \qquad \mathscr{E}(x) = c(d) \frac{x}{|x|^{d+1}}.$$

Here $c(d)$ is a nonzero real number which is chosen so that $\mathscr{E}(\cdot)$ is the fundamental solution of δ. (In particular, it is Clifford holomorphic away from 0.) According to (3.3), this means that $\mathscr{E}(\cdot)$ should be equal to -1 times δ applied to the fundamental solution of the Laplacian. We call $\mathscr{E}(\cdot)$ the Cauchy-Clifford kernel.

There are natural versions of the Cauchy formulas in this context. Let Ω be a bounded domain in $\mathbf{R}^{d+1}$ with reasonably smooth boundary, and let $f : \overline{\Omega} \to \mathscr{C}(d+1)$ be a continuous function which is Clifford holomorphic on Ω. Let $\nu(y)$ denote the inward-pointing unit normal on $\partial \Omega$. Then

$$(3.5) \qquad \int_{\partial\Omega} \mathscr{E}(x-y)\nu(y)f(y)\,dy = \begin{cases} f(x) & \text{when } x \in \Omega, \\ 0 & \text{when } x \notin \overline{\Omega}. \end{cases}$$

Here dy denotes the usual surface measure.

This is all the background information about Clifford analysis that we are going to need.

For our purposes there are two crucial features of Clifford analysis:

(3.6)   there are plenty of Clifford holomorphic functions;

(3.7)   if $f(x)$ is Clifford holomorphic, then the derivative of f in any given direction can be computed in terms of the derivatives in the orthogonal directions.

The first one, (3.6), is manifested in practice by the existence of the Cauchy kernel. (One could acknowledge also the importance of the linearity of δ.) The second feature (3.7) is made precise by the following.

LEMMA 3.8. *If* $\{v_j\}_{j=1}^{d+1}$ *is any orthonormal basis for* $\mathbf{R}^{d+1}$, *then*

$$(3.9) \qquad \delta = \sum_{j=1}^{d+1} v_j D_{v_j}.$$

Here D_v denotes the operator of differentiation in the direction v (so that $D_v f(x) = \frac{d}{dt}\big|_{t=0} f(x + tv)$).

Lemma 3.8 is the sort of simple little fact for which the proving of it can inspire heated arguments between proponents of different philosophies, and therefore we omit it. [Notice, however, what happens when you test (3.9) on $x \mapsto \langle x, w \rangle$ for $w \in \mathbf{R}^{d+1}$.]

COROLLARY 3.10. *If* $\alpha : \mathbf{R}^{d+1} \to \mathscr{C}(d+1)$ *is affine and Clifford holomorphic, then it is uniquely determined by its restriction to any hyperplane.*

3.2. Clifford analysis and square function estimates.

Let E be a d-dimensional regular set in $\mathbf{R}^{d+1}$, and set $N(x) = |x|^{1-d}$.

LEMMA 3.11. *The following are equivalent*:

(3.12) *there is a* $C > 0$ *so that*

$$\iint_{\mathbf{R}^{d+1}} \Big| \int_E \nabla^2 N(x-y) f(y) dy \Big|^2 \operatorname{dist}(x, E) dx \leq C \int_E |f|^2$$

(3.13) *for all* $f \in L^2(E)$ (*i.e.*, E *satisfies the USFE—see Definition I.2.38*);

$$\iint_{\mathbf{R}^{d+1}} \Big| \int_E \nabla \mathscr{E}(x-y) f(y) dy \Big|^2 \operatorname{dist}(x, E) dx \leq C \int_E |f|^2$$

for all $f \in L^2(E)$.

Here the integrals over E are taken with respect to $H^d\big|_E$, while the integrals over $\mathbf{R}^{d+1}$ use Lebesgue measure. We shall continue to follow the convention of the preceding chapter of using double integrals to emphasize that we are integrating with respect to Lebesgue measure rather than d-dimensional Hausdorff measure. Also, we let ∇g denote the gradient of g even if it is $\mathscr{C}(d+1)$-valued, in which case $\nabla g(x)$ should be viewed as a vector of $d+1$ elements of $\mathscr{C}(d+1)$.

The lemma is trivial, since $\mathscr{E}$ is essentially the same as ∇N. The point of the lemma is simply to record the fact that the USFE condition from Definition I.2.38 has this simple reformulation (3.13) which is related to Clifford analysis.

There are some other reformulations of (3.12) and (3.13) which are not quite as trivial, just as in §2.1. We omit the details.

Now let us turn to the "only if" part of Theorem I.2.41.

THEOREM 3.14. *If E is uniformly rectifiable, then E satisfies* (3.13).

As in the $d = 1$ case, this is basically known, but there does not seem to be a good reference. The argument that was sketched in §2.2 also works here with only cosmetic changes to account for the different dimension. We might mention that there are again many proofs known of the analogue of Lemma 2.10 in this case and that the method of [S2] still applies. This uses the Clifford-Cauchy integral formula (3.5).

3.3. From square functions to uniform rectifiability: Preliminary reductions.
This section closely parallels §2.3.
As before we prefer to work with

$$e(x) = \left| \int_E \nabla \mathscr{E}(x - y)\, dy \right| \operatorname{dist}(x, E), \qquad x \in \mathbf{R}^{d+1} \backslash E,$$

and we have that (3.13) implies (and is even equivalent to) the Carleson measure estimate

(3.15)
$$\iint_{B(y,R)} e(x)^2 \operatorname{dist}(x, E)^{-1}\, dx \leq CR^d$$

for all $y \in E$ and $R > 0$. This condition trivially implies the a priori weaker condition
(3.16)
$$\{x \in \mathbf{R}^{d+1} \backslash E : e(x) > \epsilon\} \text{ is a Carleson set in } \mathbf{R}^{d+1} \backslash E \text{ for all } \epsilon > 0.$$

[A set $U \subseteq \mathbf{R}^{d+1} \backslash E$ is said to be a Carleson set if

$$\iint_{U \cap B(y, R)} \operatorname{dist}(x, E)^{-1}\, dx \leq CR^d$$

for all $y \in E$ and $R > 0$.] Recall from Definition I.2.43 that E is said to satisfy the WUSFE if (3.16) is true. Thus the following is just a restatement of Theorem I.2.45 (when $d > 1$), and it implies the "if" part of Theorem I.2.41 trivially.

THEOREM 3.17. *If E is a d-dimensional regular set in $\mathbf{R}^{d+1}$ and* (3.16) *holds, then E satisfies the WEC and hence is uniformly rectifiable.*

Exactly as in §§2.3 and 2.4, Theorem 3.17 can be reduced to proving that the analogue of Cauchy flatness implies that each complementary component is convex. Let us state this more precisely.

DEFINITION 3.18. We denote by CF ("Cauchy flat") the set of closed sets A in $\mathbf{R}^{d+1}$ such that A is a d-dimensional regular set and such that there is a measure α on $\mathbf{R}^{d+1}$ which satisfies the following three conditions:

(3.19) $\text{supp}\,\alpha = A$;

(3.20) α is (Ahlfors) regular, i.e., there is a $C > 0$ so that
$$C^{-1}R^d \leq \alpha(B(a,R)) \leq CR^d \text{ for all } a \in A \text{ and } R > 0;$$

(3.22) $\int_A \nabla \mathscr{E}(x-y) d\alpha(y) = 0$ for all $x \in \mathbf{R}^{d+1}\backslash A$.

Of course, (3.21) is equivalent to

(3.22) $$\int_A \nabla^2 N(x-y)\,d\alpha(y) = 0 \quad \text{for all } x \in \mathbf{R}^{d+1}\backslash A.$$

PROPOSITION 3.23. *If $A \in CF$, then each component of $\mathbf{R}^{d+1}\backslash A$ is convex.*

We shall prove this in the next four sections. This proposition implies Theorem 3.17, because the obvious analogues of Lemma 2.28 and 2.36 are true in this context, with essentially the same proofs.

The proof of Proposition 3.23 has pretty much the same structure as the proof of Proposition 2.38, but there are some nontrivial differences. For instance, we cannot use connectedness in the same way as we did in the $d = 1$ case.

3.4. Cauchy flatness implies rectifiability.

In this section we shall rely somewhat heavily on some well-known results about the rectifiability of sets. The basic reference for these results is [**Fe**] (see also [**Fl1**], [**Ma1**], [**Ma3**], [**Si**]). Remember that we use the term "rectifiable" for sets which would be called "countably rectifiable" in [**Fe**].

LEMMA 3.24. *If $A \in CF$, then A is rectifiable. In particular, almost all points of A have tangent hyperplanes.*

The rest of this section is devoted to the proof of this.

The second part of the lemma is a consequence of the first part. The only issue is to notice that approximate tangent hyperplanes to A must be true tangents, because A is regular. To prove the first part we begin with some general facts about rectifiability.

LEMMA 3.25. *Let A be a d-dimensional regular set in $\mathbf{R}^{d+1}$. Suppose that there is an $\eta > 0$ so that for each $a \in A$ and $t > 0$ there is a rectifiable set $R \subseteq A \cap B(a,t)$ such that $|A \cap R| \geq \eta t^d$. Then A is rectifiable.*

Let R_0 be the union of the R's which correspond to a countable dense set in $A \times \mathbf{R}_+$. It is easy to check that we can arrange this so that R_0 is a Borel set. We want to show that $|A\backslash R_0| = 0$.

Suppose not. Then there is a point of density of $A\backslash R_0$, i.e., an $x \in A\backslash R_0$ such that
$$\lim_{t \to 1} \frac{|B(x,t) \cap (A\backslash R_0)|}{|B(x,t) \cap A|} = 1.$$

This is impossible under the hypotheses of the lemma.

Before we state the next lemma we need some more notation. Given v in $\mathbf{R}^{d+1}$, $|v| = 1$, and a set $X \subseteq \mathbf{R}^{d+1}$, let $\rho(v, X)$ denote the d-dimensional Lebesgue measure of the projection of X onto a hyperplane that is orthogonal to v. (It does not matter which one.)

LEMMA 3.26. *Let A be a d-dimensional regular set in $\mathbf{R}^{d+1}$. Suppose that there exists $\epsilon > 0$ so that for each $a \in A$ and $t > 0$ we have*

$$\rho(v, A \cap B(a, t)) \geq \epsilon t^d$$

for a set of $v \in S^d$ with positive Lebesgue measure. Then A is rectifiable.

We are going to reduce this to the previous lemma. Fix $a \in A$ and $t > 0$. Let R be a rectifiable (Borel) subset of $A \cap B(a, t)$ which has maximal measure (the existence of which is easy and well known). Then

$$A \cap B(a, t) \setminus R$$

is totally unrectifiable, and, by a result of Federer (Besicovitch when $d = 1$),

$$\rho(v, A \cap B(a, t) \setminus R) = 0$$

for almost all $v \in S^d$. Our hypotheses now imply the necessary lower bound on $|R|$.

In our case it will be convenient to use a topological criterion for the assumptions of Lemma 3.26 to hold. Let A be a d-dimensional regular set in $\mathbf{R}^{d+1}$. Given $(a, t) \in A \times \mathbf{R}_+$, let $y(a, t) \in \mathbf{R}^{d+1} \setminus A$ be any point which satisfies

(3.27) $\qquad\qquad |y(a, t) - a| \leq \dfrac{t}{2}$, and

(3.28) $\qquad\qquad \mathrm{dist}(y(a, t), A) \geq \beta t$.

Such a point exists if β is small enough, and we can take $\beta \in (0, \frac{1}{2})$ to depend only on d and the regularity constant for A. Let $U(a, t)$ be the connected component of $\mathbf{R}^{d+1} \setminus A$ which contains $y(a, t)$.

LEMMA 3.29. *With the notation and assumptions of the preceding paragraph, A satisfies the hypotheses of Lemma 3.26 if*

(3.30) $\qquad\qquad H^{d+1}(B(a, t) \setminus U(a, t)) \geq \theta t^{d+1}$

for some $\theta > 0$ and all $(a, t) \in A \times \mathbf{R}_+$.

Let $(a, t) \in A \times \mathbf{R}_+$ be given. Because of (3.30) and (3.28) we can find $z \in B(a, t)$ such that $|z - y(a, t)| \geq \beta t$ and

(3.31) $\qquad H^{d+1}(B(z, \beta t/10) \cap B(a, t) \setminus U(a, t)) \geq C^{-1} \theta t^{d+1}$.

It is not too hard to check that (3.31) implies that

$$\rho(v, A \cap B(a, t)) \geq C^{-1} \theta t^d$$

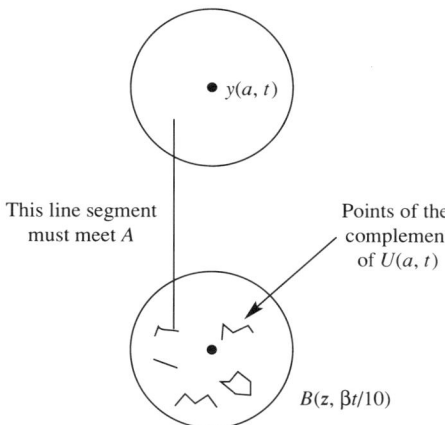

FIGURE 3.1

when v is sufficiently close to $(z - y(a, t))/|z - y(a, t)|$, using the definition of $U(a, t)$. [The main point is that if $p \in B(z, \beta t/10) \cap B(a, t)$ and $p \notin U(a, t)$, then any line segment which connects p to $B(y(a, t), \beta t)$ must intersect A. See Figure 3.1.] This proves Lemma 3.29.

Let us now prove Lemma 3.24. Let $A \in CF$ be given, and let the measure α be as in Definition 3.18. Let $y(a, t)$, β, and $U(a, t)$ be as above. We want to show that (3.30) is satisfied.

Fix $(a, t) \in A \times \mathbf{R}_+$, and set $y_0 = y(a, t)$. Define $F : \mathbf{R}^{d+1} \to \mathbf{R}$ by

$$(3.32) \quad F(x) = \int_A [N(x - u) - N(y_0 - u) - \nabla N(y_0 - u) \cdot (x - y_0)] \, d\alpha(u).$$

The integral converges at ∞ because

$$|N(x - u) - N(y_0 - u) - \nabla N(y_0 - u) \cdot (x - y_0)| = O(|u|^{-d-1})$$

as $|u| \to \infty$ (for each fixed x). Notice that F is continuous, even on A, because the singularity of $N(x - u)$ is sufficiently mild.

The Cauchy flatness of A implies that $\nabla^2 F = 0$ on $\mathbf{R}^{d+1} \backslash A$. By construction we have that $F(y_0) = \nabla F(y_0) = 0$, and hence

$$(3.33) \quad F(x) = 0 \quad \text{for all } x \in U(a, t).$$

On the other hand $\Delta F = \alpha$ in the sense of distributions on $\mathbf{R}^{d+1}$. Thus if ϕ is a smooth bump function on $\mathbf{R}^{d+1}$ such that $\operatorname{supp} \phi \subseteq B(a, t)$, $\phi \geq 0$, $\phi = 1$ on $B(a, t/2)$, and $|\Delta \phi| \leq Ct^{-2}$, then we have

$$(3.34) \quad \begin{aligned} \alpha(B(a, t/2)) &\leq \int \phi \, d\alpha = \int (\Delta \phi) F \\ &\leq Ct^{-2} \Big(\sup_{x \in B(a, t)} |F(x)| \Big) H^{d+1}(\operatorname{supp} F \cap B(a, t)). \end{aligned}$$

Observe that

$$(3.35) \quad |F(x)| \leq Ct \quad \text{when } x \in B(a, t).$$

This is not hard to check, using the easy fact that

$$|N(x-u) - N(y_0 - u) - \nabla N(y_0 - u) \cdot (x - y_0)| \leq Ct^2(|x-u|+t)^{-d-1}$$

when $x \in B(a, t)$ and $u \notin B(a, 2t)$ and brutal estimates when u lies in $B(a, 2t)$.

Once you have (3.35), (3.34) yields

$$t^{d+1} \leq CH^{d+1}(\operatorname{supp} F \cap B(a, t)).$$

Combining this with (3.33) we get that $H^{d+1}(B(a, t) \backslash U(a, t)) \geq C^{-1} t^{d+1}$. The rectifiability of E now follows from Lemmas 3.29 and 3.26. This proves Lemma 3.24.

3.5. The analogue of Proposition 2.59.

We want to give a reformulation of the Cauchy flatness condition (3.21) that applies to all C^1 functions with compact support, just as Proposition 2.59 does when $d = 1$. We do this by following the obvious analogy with §2.6, although the algebra and the notation will become a little more complicated.

Let A be Cauchy flat, and let α be as in Definition 3.18. Let A' denote the set of $a \in A$ for which there is a tangent hyperplane P_a at a. Thus $A \backslash A'$ has measure zero and

$$(3.36) \qquad \lim_{r \to 0} \sup_{x \in B(a, r)} \left(r^{-1} \operatorname{dist}(x, P_a) \right) = 0$$

for all $a \in A'$. Let $T_a A$ denote the translation of P_a to a hyperplane through the origin (when $a \in A'$).

We want to define a differential operator H associated to A as follows. Let $U \subseteq \mathbf{R}^{d+1}$ be open, and let $f : U \to \mathscr{C}(d+1)$ be a C^1 function. Given $a \in A' \cap U$, let $Hf(a)$ be the unique linear mapping from $\mathbf{R}^{d+1}$ into $\mathscr{C}(d+1)$ with the following properties:

$$(3.37) \qquad Hf(a)(v) = D_v f(a) \text{ when } v \in T_a A;$$

(3.38)

$$Hf(a) : \mathbf{R}^{d+1} \to \mathscr{C}(d+1) \text{ is a Clifford holomorphic linear mapping.}$$

[Remember that D_v denotes directional differentiation in the direction v.] The existence and uniqueness of $Hf(a)$ follows from Lemma 3.8 and Corollary 3.10. It is also easy to see that $Hf(a)$ is a measurable function of a. Notice that $Hf(a)(v) = D_v f(a)$ for all $v \in \mathbf{R}^{d+1}$ if f is Clifford holomorphic at a.

PROPOSITION 3.39. *Assume, as above, that A is Cauchy flat and that α is as in Definition 3.18. Then*

$$(3.40) \qquad \int_A Hg(a)(v) \, d\alpha(a) = 0$$

for all C^1 functions $g : \mathbf{R}^{d+1} \to \mathscr{C}(d+1)$ with compact support and all $v \in \mathbf{R}^{d+1}$.

Notice that (3.40) implies (3.21), as you see by taking $g(y) = \mathscr{E}(x - y)$ for any fixed $x \in \mathbf{R}^{d+1} \backslash A$ and using a simple approximation argument.

In the next section we take up the issue of putting (3.40) in a form which is more convenient for application.

The proof of Proposition 3.39 is essentially the same as that of Proposition 2.59. We shall not repeat it in detail, but we shall walk through the main steps.

Let $g : \mathbf{R}^{d+1} \to \mathscr{C}(d+1)$ be C^1, compactly supported, and given. The first step is to construct $G : \mathbf{R}^{d+1} \to \mathscr{C}(d+1)$ in such a way that $G = g$ on A, G is Lipschitz of order 1, compactly supported, and satisfies

$$(3.41) \qquad \lim_{r \to 0} \sup\{|\delta G(x)| : x \in B(a, r) \text{ and } \operatorname{dist}(x, A) \geq \eta r\} = 0$$

for all $a \in A'$ and $\eta > 0$. [Recall that the differential operator δ was defined in (3.2).]

The construction of G and the verification of these properties is completely analogous to the corresponding steps in §2.6. You take a Whitney decomposition $\{Q_i\}$ of $\mathbf{R}^{d+1} \backslash A$, you choose d-planes P_i which usually approximate A well near Q_i, you choose Clifford holomorphic affine functions $h_i : \mathbf{R}^{d+1} \to \mathscr{C}(d+1)$ that approximate g well on P_i near Q_i, and you combine the h_i's using a partition of unity associated to the Q_i's. The necessary changes are mostly cosmetic, although occasionally annoying. [When you write down the various formulas, you must sometimes be a little more careful because of the noncommutativity of the Clifford algebra. This is true, for instance, with (2.67).]

There is one part that is a little more complicated in the present situation, namely, the analogue of the estimation of $|h_i(z) - h_j(z)|$ which culminates just before (2.73). For the sake of clarity let us state a lemma which incorporates this issue in an explicit and precise manner.

LEMMA 3.42. *Let $\delta \in (0, 1)$ be given. Let f_1, $f_2 : \mathbf{R}^{d+1} \to \mathscr{C}(d+1)$ be a pair of affine mappings such that*

$$\sup_{B(0,1)} |f_i| \leq 1 \quad \text{and} \quad \sup_{B(0,1)} |f_1 - f_2| \leq \delta.$$

Let P_1 and P_2 be two d-planes in $\mathbf{R}^{d+1}$, with P_1 passing through the origin, such that

$$\sup\{\operatorname{dist}(p, P_2) : p \in P_1 \cap B(0, 1)\} + \sup\{\operatorname{dist}(q, P_1) : q \in P_2 \cap B(0, 1)\} \leq \delta.$$

Let h_1, $h_2 : \mathbf{R}^{d+1} \to \mathscr{C}(d+1)$ be the holomorphic affine functions which satisfy $h_i = f_i$ on P_i, $i = 1, 2$. Then

$$\sup_{B(0,1)} |h_1 - h_2| \leq C(d)\delta.$$

The proof of this is rather straightforward. Here is an outline.
It is easy to see that

$$\left|\frac{\partial}{\partial x_j}(f_1 - f_2)\right| \leq C\delta, \qquad 1 \leq j \leq d+1.$$

A slightly trickier fact is that

$$\left|\frac{\partial}{\partial x_j}(h_1 - h_2)\right| \leq C\delta, \qquad 1 \leq j \leq d+1.$$

To prove this you need to be able to compute the derivatives of the h_i's in terms of the P_i's and the derivatives of the f_i's. This can be accomplished using Lemma 3.8, for instance.

It remains to check that

$$|h_1(0) - h_2(0)| \leq C\delta.$$

Let z be the point on P_2 which is as close to 0 as possible, so that $|z| \leq \delta$. Then

$$\begin{aligned}
|h_1(0) - h_2(0)| &= |f_1(0) - h_2(0)| \\
&\leq |f_1(0) - f_2(0)| + |f_2(0) - f_2(z)| + |f_2(z) - h_2(0)| \\
&\leq \delta + |\nabla f_2|\delta + |h_2(z) - h_2(0)| \leq C\delta,
\end{aligned}$$

as desired. [We are using here the fact that $|\nabla f_2| \leq C$ and $|\nabla h_2| \leq C$.]

Now let us explain how Lemma 3.42 should be used in our situation. Actually, we do not use it but rather its obvious extension to all balls (and not just the unit ball).

Let h_i and h_j be as before, let Q_i and Q_j be the associated Whitney cubes, and let P_i and P_j be the associated d-planes. We are supposed to have chosen h_i so that it approximates g well on P_i near Q_i, and similarly for h_j. In practice, we usually choose h_i so that

$$h_i|_{P_i} = f_i|_{P_i},$$

where f_i is the affine Taylor approximation to g at some point on P_i near Q_i, and similarly for h_j and f_j.

In the relevant cases Q_i and Q_j are small cubes of roughly the same size which are not too far from each other. This allows us to control $|f_1 - f_2|$ using the assumption that g is C^1. In general, we also have that P_i and P_j are very close together, because they are both trying to approximate the same piece of A. The extended version of the lemma can then be applied to get suitable estimates for $|h_i - h_j|$.

Once we have G the next step is to use the formula

$$G(x) = \iint_{\mathbf{R}^{d+1}} \mathscr{E}(x-y)\delta G(y)\, dy,$$

3.5. THE ANALOGUE OF PROPOSITION 2.59

which holds for all $x \in \mathbf{R}^{d+1}$. In particular we have

(3.43) $$g(x) = \iint_{\mathbf{R}^{d+1}} \mathscr{E}(x-y)\delta G(y)\,dy$$

for all $x \in A$. To make use of this we need the following version of Lemma 2.76.

LEMMA 3.44. *For each $a \in A'$ and $v \in \mathbf{R}^{d+1}$ we have that*

$$\lim_{r \to 0} \left\{ \frac{1}{r^{d+1}} \iint_{B(a,r)} |D_v G(x) - Hg(a)(v)|\,d\alpha(x) \right\} = 0.$$

This is superficially somewhat different from Lemma 2.76; in our story we are taking δ to be the analogue in $\mathbf{R}^{d+1}$ of $\overline{\partial}$ for $\mathbf{C}$, and there is no proper analogue of ∂ for this higher-dimensional business. This apparent difference does not really matter, though, because we constructed G to be "asymptotically holomorphic" at A'.

At any rate one can prove Lemma 3.44 using the same techniques as for the proof of Lemma 2.76. The necessary modifications are pretty straightforward; the main point is to use Lemma 3.8 sometimes, like we did for Lemma 3.42. We should point out that Lemma 3.44 is not supposed to be derived from the properties of G that we listed before but rather from the construction of G.

Although the last part of the proof of Proposition 3.39 is quite similar to that of Proposition 2.59, there are some slightly nontrivial changes which are needed, so we go through it in a little more detail.

Let θ be a smooth real-valued radial bump function on $\mathbf{R}^{d+1}$ such that $\operatorname{supp} \theta \subseteq B(0, 1)$ and $\iint_{\mathbf{R}^{d+1}} \theta(x)\,dx = 1$. Define θ_t on $\mathbf{R}^{d+1}$ by $\theta_t(x) = t^{-d-1}\theta(t^{-1}x)$. Then Lemma 3.44 implies that

$$\lim_{t \to 0} \theta_t * (D_v G)(a) = Hg(a)(v)$$

for all $a \in A'$ and $v \in \mathbf{R}^{d+1}$. Hence

(3.45) $$\lim_{t \to 0} \int_A |\theta_t * (D_v G)(a) - Hg(a)(v)|\,d\alpha(a) = 0$$

for all $v \in \mathbf{R}^{d+1}$, because of the dominated convergence theorem. (Remember that G is Lipschitz and has compact support.) In order to verify (3.40) it is therefore enough to show that

(3.46) $$\lim_{t \to 0} \int_A \theta_t * (D_v G)(a)\,d\alpha(a) = 0$$

for all $v \in \mathbf{R}^{d+1}$.

Fix $v \in \mathbf{R}^{d+1}$. Let T denote the linear operator which satisfies

(3.47) $$T(f) = D_v(\mathscr{E} * f)$$

for, say, all $f \in L^2(\mathbf{R}^{d+1})$. [The right-hand side should be interpreted in the sense of distributions.] Standard results in Fourier analysis imply that T can be expressed as a sum of two pieces, one of which is a Calderón-Zygmund convolution singular integral operator, and the other being a constant multiple of the identity. (See Theorem 6 on p. 75 of [St], for instance. Note that the "constant" which was just mentioned will lie in the Clifford algebra.)

We know that G is a compactly supported Lipschitz function, so $G = \mathscr{E} * (\delta G)$ and hence

$$D_v G = T(\delta G)$$

a.e. on $\mathbf{R}^{d+1}$ and in the sense of distributions. Set $k_t = T(\theta_t)$, so that k_t is a smooth function on $\mathbf{R}^{d+1}$ which satisfies

(3.48) $$|k_t(x)| \leq C(t + |x|)^{-d-1}.$$

[This too follows from standard results in Fourier analysis.] We have that

$$\theta_t * (D_v G) = k_t * (\delta G) \quad \text{on } \mathbf{R}^{d+1},$$

and hence

(3.49) $$\int_A \theta_t * (D_v G) \, d\alpha = \int_A k_t * (\delta G) \, d\alpha = \iint_{\mathbf{R}^{d+1}} (k_t * \alpha)(x) \delta G(x) \, dx.$$

[We are also using here the fact that k_t is even. Note that k_t takes values in $\mathscr{C}(d+1)$, so one must be a little careful here, because of the noncommutativity. Notice also that all the integrals in (3.49) converge absolutely.]

LEMMA 3.50. *The support of $k_t * \alpha$ is contained in $A_t = \{x \in \mathbf{R}^{d+1} : \text{dist}(x, A) \leq t\}$, and*

$$|k_t * \alpha(x)| \leq Ct^{-1} \quad \text{for all } x \in \mathbf{R}^{d+1}.$$

The second part follows from (3.48) and the regularity condition (3.20). Let us prove the first part. Let $x \in \mathbf{R}^{d+1} \setminus A_t$ be given, and consider

(3.51) $$k_t * \alpha(x) = \int_A k_t(x - y) \, d\alpha(y).$$

For $y \in A$ we have

(3.52) $$k_t(x - y) = \iint_{\mathbf{R}^{d+1}} \theta_t(x - u)(D_v \mathscr{E})(u - y) \, du,$$

by definition of k_t and T. This is a true formula in the most classical sense, i.e., there are no singularities or anything like that: if $u \in \mathbf{R}^{d+1}$ satisfies $\theta_t(x - u) \neq 0$, then $\text{dist}(u, A) \geq \eta t$ for some $\eta > 0$ (which depends only on θ), so $|u - y| \geq \eta t$ for all $y \in A$. Substituting (3.52) into (3.51) and applying Fubini we get that $k_t * \alpha(x) = 0$ by (3.21), as desired.

Combining Lemma 3.50 with (3.49) we get that

$$\left| \int_A \theta_t * (D_v G) \, d\alpha \right| \leq C t^{-1} \iint_{A_t} |\delta G(x)| \, dx. \tag{3.53}$$

The proof that this tends to 0 as $t \to 0$ is practically the same as in the $d = 1$ case (given right after (2.88)). We omit the details.

This completes our discussion of the proof of Proposition 3.39.

3.6. Cauchy flatness implies weak flatness.

We want to reformulate the conclusion of Proposition 3.39 to get a condition that is easier to use. As a practical matter that means getting rid of the Clifford algebras. That is what we shall do in this section.

Let A be Cauchy flat, and let α be as in Definition 3.18. We shall continue to use the notation established in the preceding section.

For each $a \in A'$ let $\nu(a)$ be a choice of unit normal to $T_a A'$. This is only determined up to multiplication by -1, but in our most important formulae there will be an even number of ν's and so this will not matter. It is convenient, though, to require that $\nu(a)$ be a measurable function of a.

Let g be a real-valued C^1 function on $\mathbf{R}^{d+1}$ which has compact support. It is easy to see that we suffer no loss in generality by restricting our attention to real-valued functions, in the sense that (3.40) must be true in general if it is true in the real-valued case.

Define $Ng : A' \to \mathscr{C}(d+1)$ by $Ng(a) = Hg(a)(\nu(a))$. The definition of Hg implies that $Ng(a)$ can be computed in terms of the tangential gradient of g at a, and we want to calculate it explicitly.

Fix $a \in A'$ for the moment, and let $\{w_j\}_{j=1}^d$ be an orthonormal basis for $T_a A$. By definitions,

$$Hg(a)(u) = Ng(a) \langle u, \nu(a) \rangle + \sum_{j=1}^d D_{w_j} g(a) \langle u, w_j \rangle. \tag{3.54}$$

We also have that $Hg(a)(u)$ is Clifford holomorphic in u, and that

$$\delta = \nu(a) D_{\nu(a)} + \sum_{j=1}^d w_j D_{w_j},$$

by Lemma 3.8. [This might be a good time to remind the reader about Convention 3.1.] Hence

$$\nu(a) Ng(a) + \sum_{j=1}^d w_j D_{w_j} g(a) = 0,$$

and therefore

$$Ng(a) = \nu(a) \sum_{j=1}^d w_j D_{w_j} g(a) \tag{3.55}$$

since $\nu(a)^2 = -1$.

In order to make use of this we need some simple algebraic observations. For each l in $\{0, 1, \ldots, d+1\}$ let $\mathscr{E}_l(d+1)$ denote the vector subspace of $\mathscr{E}(d+1)$ of elements of degree l, i.e.,

$$\mathscr{E}_l(d+1) = \mathrm{span}\{e_{i_1} e_{i_2} \cdots e_{i_l} : i_1 < i_2 < \cdots < i_l\}.$$

Thus $\mathscr{E}_0(d+1) = \mathbf{R}$, $\mathscr{E}_1(d+1) = \mathbf{R}^{d+1}$, and $\mathscr{E}(d+1)$ is the vector space direct sum of $\mathscr{E}_l(d+1)$, $l = 0, 1, \ldots, d+1$.

LEMMA 3.56. *If $x, y \in \mathscr{E}_1(d+1) = \mathbf{R}^{d+1}$ and if $\langle x, y \rangle = 0$, then $xy \in \mathscr{E}_2(d+1)$.*

This is trivial but also very useful. From (3.55) we see that $Ng(a)$ lies in $\mathscr{E}_2(d+1)$, while the second piece on the right side of (3.54) is real. Thus if $u \in \mathbf{R}^{d+1}$ is arbitrary and $u_n(a)$ and $u_t(a)$ denote its normal and tangential parts at a, then

$$Hg(a)(u_n(a)) \in \mathscr{E}_2(d+1) \quad \text{and} \quad Hg(a)(u_t(a)) \in \mathbf{R}.$$

This implies the following.

LEMMA 3.57. *Let A be Cauchy flat, and let α be as in Definition 3.18. Then*

$$(3.58) \qquad \int_A Hg(a)(u_n(a))\, d\alpha(a) = 0$$

and

$$(3.59) \qquad \int_A Hg(a)(u_t(a))\, d\alpha(a) = 0$$

for all $u \in \mathbf{R}^{d+1}$ and all C^1 functions $g : \mathbf{R}^{d+1} \to \mathbf{R}$ with compact support.

Of course,

$$Hg(a)(u_t(a)) = (D_{u_t} g)(a),$$

so (3.59) becomes

$$(3.60) \qquad \int_A D_{u_t} g\, d\alpha = 0$$

for all $u \in \mathbf{R}^{d+1}$ and all g as above. This is equivalent to saying that the rectifiable varifold which is associated to A and α is stationary. We will not need the precise statement, but, roughly speaking, this says that A and α define a kind of generalized minimal surface.

This is not as strong a condition as in the $d = 1$ case. There are plenty of examples of minimal surfaces for which it is not true that each component of the complement is convex, e.g., a catenoid. In our case, however, we have more information, coming from (3.58).

Let us calculate (3.58) more explicitly. We have that

$$Hg(a)(u_n(a)) = Ng(a)\langle u, \nu(a)\rangle = \nu(a)(\delta g(a) - \nu(a)D_{\nu(a)}g(a))\langle u, \nu(a)\rangle,$$

because of (3.55) and Lemma 3.8. Hence (3.58) implies that

(3.61) $$\int_A (\nu\delta g + D_\nu g)\langle u, \nu\rangle\,d\alpha = 0.$$

Let us write this out in terms of the e_j's. Let ν_j, $1 \le j \le d+1$, denote the components of ν. Then

(3.62) $$\nu\delta g + D_\nu g = \left\{\sum_{j=1}^{d+1}\nu_j e_j\right\}\left\{\sum_{k=1}^{d+1}e_k D_{e_k}g\right\} + \sum_{j=1}^{d+1}\nu_j D_{e_j}g$$
$$= \sum_{1 \le j < k \le d+1}\{\nu_j D_{e_k}g - \nu_k D_{e_j}g\}e_j e_k.$$

[The $j = k$ terms from the product cancel with the last sum on the first line.] Combining this with (3.61) and taking $u = e_l$ we get the following.

PROPOSITION 3.63. *Suppose that A is Cauchy flat, and let α be as in Definition 3.18. Let ν denote the (almost everywhere defined) unit normal vector on A. Then*

(3.64) $$\int_A \nu_l\{\nu_j D_{e_k}g - \nu_k D_{e_j}g\}\,d\alpha = 0$$

for all C^1 functions $g: \mathbf{R}^{d+1} \to \mathbf{R}$ with compact support and all j, k, l in $\{1, 2, \ldots, d+1\}$.

[To be fastidious, our argument gives this only when $j \ne k$, but the $j = k$ case is trivial.]

Proposition 3.63 is the main result of this section. We shall show in the next section how it implies that each component of $\mathbf{R}^{d+1}\setminus A$ is convex. We shall spend the rest of this section making some remarks about it.

Let us first show that (3.64) implies (3.60) and, hence, (3.59). If you set $l = j$ in (3.64) and sum in j, you get

(3.65) $$\int_A \{D_{e_k}g - \nu_k D_\nu g\}\,d\alpha = 0.$$

This is simply (3.60) with $u = e_k$.

Observe next that (3.64) really involves only tangential derivatives of g, because the total contribution of the normal derivatives is

$$\nu_l\{\nu_j\nu_k D_\nu g - \nu_k\nu_j D_\nu g\},$$

which is zero. Of course, this had to be true, because it was true of Hg.

We shall refer to the conclusion of Proposition 3.63 as weak flatness. It has much the same relationship with truly flat sets (i.e., d-planes) as (3.60) has with minimal surfaces. To make this clearer let us see how weak flatness implies that A is a d-plane if you assume a priori that A is smooth (and connected). In this case we can take ν to be a smooth choice of unit normal.

Define the first-order differential operator τ_k by

(3.66) $$\tau_k g(a) = D_{e_k} g(a) - \nu_k(a) D_{\nu(a)} g(a)$$

for $a \in A'$. Thus τ_k is just the tangential part of D_{e_k}. As noted above, (3.64) is equivalent to

(3.67) $$\int_A \nu_l \{\nu_j \tau_k g - \nu_k \tau_j g\} \, d\alpha = 0.$$

We want to show that this implies that

(3.68) $$\tau_a(\nu_b) = 0 \quad \text{for } a, b \in \{1, \ldots, d+1\}$$

(so that ν is constant).

Rewrite (3.65) as

(3.69) $$\int_A \tau_k g \, d\alpha = 0.$$

Take $g = \nu_k \phi$, where $\phi : \mathbf{R}^{d+1} \to \mathbf{R}$ is C^1 and compactly supported, and sum in k. This gives

$$0 = \int_A \sum_{k=1}^{d+1} \tau_k(\nu_k \phi) \, d\alpha = \int_A \phi \left(\sum_{k=1}^{d+1} \tau_k(\nu_k) \right) d\alpha + \int_A \sum_{k=1}^{d+1} \nu_k \tau_k(\phi) \, d\alpha.$$

This last piece vanishes, because $\sum_{k=1}^{d+1} \nu_k \tau_k = 0$, as one can easily verify using (3.66). Hence

$$\int_A \phi \left(\sum_{k=1}^{d+1} \tau_k(\nu_k) \right) d\alpha = 0,$$

and since this is true for all ϕ, we get

(3.70) $$\sum_k \tau_k(\nu_k) = 0.$$

In other words, we have just given the standard calculation for showing that (3.65) implies that the mean curvature vanishes.

Now take $g = \nu_k \phi$ in (3.67) and sum in k. This gives

$$0 = \int_A \nu_l \sum_{k=1}^{d+1} \left\{ \nu_j \phi \tau_k(\nu_k) + \nu_j \nu_k \tau_k(\phi) - \nu_k \phi \tau_j(\nu_k) - \nu_k^2 \tau_j(\phi) \right\} d\alpha.$$

The first term vanishes because of (3.70), while the second term vanishes because $\sum_{k=1}^{d+1} \nu_k \tau_k = 0$. The third term is also zero, because

$$\sum_{k=1}^{d+1} \nu_k \tau_j(\nu_k) = \frac{1}{2} \sum_{k=1}^{d+1} \tau_j(\nu_k^2) = \frac{1}{2} \tau_j(1) = 0.$$

Thus we get that $\int_A \nu_l \tau_j(\phi)\, d\alpha = 0$. Because of (3.69) this implies that $\int_A \phi \tau_j(\nu_l)\, d\alpha = 0$, and since this is true for all ϕ, we get (3.68), as desired.

Although we have only been concerned with flatness here, we could also use the left side of (3.67) to define a generalized second fundamental form for (A, α) even when it is not supposed to vanish, in much the same way that (3.69) is used to define a generalized mean curvature. A notion of generalized second fundamental form has been given by Hutchinson [H], but it is different from this one. Actually, ours is really a piece of his, and in particular ours vanishes when his does. The converse is not true, however, because our notion of weak flatness allows sets which have "propellors" without being a finite union of planes.

3.7. Weak flatness implies exterior convexity. Let A be Cauchy flat, and let α be as in Definition 3.18. We shall continue to use the same notation as in the preceding section. We want to use Proposition 3.63 to prove that every component of $\mathbf{R}^{d+1}\setminus A$ is convex.

Lemma 2.92 generalizes in the obvious way to reduce the problem to the following. Let $p, q \in \mathbf{R}^{d+1}\setminus A$ and $\epsilon > 0$ be given, and suppose that

(3.71) $$\operatorname{dist}(p, A) > \epsilon, \quad \operatorname{dist}(q, A) > \epsilon,$$

(3.72) $$\operatorname{dist}(S(p, q), A) > \epsilon/4,$$

where $S(p, q)$ is the line segment which joins p to q. If we can prove that we must have

(3.73) $$\operatorname{dist}(S(p, q), A) \geq \epsilon/2$$

under these circumstances, then we shall be finished.

Without loss of generality we may as well assume that $p = 0$ and $q = e_{d+1}$, because of the natural invariance under translations, dilations, and rotations in this story. Let $\pi : \mathbf{R}^{d+1} \to \mathbf{R}^d$ be the projection which forgets the last coordinate. Let Γ_0 and Γ_1 denote the open cylinders

$$\Gamma_0 = \{x \in \mathbf{R}^{d+1} : 0 < x_{d+1} < 1, \ |\pi(x)| < \epsilon/4\},$$
$$\Gamma_1 = \{x \in \mathbf{R}^{d+1} : 0 < x_{d+1} < 1, \ |\pi(x)| < \epsilon/2\}.$$

(See Figure 3.2.) Our assumption (3.72) implies that $A \cap \Gamma_0 = \varnothing$, and our goal is to prove

(3.74) $$\Gamma_1 \cap A = \varnothing,$$

since that and (3.71) imply (3.73).

LEMMA 3.75. *Under these assumptions*

$$\nu_{d+1} = 0 \quad \text{a.e. on } A \cap \Gamma_1.$$

This will of course take us a long way toward our goal of proving (3.74).

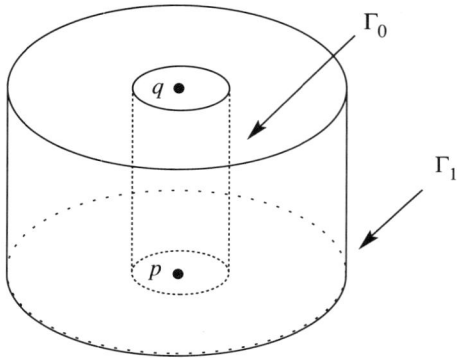

FIGURE 3.2

We want to use (3.64). Let ϕ be any C^1 function on $\mathbf{R}^d$ such that $\operatorname{supp} \phi \subseteq B(0, \epsilon/2)$. For any such function we can find a C^1 function $g : \mathbf{R}^{d+1} \to \mathbf{R}$ such that $\operatorname{supp} g \subseteq \Gamma_1$ and $g = \phi \circ \pi$ on a neighborhood of $A \cap \Gamma_1$. [We can do this because A does not touch the top or bottom of Γ_1.] Applying (3.64) with $j = l = d+1$ we get that

$$(3.76) \qquad \int_A \nu_{d+1}^2 (D_{e_k} g) \, d\alpha = 0$$

for $k = 1, 2, \ldots, d$.

Let us take ϕ to be of the form

$$\phi(u) = u_k \beta \left(\sum_{j=1}^d u_j^2 \right),$$

where $\beta(t)$ is a C^1 function on $\mathbf{R}$ which is supported in $(0, (\epsilon/2)^2)$. Then (3.76) yields

$$\int_{A \cap \Gamma_1} \nu_{d+1}(y)^2 \left\{ \beta \left(\sum_{j=1}^d y_j^2 \right) + 2 y_k^2 \beta' \left(\sum_{j=1}^d y_j^2 \right) \right\} d\alpha(y) = 0.$$

Summing in k we get that

$$\int_{A \cap \Gamma_1} \nu_{d+1}(y)^2 \{ d\beta(|\pi(y)|^2) + 2|\pi(y)|^2 \beta'(|\pi(y)|^2) \} \, d\alpha(y) = 0.$$

We can write this more conveniently as

$$(3.77) \qquad \int_{A \cap \Gamma_1} \nu_{d+1}(y)^2 \gamma(|\pi(y)|^2) d\alpha(y) = 0,$$

where $\gamma(t) = d\beta(t) + 2t\beta'(t) = 2t^{1-\frac{d}{2}} \{ t^{\frac{d}{2}} \beta(t) \}'$.

For each $s \in ((\epsilon/4)^2, (\epsilon/2)^2)$ we can choose $\beta(t)$ so that it is C^1 and supported in $(0, (\epsilon/2)^2)$ and also so that $\gamma \geq 1$ on $\left[(\epsilon/4)^2, s \right]$ and $\gamma \geq 0$

on $[s, (\epsilon/2)^2]$. Since s is arbitrary, we get $\nu_{d+1} = 0$ a.e. on $A \cap \Gamma_1$ from (3.77) and the fact that $A \cap \Gamma_0 = \varnothing$. This proves the lemma.

Now we want to use (3.64) again but with different choices. As before let ϕ be any C^1 function on $\mathbf{R}^d$ with $\operatorname{supp} \phi \subseteq B(0, \epsilon/2)$, and let $\theta : \mathbf{R} \to \mathbf{R}$ be a C^1 function with $\operatorname{supp} \theta \subseteq (0, 1)$. Define $g : \mathbf{R}^{d+1} \to \mathbf{R}$ by $g(x) = \phi(\pi(x))\theta(x_{d+1})$ so that g is C^1 and supported in Γ_1. Applying (3.64) with $k = d+1$ and $l = j$ we get that

$$\int_{A \cap \Gamma_1} \nu_j(y)^2 \phi(\pi(y)) \theta'(y_{d+1}) \, d\alpha(y) = 0.$$

Because A does not touch the top or bottom of Γ_1, we may choose θ so that it also satisfies $\theta'(y_{d+1}) = 1$ when $y \in A \cap \Gamma_1$. With this choice we get

$$\int_{A \cap \Gamma_1} \nu_j(y)^2 \phi(\pi(y)) \, d\alpha(y) = 0.$$

Summing in j gives

$$\int_{A \cap \Gamma_1} \phi(\pi(y)) \, d\alpha(y) = 0,$$

and since ϕ is arbitrary, we obtain our desired conclusion that $A \cap \Gamma_1 = \varnothing$. [This uses also the fact that Γ_1 is open and A is regular.]

This completes the proof of Proposition 3.23.

3.8. Some remarks about the higher-codimension case.

It is natural to ask whether there is a reasonable version of Theorems I.2.41 and I.2.45 for the case where $d < n - 1$. This question leads immediately to the problem of formulating a version of the square function estimates when $d < n - 1$. An obvious guess would be to look at

$$(3.78) \qquad \int_E \nabla^2 N_d(x - y) \, dy, \qquad x \in \mathbf{R}^n \setminus E,$$

where E is a d-dimensional regular set in $\mathbf{R}^n$, and $N_d(x) = |x|^{1-d}$ when $d > 1$, $N_1(x) = \log |x|$. In order for this to work, though, one really should have that (3.78) vanishes when E is a d-plane. It is easy to see that this is not true when $d < n - 1$.

There are a couple of ways to try to find a reasonable substitute for the square function estimates in higher codimensions. One approach would be to try to work with differential forms. This is natural for the following topological reason. In the codimension 1 case one is interested in pairs of points in the complement and knowing when they lie in the same complementary component, or, equivalently, when they can be joined by a curve which does not touch the set. When the codimension is larger than 1 one should be interested in topological spheres (of the correct dimension) in the complement and in whether they can be filled in without touching the set. In the codimension 1 situation Cauchy integrals can be used to detect whether two points lie in the same complementary component, while in higher codimensions there are similar analytic objects given in terms of differential forms which can be used to detect the linking of a complementary sphere with the given set. Perhaps there is a reasonable version of "square function estimates" in terms of these analytic objects.

An aspect of this approach which is attractive is that the algebra associated to differential forms and exterior differentiation may permit you to avoid the problem that we ran into with (3.78), namely, the fact that it was nonvanishing even when E is a d-plane. More precisely, there might be some natural quantity with the property that its gradient is not zero but its exterior derivative is.

The problem with this approach is that it does not seem to let you start off with merely a set E and Hausdorff measure on E. It needs more structure on E, so that you can integrate differential forms on it.

Another approach would be to go back to (3.78) and look at some other conditions on it. For example, although (3.78) does not vanish when E is a d-plane, its rank—as a matrix-valued function—is always at most $n-d$, and it has other rather special features. You could look at versions of "square function estimates" on a general set E which imposes approximate versions of these conditions, e.g., estimates on the determinants of minors of (3.78) (viewed as a matrix-valued function). It is easy to imagine how the ensuing nonlinearities could be difficult to manage.

At this point it is probably a good idea to look again at the two features of Clifford analysis—(3.6) and (3.7) in §1—which we identified as being crucial. It is not so easy to see how to build a version of Clifford analysis which is good for the higher-codimension case and which satisfies appropriate versions of these properties. It could well not exist.

Notice that many of the methods that we have used in this and the preceding chapter do lend themselves reasonably well to higher codimensions even though we lack a statement for the result we want. If you had a good notion of square function estimates when $d < n-1$, then you ought to be able to show that uniformly rectifiable sets satisfy them by using corona decompositions. On the other hand, the results of Chapter 3 in Part II include criteria for uniform rectifiability which provide good generalizations of the WEC to higher codimensions.

CHAPTER 4

Approximating Lipschitz Functions by Affine Functions

This chapter is devoted to the proof of Theorem 2.49 in Part I. The first half of that result was noted previously in §4 of [**DS3**], but only a rough outline of the argument was given there, so we provide a fairly detailed proof in §4.1. The second half of Theorem I.2.49 is taken up in the second section of this chapter. In the third section we discuss a more abstract version of the WALA.

4.1. The direct estimates.

Let E be a d-dimensional regular set in $\mathbf{R}^n$. We want to show how the uniform rectifiability of E has certain consequences for the behavior of Lipschitz functions on E which are analogous to known results for Lipschitz functions on $\mathbf{R}^d$.

We first need to introduce some notation. Given $f : E \to \mathbf{R}$, $1 \leq q \leq \infty$, and $K > 0$, define $\gamma_q^{(K)}(x, t)$ on $E \times \mathbf{R}_+$ by

$$(4.1) \qquad \gamma_q^{(K)}(x, t) = t^{-1} \inf_a \left\{ \left(t^{-d} \int_{B(x,t) \cap E} |f - a|^q \right)^{1/q} \right\},$$

where the infimum is taken over the set of all affine functions $a : \mathbf{R}^n \to \mathbf{R}$ such that $|\nabla a| \leq K$. When $q = \infty$ (4.1) should be interpreted in the usual way, so that $\gamma_\infty^{(K)}$ is the same as the function $\gamma^{(K)}$ which is defined in (I.2.46).

PROPOSITION 4.2. *Suppose that E is uniformly rectifiable. Then there is a $K > 0$ so that if $f : E \to \mathbf{R}$ is Lipschitz with norm ≤ 1 and if $1 \leq q < \frac{2d}{d-2}$ ($1 \leq q \leq \infty$ when $d = 1$), then*

$$(4.3) \qquad \gamma_q^{(K)}(x, t)^2 dx \frac{dt}{t} \text{ is a Carleson measure on } E \times \mathbf{R}_+,$$

with norm bounded independently of f.

In fact, any $K > 1$ will work.

The first part of Theorem I.2.49 is an easy consequence of Proposition 4.2 and the fact that

$$(4.4) \qquad \gamma_\infty^{(K)}(x, t) \leq C(K, q)\gamma_q^{(K)}(x, 2t)^{q/(d+q)}$$

holds for Lipschitz functions f with norm ≤ 1. Let us quickly run through the proof of (4.4).

Fix x and t, and let a be the affine function with $|\nabla a| \leq K$ which attains the infimum in the definition of $\gamma_q^{(K)}(x, 2t)$. Let $y \in B(x, t) \cap E$ be given. For each $\lambda \in (0, 1)$ and $u \in B(y, \lambda t) \cap E$ we have that

$$\begin{aligned} |f(y) - a(y)| &\leq |f(y) - a(y) - f(u) + a(u)| + |f(u) - a(u)| \\ &\leq (K + 1)\lambda t + |f(u) - a(u)|. \end{aligned}$$

Taking the L^q-average over $u \in B(y, \lambda t) \cap E$ we get

$$\begin{aligned} |f(y) - a(y)| &\leq (K + 1)\lambda t + C\left((\lambda t)^{-d} \int_{B(y, \lambda t) \cap E} |f - a|^q\right)^{1/q} \\ &\leq (K + 1)\lambda t + C\lambda^{-d/q}\gamma_q^{(K)}(x, 2t)t. \end{aligned}$$

If $\gamma_q^{(K)}(x, 2t) \leq 1$, then we take $\lambda = \gamma_q^{(K)}(x, 2t)^{q/(d+q)}$ and get

$$t^{-1}|f(y) - a(y)| \leq ((K + 1) + C)\gamma_q^{(K)}(x, 2t)^{q/(d+q)},$$

which is fine. If $\gamma_q^{(K)}(x, 2t) \geq 1$, then (4.4) follows from the trivial estimate $\gamma_\infty^{(K)}(x, 2t) \leq C$.

Let us now proceed to the proof of Proposition 4.2. We begin with a couple of special cases.

LEMMA 4.5. *Proposition 4.2 holds when E is a d-plane.*

This is essentially the same as Theorem I.1.42, but there is a difference which is not entirely insignificant: Theorem I.1.42 says that (4.3) is true if we are willing to replace $\gamma_q^{(K)}$ by $\gamma_q = \gamma_q^{(\infty)}$. However, it is an easy exercise to show that if E is a d-plane and if f has Lipschitz norm ≤ 1, then, for each $\eta > 0$, either $\gamma_q^{(1+\eta)} = \gamma_q$ or $\gamma_q \geq C(\eta)^{-1}$, in which case $\gamma_q^{(1+\eta)} \leq C(\eta)\gamma_q$. [In other words, if an affine function has a directional derivative which is $\geq 1 + \eta$ for some direction tangent to E, then that affine function cannot approximate f very well.]

LEMMA 4.6. *Proposition 4.2 holds when E is a Lipschitz graph, with estimates that depend only on the Lipschitz constant.*

This reduces to the preceding lemma for rather dumb reasons. By hypothesis we can represent E as

$$E = \{p + A(p) : p \in P\},$$

4.1. DIRECT ESTIMATES

where P is a d-plane in $\mathbf{R}^n$, $A: P \to Q$ is a Lipschitz mapping with norm M, say, and Q is an $(n-d)$-plane in $\mathbf{R}^n$ which is orthogonal to P. Let π denote the orthogonal projection of E onto P. Given $f: E \to \mathbf{R}$ which is Lipschitz with norm ≤ 1, define $F: P \to \mathbf{R}$ by $F \circ \pi = f$ so that F has norm $\leq 1 + M$. Let $\gamma_q^{(K)}(x, t)$ be as in (4.1), and let $\tilde{\gamma}_q^{(\widetilde{K})}(p, t)$ be the corresponding object for F and P instead of f and E, with $\widetilde{K} = (1+M)K$. Then

$$\gamma_q^{(\widetilde{K})}(x, t) \leq C \tilde{\gamma}_q^{(\widetilde{K})}(\pi(x), t)$$

for all $x \in E$ and $t > 0$. This is not hard to check. [Roughly speaking, we are simply saying that if $a(p)$ is an affine function on P that approximates F well on $B(\pi(x), t) \cap P$, then $a \circ \pi$ is an affine function on $\mathbf{R}^n$ which approximates f well on $E \cap B(x, t)$.] From here Lemma 4.6 follows easily.

Let us come back now to Proposition 4.2 itself. So let E be given, with E uniformly rectifiable. We are going to use the uniform rectifiability in the form of the existence of a corona decomposition (see Definition I.3.19) to reduce to the case of Lipschitz graphs. [The argument could be simplified substantially if we were merely going to prove that E satisfies the WALA, rather than the conclusion of Proposition 4.2.]

Let Δ be a family of cubes on E, as in §I.3.1, and let $\theta \in (0, 1)$ be small, to be chosen later. By definition of a corona decomposition there is a coronization $(\mathscr{B}, \mathscr{G}, \mathscr{F})$ of E and a family of d-dimensional Lipschitz graphs $\Gamma(S)$, $S \in \mathscr{F}$, with constant $\leq \eta$, such that

(4.7) $\quad \operatorname{dist}(x, \Gamma(S)) \leq \theta \operatorname{diam} Q$ whenever $x \in 2Q$ and $Q \in S$.

Here η can be taken to be any positive number. We may as well take $\eta = 1$, but to get the optimal choices for K we should take η to be arbitrarily small.

Let $f: E \to \mathbf{R}$ be given, with Lipschitz norm ≤ 1. Let $\phi: \mathbf{R}^n \to \mathbf{R}$ be a Lipschitz extension of f, e.g.,

$$\phi(x) = \inf_{y \in E}(f(y) + |x - y|).$$

This choice of ϕ also has Lipschitz norm ≤ 1. For each $S \in \mathscr{F}$ let $\sigma_S(x, t)$ be the function on $\Gamma(S) \times \mathbf{R}_+$ which is the analogue of $\gamma_q^{(K)}(x, t)$, but for $\Gamma(S)$ and $\phi|_{\Gamma(S)}$ instead of E and f. Thus σ_S also depends on K and q, but we shall suppress that dependence in the notation for the convenience of all parties. From Lemma 4.6 it follows that if K is sufficiently large ($K = 1 + 2\eta$ will do) and if q lies in the same range as in Proposition 4.2, then

(4.8) $\quad \sigma_S(x, t)^2 dx \dfrac{dt}{t}$ is a Carleson measure on $\Gamma(S) \times \mathbf{R}_+$,

with norm $\leq C(n, d)$.

We shall assume from now on that K has been chosen and fixed so that this is true and that q lies in the usual range and is fixed.

We want to derive (4.3) from (4.8) and the definition of a corona decomposition. To do this we first set some more notation, and then we get rid of some garbage terms.

Given $Q \in \Delta$, set

$$\hat{Q} = \{(x, t) \in E \times \mathbf{R}_+ : x \in Q \text{ and } \operatorname{diam} Q' < t \leq \operatorname{diam} Q,$$
$$\text{where } Q' \text{ is the child of } Q \text{ containing } x\}.$$

Thus $\bigcup_{Q \in \Delta} \hat{Q} = E \times \mathbf{R}_+$, and the $\hat{Q}$'s are pairwise disjoint. If $\mathscr{A} \subseteq \Delta$, set

$$\hat{\mathscr{A}} = \bigcup_{Q \in \mathscr{A}} \hat{Q}.$$

LEMMA 4.9. *If $\mathscr{A} \subseteq \Delta$ satisfies a Carleson packing condition (i.e., if there is a $C > 0$ so that*

$$\sum_{\substack{Q \in \mathscr{A} \\ Q \subseteq R}} |Q| \leq C|R| \quad \text{for all } R \in \Delta),$$

then $\gamma_q^{(K)}(x, t)^2 \chi_{\hat{\mathscr{A}}}(x, t) dx \frac{dt}{t}$ is a Carleson measure on $E \times \mathbf{R}_+$.

This is an immediate consequence of the definitions and the trivial estimate $\gamma_q^{(K)}(x, t) \leq C$.

This lemma allows us to throw away some otherwise annoying parts of $E \times \mathbf{R}_+$. More precisely we can use it to throw away $\hat{\mathscr{B}}$ and also $\hat{\mathscr{A}_0}$, where

$$\mathscr{A}_0 = \{Q \in \mathscr{G} : 10Q \not\subseteq Q(S), \text{ where } S \in \mathscr{F} \text{ satisfies } Q \in S\}.$$

[Remember that $Q(S)$ is the maximal element of the stopping-time region $S \in \mathscr{F}$.] The fact that $\mathscr{A}_0$ satisfies a Carleson packing condition follows from Lemma I.3.29.

Thus we are reduced to showing that

$$(4.10) \qquad \gamma_q^{(K)}(x, t)^2 \chi_{(\Delta \setminus (\mathscr{B} \cup \mathscr{A}_0))^{\wedge}}(x, t) dx \frac{dt}{t}$$

is a Carleson measure on $E \times \mathbf{R}_+$. Let us make one more reduction.

LEMMA 4.11. *To prove (4.10) it suffices to show that*

$$(4.12) \qquad \gamma_q^{(K)}(x, t)^2 \chi_{(S \setminus \mathscr{A}_0)^{\wedge}}(x, t) dx \frac{dt}{t}$$

is a Carleson measure on $E \times \mathbf{R}_+$ for each $S \in \mathscr{F}$, with uniformly bounded norm.

Fix $u \in E$ and $R > 0$, and consider

$$(4.13) \qquad \sum_{S \in \mathscr{F}} \int_0^R \int_{B(u,R) \cap E} \gamma_q^{(K)}(x, t)^2 \chi_{(S \setminus \mathscr{A}_0)^{\wedge}}(x, t) \, dx \, \frac{dt}{t}.$$

To prove (4.10), we are supposed to show that this is $\leq C R^d$.

Let $I(S)$ denote the integral in the sum (4.13), and let $\mathscr{F}_1$ denote the class of $S \in \mathscr{F}$ such that $I(S) \neq 0$ and $\operatorname{diam} Q(S) \leq R$. Set $\mathscr{F}_2 = \{S \in \mathscr{F} \setminus \mathscr{F}_1 : I(S) \neq 0\}$.

If $S \in \mathscr{F}_1$, then $Q(S)$ intersects $B(u, R)$ and so $Q(S) \subseteq B(u, 2R)$. Also,

$$I(S) \leq \int_0^{\operatorname{diam} Q(S)} \int_{Q(S)} \gamma_q^{(K)}(x, t)^2 \chi_{(S \setminus \mathscr{A}_0)^\wedge}(x, t)\, dx\, \frac{dt}{t}$$

by definitions, and hence

$$I(S) \leq C|Q(S)|$$

by hypothesis. Thus

$$\sum_{S \in \mathscr{F}_1} I(S) \leq C \sum_{S \in \mathscr{F}_1} |Q(S)| \leq CR^d,$$

since $\{Q(S) : S \in \mathscr{F}\}$ satisfies a Carleson packing condition, by the definition of a coronization.

If $S \in \mathscr{F}_2$, then there must be a $Q \in S$ such that Q intersects $B(u, R)$ and satisfies $\operatorname{diam} Q \leq CR$. Because $\operatorname{diam} Q(S) \geq R$, there is an ancestor Q' of Q in S such that Q' intersects $B(u, R)$ and $C^{-1}R \leq \operatorname{diam} Q' \leq CR$. This implies that $\mathscr{F}_2$ has a bounded number of elements, because there is only a bounded number of cubes in Δ which intersect $B(u, R)$ and whose diameter lies in $[C^{-1}R, CR]$. Hence

$$\sum_{S \in \mathscr{F}_2} I(S) \leq C \left(\sup_{S \in \mathscr{F}_2} I(S) \right) \leq CR^d,$$

by hypothesis. This proves Lemma 4.11.

With these reductions out of the way we can get to the heart of the argument, which is to control $\gamma_q^{(K)}(x, t)$ for $(x, t) \in (S \setminus \mathscr{A}_0)^\wedge$ in terms of σ_S. We first introduce some more notation.

Fix $S \in \mathscr{F}$. Given $Q \in S$ define $\widetilde{Q} \subseteq \Gamma(S)$ by

(4.14) $\qquad \widetilde{Q} = \{y \in \Gamma(S) : \operatorname{dist}(y, Q) \leq 10 \operatorname{diam} Q\}$.

Notice that $|\widetilde{Q}| \approx |Q|$, because of (4.7). We want to control $\gamma_q^{(K)}(x, t)$ for $(x, t) \in \widehat{Q}$, $Q \in S$, in terms of $\sigma_S(y, \tau)$, where $y \in \widetilde{Q}$ and $25 \operatorname{diam} Q \leq \tau \leq 26 \operatorname{diam} Q$.

For $Q \in S$ set

(4.15) $\qquad \sigma(Q) = \inf_a \left\{ (\operatorname{diam} Q)^{-1} \left(\frac{1}{|Q|} \int_{\widetilde{Q}} |\phi - a|^q \right)^{1/q} \right\}$,

where the infimum is taken over all affine functions a on $\mathbf{R}^n$ such that $|\nabla a| \leq K$. By definitions we have that

$$\sigma(Q) \leq C\sigma_S(y, \tau)$$

whenever $y \in \widetilde{Q}$ and $25 \operatorname{diam} Q \leq \tau \leq 26 \operatorname{diam} Q$. Using this and (4.8) it is not hard to prove the following.

LEMMA 4.16.
$$\sum_{Q \in S} \sigma(Q)^2 \chi_{\widehat{Q}}(x, t) dx \frac{dt}{t}$$
is a Carleson measure on $E \times \mathbf{R}_+$, with uniformly bounded norm (independently of S).

We leave the details of the proof to the reader. It is helpful to notice that the subsets of $\Gamma(S) \times \mathbf{R}_+$ which are given by

$$\{(y, \tau) \in \Gamma(S) \times \mathbf{R}_+ : y \in \widetilde{Q}, \ 25 \operatorname{diam} Q \leq \tau \leq 26 \operatorname{diam} Q\}$$

have bounded overlap.

For each $Q \in S \setminus \mathscr{A}_0$ set

$$\gamma(Q) = \inf_a \left\{ (\operatorname{diam} Q)^{-1} \left(\frac{1}{|Q|} \int_{2Q} |f - a|^q \right)^{1/q} \right\},$$

where the infimum is taken over all affine functions a on $\mathbf{R}^n$ with $|\nabla a| \leq K$. Clearly

(4.17) $\quad \gamma_q^{(K)}(x, t) \leq C\gamma(Q) \quad \text{for all } (x, t) \in \widehat{Q}.$

We are going to estimate $\gamma(Q)$ in terms of $\sigma(Q)$ and an error term that reflects the extent to which E is well approximated by $\Gamma(S)$ on $2Q$. This of course involves the minimal cubes of S in a natural way, because of (4.7). Let us make this precise.

Define $\mu : E \to \mathbf{R}$ by $\mu(x) = 0$ if $x \in E \setminus Q(S)$ and

$$\mu(x) = \inf\{\operatorname{diam} Q : Q \in S, x \in Q\}$$

when $x \in Q(S)$. Set $Z = \{x \in Q(S) : \mu(x) = 0\}$, and let $m(S)$ denote the set of minimal cubes in S. Then

(4.18) $\quad Q(S) = Z \cup \bigcup_{Q \in m(S)} Q,$

and $\mu(x) = \operatorname{diam} Q$ if $x \in Q$ and $Q \in m(S)$. Also,

(4.19) $\quad Z \subseteq E \cap \Gamma(S),$

by (4.7).

LEMMA 4.20. For each $Q \in S \setminus \mathscr{A}_0$ we have

$$\gamma(Q) \leq C(\sigma(Q) + M_q(Q)),$$

where $M_q(Q) = (\operatorname{diam} Q)^{-1} \left(\frac{1}{|Q|} \int_{2Q} \min(\operatorname{diam} Q, \mu)^q \right)^{1/q}$.

Let $Q \in S \setminus \mathscr{A}_0$ be given. Let a be an affine function on $\mathbf{R}^n$ with $|\nabla a| \leq K$ which achieves the infimum in the definition of $\sigma(Q)$. Since we have $\sigma(Q) \leq C$ trivially, we must have in particular that $\inf_{\widetilde{Q}} |\phi - a| \leq C \operatorname{diam} Q$.

Since $\phi - a$ is a Lipschitz function on $\mathbf{R}^n$ with norm $\leq K+1$, we conclude that

(4.21) $\quad |\phi(z) - a(z)| \leq C \operatorname{diam} Q \quad \text{for all} \quad z \in \mathbf{R}^n$
$\quad\quad\quad\quad\quad\quad\text{such that}\quad \operatorname{dist}(z, Q) \leq 100 \operatorname{diam} Q.$

In particular,

(4.22) $\quad\quad\quad \sup_{2Q} |f - a| \leq C \operatorname{diam} Q.$

For the record, let us mention the trivial facts that

(4.23) $\quad\quad \sigma(Q) = (\operatorname{diam} Q)^{-1} \left(\frac{1}{|Q|} \int_{\widetilde{Q}} |\phi - a|^q \right)^{1/q},$

by the choice of a, and

(4.24) $\quad\quad \gamma(Q) \leq (\operatorname{diam} Q)^{-1} \left(\frac{1}{|Q|} \int_{2Q} |f - a|^q \right)^{1/q}.$

To control $\gamma(Q)$ by $\sigma(Q)$ we begin by observing that

(4.25) $\quad (\operatorname{diam} Q)^{-1} \left(\frac{1}{|Q|} \int_{2Q \cap Z} |f - a|^q \right)^{1/q} \leq \sigma(Q),$

which follows from (4.19).

Now we need to control the contribution to the right side of (4.24) that comes from $2Q \setminus Z$. Keep in mind that $2Q \subseteq Q(S)$, since $Q \notin \mathscr{A}_0$, so $2Q \setminus Z$ is covered by minimal cubes of S.

Set $Y = \bigcup \{ R \in m(S) : \operatorname{diam} R \geq \operatorname{diam} Q \}$. Then (4.22) yields

(4.26) $\quad (\operatorname{diam} Q)^{-1} \left(\frac{1}{|Q|} \int_{2Q \cap Y} |f - a|^q \right)^{1/q} \leq C M_q(Q),$

by definition of $M_q(Q)$.

We are left with $2Q \cap X$, where $X = \bigcup \{ R \in m(S) : \operatorname{diam} R < \operatorname{diam} Q \}$. This piece is a little more complicated. Let $\{T_i\}$ be an enumeration of the elements of $m(S)$ which have diameter $< \operatorname{diam} Q$ and which intersect $2Q$. Let $c(T_i)$ (the "center" of T_i) be as in Lemma I.3.5, so that

(4.27) $\quad\quad\quad \operatorname{dist}(c(T_i), E \setminus T_i) \geq C^{-1} \operatorname{diam} T_i.$

Set $B_i = B(c(T_i), 2\theta \operatorname{diam} T_i)$. If θ is small enough, then (4.27) ensures that the B_i's are pairwise disjoint. We choose $\theta \in (0, 1)$ now, once and for all, so that this is true. [Remember that θ is the parameter from the corona decomposition which appears in (4.7) and which we get to choose.] Notice that

(4.28) $\quad\quad\quad |B_i \cap \Gamma(S)| \geq C^{-1} (\operatorname{diam} T_i)^d,$

by (4.7).

Since $\phi - a$ is a Lipschitz function on $\mathbf{R}^n$ with bounded norm, we have that
$$|f(x) - a(x)| \leq |\phi(y) - a(y)| + C \operatorname{diam} T_i$$
for all $x \in T_i$ and all $y \in B_i \cap \Gamma(S)$. This implies that
$$\int_{2Q \cap T_i} |f - a|^q \leq C \int_{B_i \cap \Gamma(S)} |\phi - a|^q + C \int_{2Q \cap T_i} (\operatorname{diam} T_i)^q.$$
(This uses also (4.28).) Hence

(4.29)
$$\begin{aligned}
(\operatorname{diam} Q)^{-1} &\left(\frac{1}{|Q|} \int_{2Q \cap X} |f - a|^q \right)^{1/q} \\
= (\operatorname{diam} Q)^{-1} &\left(\frac{1}{|Q|} \sum_i \int_{2Q \cap T_i} |f - a|^q \right)^{1/q} \\
\leq C(\operatorname{diam} Q)^{-1} &\left(\frac{1}{|Q|} \int_{(\cup B_i) \cap \Gamma(S)} |\phi - a|^q \right)^{1/q} \\
+ C(\operatorname{diam} Q)^{-1} &\left(\frac{1}{|Q|} \int_{2Q \cap X} \mu^q \right)^{1/q} \\
\leq C\sigma(Q) + CM_q(Q).
\end{aligned}$$

For this last we used the fact that $\bigcup B_i \subseteq \tilde{Q}$, where $\tilde{Q}$ is as in (4.14). [As usual, when $q = \infty$ this argument should be written differently, but it works out the same way.]

Lemma 4.20 now follows from (4.24)–(4.26) together with (4.29).

We are nearly finished now with the proof of Proposition 4.2. We had previously reduced the problem to proving (4.12), with a uniform bound. Because of (4.17), it is enough to show that
$$\sum_{Q \in S \setminus \mathscr{A}_0} \gamma(Q)^2 \chi_{\hat{Q}}(x, t) dx \frac{dt}{t}$$
is a Carleson measure on $E \times \mathbf{R}_+$, with norm bounded independently of S. This Carleson measure estimate follows from Lemmas 4.16 and 4.20 once we have the next result.

LEMMA 4.30. *If $1 \leq q < \frac{2d}{d-2}$ ($1 \leq q \leq \infty$ when $d = 1$), then*
$$\sum_{Q \in S \setminus \mathscr{A}_0} M_q(Q)^2 \chi_{\hat{Q}}(x, t) dx \frac{dt}{t}$$
is a Carleson measure on $E \times \mathbf{R}_+$, with norm bounded independently of S.

The type of quantity which appears in Lemma 4.30 arises frequently when controlling error terms in this sort of situation (i.e., in deriving consequences of the existence of a corona decomposition), and similar estimates have been given in [DS2], for instance. We are including a proof here for the reader's convenience.

To prove Lemma 4.30, it is helpful to notice that its conclusion is equivalent to the following discrete version:

$$\text{(4.31)} \qquad \sum_{\substack{Q \in S \setminus \mathscr{A}_0 \\ Q \subseteq R}} M_q(Q)^2 |Q| \leq C|R|$$

for all $R \in \Delta$. This is easily verified.

Notice that $M_q(Q)$ becomes larger as q increases, modulo a bounded factor, anyway. Since the range of q's in which we are interested contains $q = 2$, we may as well restrict our attention to $q \geq 2$. The special case of $q = 2$ is contained in the next result.

LEMMA 4.32. *For each* $s \in (0, \infty)$,

$$\sum_{\substack{Q \in S \\ Q \subseteq R}} M_s(Q)^s |Q| \leq C(s)|R|$$

for all $R \in \Delta$.

Fix $R \in \Delta$. By definition of $M_s(Q)$ and $\mu(x)$ we have that

$$\sum_{\substack{Q \in S \setminus \mathscr{A}_0 \\ Q \subseteq R}} M_s(Q)^s |Q| = \sum_{\substack{Q \in S \setminus \mathscr{A}_0 \\ Q \subseteq R}} \sum_{T \in m(S)} \min\left(1, \left(\frac{\operatorname{diam} T}{\operatorname{diam} Q}\right)^s\right) |T \cap 2Q|.$$

For simplicity let us call this whole thing $\Sigma(R)$. Let us also split $\Sigma(R)$ into $\Sigma_1(R) + \Sigma_2(R)$, where $\Sigma_1(R)$ corresponds to the part where $\operatorname{diam} Q < \operatorname{diam} T$ and $\Sigma_2(Q)$ corresponds to $\operatorname{diam} Q \geq \operatorname{diam} T$.

To control $\Sigma_1(R)$, we begin with the observation that if $Q \in S$, $T \in m(S)$, and $\operatorname{diam} Q < \operatorname{diam} T$, then $Q \cap T = \emptyset$. [Clearly $T \subseteq Q$ and $Q \subseteq T$ are not possible.] Hence

$$\Sigma_1(R) \leq \sum_{T \in m(S)} \sum_{\substack{Q \subseteq R \\ Q \cap T = \emptyset \\ \operatorname{diam} Q < \operatorname{diam} T}} |T \cap 2Q|.$$

It is not so hard to prove that this is

$$\leq C \sum_{T \in m(S)} |T \cap 2R|,$$

using (I.3.4). [Consider separately the sums over Q in each Δ_j and then the sum in j. The sum in j is dominated by a geometric series because of (I.3.4).] Thus

$$\Sigma_1(R) \leq C|R|,$$

since the elements of $m(S)$ are pairwise disjoint.

For $\Sigma_2(R)$ we have

$$\Sigma_2(R) \leq \sum_{T \in m(S)} \sum_{\substack{Q \subseteq R \\ \operatorname{diam} Q \geq \operatorname{diam} T}} \left(\frac{\operatorname{diam} T}{\operatorname{diam} Q}\right)^s |T \cap 2Q|$$

$$\leq \sum_{T \in m(S)} \sum_{\substack{Q \in \Delta \\ \operatorname{diam} Q \geq \operatorname{diam} T \\ 2Q \cap T \neq \emptyset}} \left(\frac{\operatorname{diam} T}{\operatorname{diam} Q}\right)^s |T \cap 2R|$$

$$\leq C \sum_{T \in m(S)} |T \cap 2R| \leq C|R|.$$

For the third inequality we used the fact that for each $j \in \mathbf{Z}$ there are only a bounded number of $Q \in \Delta_j$ such that $\operatorname{diam} Q \geq \operatorname{diam} T$ and $2Q \cap T \neq \emptyset$ to reduce to a geometric series. This proves Lemma 4.32.

It remains to deal with the case where $q > 2$ in Lemma 4.30. We are going to derive this from Lemma 4.32, but it is a little bit trickier.

Let $\widetilde{M}_q(Q)$ be defined in the same way that $M_q(Q)$ was, except that you integrate over $3Q$ instead of $2Q$. Then it is not hard to check that

(4.33) $$M_\infty(Q) \leq C(r) \widetilde{M}_r(Q)^{r/(r+d)}$$

for all $r \in (0, \infty)$ and all $Q \in S$. Indeed, this amounts to showing that

(4.34) $$\min\left(1, \frac{\operatorname{diam} T}{\operatorname{diam} Q}\right) \leq C(r) \widetilde{M}_r(Q)^{r/(r+d)}$$

whenever $T \in m(S)$ satisfies $T \cap 2Q \neq \emptyset$. When $\operatorname{diam} T \leq \operatorname{diam} Q$ we have $T \subseteq 3Q$ and hence

$$\left(\widetilde{M}_r(Q)\right)^{r/(r+d)} \geq \left(\frac{1}{|Q|} \int_T \left(\frac{\operatorname{diam} T}{\operatorname{diam} Q}\right)^r\right)^{1/(r+d)} \geq C^{-1} \frac{\operatorname{diam} T}{\operatorname{diam} Q},$$

which implies (4.34). If $\operatorname{diam} T \geq \operatorname{diam} Q$ then you use instead the fact that

$$|T \cap 3Q| \geq C^{-1}|Q|,$$

which is itself a simple consequence of the properties satisfied by the cubes (i.e., (3.1)–(3.3) in §I.3.1). This implies that $\widetilde{M}_r(Q) \geq C^{-1}$, which is what we wanted in this case.

If $q \geq r$ then we have

$$M_q(Q) \leq M_r(Q)^{r/q} M_\infty(Q)^{1-r/q} \leq C(r) \widetilde{M}_r(Q)^s$$

where $s = r/q + (1 - r/q) r/(r+d)$. Thus

$$M_q(Q)^2 \leq C(r) \widetilde{M}_r(Q)^{2s}.$$

If we can choose r so that $r \leq 2s$, with s given as above, then we shall be done. Indeed, in that case we would have

$$M_q(Q)^2 \leq C(r) \widetilde{M}_r(Q)^r,$$

since $\widetilde{M}_r(Q) \leq C$, and so we could derive the desired estimate (4.31) from the version of Lemma 4.32 with $M_s(Q)$ replaced by $\widetilde{M}_s(Q)$. [Of course, the same proof still works in that case.]

Thus it all comes down to choosing $r > 0$ so that

$$r \leq 2s = 2[r/q + (1 - r/q)r/(r+d)].$$

This is equivalent to

$$\frac{1}{2} \leq \frac{1}{q} + (1 - r/q)\frac{1}{r+d},$$

which is itself equivalent to

$$\frac{1}{q} - \frac{1}{2} + \frac{1}{d} \geq \frac{1}{d} + (r/q - 1)\left(\frac{1}{r+d}\right).$$

Clearly the right side tends to zero as $r \to 0$, so we are in business if

$$\frac{1}{q} - \frac{1}{2} + \frac{1}{d} > 0.$$

This condition is the same as $q < \frac{2d}{d-2}$ when $d \geq 2$, while all $q \in [1, \infty]$ are allowed when $d = 1$. This completes the proof of Lemma 4.30, and hence of Proposition 4.2 as well.

4.2. The converse when $d = 1$.

For the next result we are going to be able to work with a weaker version of the WALA.

PROPOSITION 4.35. *Let E be a d-dimensional set in $\mathbf{R}^n$. Assume that there is a $q > 0$ so that for every Lipschitz function $f : E \to \mathbf{R}$ and every $\alpha > 0$ we have that*

$$\{(x, t) \in E \times \mathbf{R}_+ : \gamma_q(x, t) > \alpha\}$$

is a Carleson set, where $\gamma_q(x, t)$ is defined to be as in (4.1), but with $K = \infty$ (i.e., with no restriction on $|\nabla a|$). Then E satisfies the WNB (see Definition II.3.44).

Recall that the WNB implies uniform rectifiability when $d = 1$, by Proposition II.3.45. Thus the second half of Theorem I.2.49 follows from this proposition. [We could, incidentally, just as well have used the WNM.]

Notice that the hypothesis of Proposition 4.35 becomes weaker as q gets smaller, so we may as well ignore the case where $q = \infty$.

Let us recall some of the terminology which is relevant for the WNB. Let n and d be fixed. Given $\epsilon \in (0, 1)$, the "standard ϵ-box" is the subset $M'(\epsilon)$ of $\mathbf{R}^n$ which is given by $M'(\epsilon) = B' \cup S'$, where $B' = \{(x_1, \ldots, x_n) \in \mathbf{R}^n : 0 \leq x_n \leq \epsilon$ and $\sup_{1 \leq i \leq n-1} |x_i| \leq 1\}$ is the bottom of the box and $S' = \{(x_1, \ldots, x_n) \in \mathbf{R}^n : 0 \leq x_n \leq 10\epsilon$ and $1 - \epsilon \leq \sup_{1 \leq i \leq n-1} |x_i| \leq 1\}$ is the side of the box. We call the set $C' = \{(x_1, \ldots, x_n) \in \mathbf{R}^n : \epsilon < x_n < 2\epsilon, \sup_{1 \leq i \leq n-1} |x_i| < 1 - \epsilon\}$ the content of the box $M'(\epsilon)$.

A subset of $\mathbf{R}^n$ is called an ϵ-box if it can be obtained from the standard ϵ-box by a combination of translations, dilations, and rotations. The notions of "bottom", "side", and "content" make sense for general boxes, in the obvious way.

Roughly speaking a regular set E in $\mathbf{R}^n$ satisfies the WNB if for each $\epsilon > 0$ there are not too many ϵ-boxes M in $\mathbf{R}^n \setminus E$ such that E intersects the content of M. This is made precise using Carleson sets, as in Definition II.3.44. With this in mind let us think about how we are going to prove Proposition 4.35. The idea is to try to build a finite family of Lipschitz functions with the property that every time you have a box as above something bad happens to one of these Lipschitz functions. This "something bad" should be something which is bad for the WALA, i.e., it should prevent the Lipschitz function from being well approximated by an affine function near the given box. For instance, you can think in terms of building the Lipschitz functions so that there will be some kind of folding near the box. It is not hard to see why this should be bad for affine approximation, or why the presence of the box is relevant for putting in a fold.

One should think about this approach in light of the fact that we only know how to prove that the WALA implies uniform rectifiability when $d = 1$. Although we know criteria for uniform rectifiability which work when $d > 1$ and which are similar to the WNB (see §II.3.4), they do not seem to work so well for the WALA. The problem lies in building Lipschitz functions with the right kind of behavior. We shall say a little more about this at the end of this section, in Remark 4.90.

Before beginning the proof of Proposition 4.35 in earnest we want to make a few reductions to simplify the geometry.

Let us call an ϵ-box "special" if it is obtained from the standard ϵ-box by translations and dilations. In other words, we are not permitting rotations; special boxes point in the x_n direction.

Let E be a d-dimensional regular set in $\mathbf{R}^n$. For each $\epsilon > 0$ let $\mathscr{G}_s(\epsilon)$ denote the set of points (x, t) in $E \times \mathbf{R}_+$ for which it is impossible to find a special ϵ-box $M' \subseteq B(x, t)$ such that

(4.36) $\operatorname{diam} M' \geq \epsilon t$ and

(4.37) $M' \cap E = \varnothing$, but E intersects the content of M'.

LEMMA 4.38. *If E is as in Proposition 4.35, then $E \times \mathbf{R}_+ \setminus \mathscr{G}_s(\epsilon)$ is a Carleson set for each $\epsilon > 0$.*

The difference between this lemma and Proposition 4.35 is that we are only considering special boxes here. However, this is not a crucial difference, in light of the following.

LEMMA 4.39. *Lemma 4.38 implies Proposition 4.35.*

Indeed, let E be given, and suppose that E satisfies the hypotheses of Proposition 4.35. Then $\rho(E)$ also satisfies the same conditions for all rotations ρ, so each $\rho(E)$ also satisfies the conclusion of Lemma 4.38. This does not quite imply immediately that E satisfies the WNB, since the rotation group is infinite, but it almost does. Let $\epsilon > 0$ be given, so that we want to show that $E \times \mathbf{R}_+ \setminus \mathscr{G}_b(\epsilon)$ is a Carleson set, where $\mathscr{G}_b(\epsilon)$ is as in Definition II.3.44. Let $\rho_1, \ldots, \rho_N$ be a finite set of rotations on $\mathbf{R}^n$ so that every other rotation is at distance $< \eta$ (with respect to some reasonable metric) from one of the ρ_j's. If η is small enough, then $E \times \mathbf{R}_+ \setminus \mathscr{G}_b(\epsilon)$ is a Carleson set if $E \times \mathbf{R}_+ \setminus \mathscr{G}_s(\eta)$ is and if the same is true for E replaced by $\rho_j(E)$ for $j = 1, \ldots, N$. This is not difficult to check, and we leave the details as an exercise for the devoted reader.

[Notice that the preceding point would be completely trivial if we were using the version of the WNB in which we only consider boxes with sides parallel to the coordinate hyperplanes. This restricted version of the WNB is still a sufficient condition for uniform rectifiability when $d = 1$, as we observed in §II.3.4. However, it is going to be convenient for us not to limit ourselves to such boxes.]

Thus we need only prove Lemma 4.38. To do this we need to analyze the possible behavior of E inside a box—in particular, its projection onto the x_n-axis.

Define $\pi_n : \mathbf{R}^n \to \mathbf{R}$ by $\pi_n(x) = x_n$, i.e., the projection onto the x-axis.

Let $T'(\epsilon)$ denote the "top" of standard ϵ-box $M'(\epsilon)$, i.e.,

$$T'(\epsilon) = \left\{ (x_1, \ldots, x_n) \in \mathbf{R}^n : x_n = 10\epsilon, \sup_{1 \leq i \leq n-1} |x_i| \leq 1 \right\}.$$

Similarly let $I'(\epsilon)$ denote the "inside" of $M'(\epsilon)$,

$$I'(\epsilon) = \left\{ (x_1, \ldots, x_n) \in \mathbf{R}^n : \epsilon < x_n < 9\epsilon, \sup_{1 \leq i \leq n-1} |x_i| < 1 - \epsilon \right\}.$$

[Thus the content of the box is contained in its inside part, as it should be. Notice that we are not allowing the inside of the box to get too close to the top.] The notions of "top" and "inside" of a box extend to general boxes in the obvious way.

Let E be our d-dimensional regular set again. Set

$$\mathscr{B}_s(\epsilon) = E \times \mathbf{R}_+ \setminus \mathscr{G}_s(\epsilon)$$

so that $(x, t) \in \mathscr{B}_s(\epsilon)$ if there is a special ϵ-box $M' \subseteq B(x, t)$ which satisfies (4.36) and (4.37). Given $\mu > 0$, let $\mathscr{B}_s(\epsilon, \mu)$ denote the set of (x, t) in $E \times \mathbf{R}_+$ for which there exists a special ϵ-box M', $M' \subseteq B(x, t)$, which satisfies (4.36), (4.37), and

(4.40) there exist $u^j \in E$, $1 \leq j \leq n + 2$, which lie in the inside I' of the box M', and which satisfy $\pi_n(u^j) < \pi_n(u^{j+1}) - \mu \operatorname{diam} M'$ for $j = 1, 2, \ldots, n + 1$.

Thus $\mathscr{B}_s(\epsilon, \mu) \subseteq \mathscr{B}_s(\epsilon)$ for all $\mu > 0$.

LEMMA 4.41. *If E satisfies the hypotheses of Proposition 4.35, then $\mathscr{B}_s(\epsilon, \mu)$ is a Carleson set for all $\epsilon, \mu > 0$.*

It will be easier for us to work with $\mathscr{B}_s(\epsilon, \mu)$ than $\mathscr{B}_s(\epsilon)$, because it is easier to build the Lipschitz functions with the right kind of bad behavior. Fortunately, it is sufficient to consider only $\mathscr{B}_s(\epsilon, \mu)$.

LEMMA 4.42. *Lemma 4.41 implies Lemma 4.38.*

Let us explain now why Lemma 4.42 is true and then deal with Lemma 4.41 afterward. The idea is that we can always rotate our boxes a little bit to produce the u^j's of (4.40).

Given a rotation ρ on $\mathbf{R}^n$, we say that an ϵ-box is ρ-special if it can be obtained from a special box by the action of ρ on $\mathbf{R}^n$. Thus ρ-special boxes point in the direction $v(\rho) =$ the image of $(0, \ldots, 0, 1)$ under ρ. Let $\pi_\rho : \mathbf{R}^n \to \mathbf{R}$ denote the linear functional defined by $\pi_\rho(x) = \langle x, v(\rho) \rangle$, and let $\mathscr{B}_{s,\rho}(\epsilon)$ and $\mathscr{B}_{s,\rho}(\epsilon, \mu)$ be defined as above, except that we use ρ-special boxes instead of special boxes, and we replace π_n by π_ρ.

LEMMA 4.43. *For each $\epsilon > 0$ there exist $\mu > 0$ and a finite set F of rotations $\mathbf{R}^n$ such that*

$$\mathscr{B}_s(\epsilon) \subseteq \bigcup_{\rho \in F} \mathscr{B}_{s,\rho}(\epsilon/2, \mu).$$

Of course, Lemma 4.42 follows immediately from Lemma 4.43 and the rotation-invariance of the hypotheses of Proposition 4.35.

Rather than burden the reader with a detailed proof of Lemma 4.43, we shall confine ourselves to a discussion of the main ideas. We begin with the observation that

$$\mathscr{B}_s(\epsilon) \subseteq \mathscr{B}_{s,\rho}(\epsilon/2)$$

whenever ρ is close enough to the identity. In other words, if M' is a special ϵ-box which does not intersect E but whose content does, then it is easy to perturb M' a little to get a ρ-special $\epsilon/2$-box which also has these properties, as long as $|\rho - I|$ is sufficiently small.

Now suppose that $(x, t) \in \mathscr{B}_s(\epsilon)$, so that there is a special ϵ-box $M' \subseteq B(x, t)$ which satisfies (4.36) and (4.37). From the preceding paragraph we know that for each rotation ρ which is close enough to the identity we can find a ρ-special $\epsilon/2$-box M'_ρ which is contained in $B(x, t)$ and satisfies (4.36) (with ϵ replaced by $\epsilon/2$) and (4.37). It is easy to show that you can do this in such a way that there is a $w^1 \in E$ which lies in the content of each such M'_ρ. [The point is that we require w^1 to be independent of ρ.]

The regularity of E implies that we can find $w^j \in E$, $2 \leq j \leq n+2$, such that $w^j \in B(w^1, 10^{-2} \epsilon \operatorname{diam} M')$ for all j and

$$|w^j - w^i| \geq C^{-1} \epsilon \operatorname{diam} M'$$

when $j \neq i$. We also have that the w^j's lie in the inside part of M' and each M'_ρ, as long as we did not choose the M'_ρ's stupidly. (They should be close to M' and have almost the same size, in particular.)

Lemma 4.43 now comes down to the fact that we can find a ρ which is as close as we want to the identity such that

$$|\pi_\rho(w^j) - \pi_\rho(w^i)| > \mu \operatorname{diam} M'_v$$

for all i and j, $i \neq j$, at least if μ is sufficiently small. This is not hard to check. We can even choose ρ from a finite set that does not depend on x, t, M', the w^j's, or anything like that.

We have been a little sloppy about the quantifiers here, as well as other details, but we hope that the reader will agree that it is less painful to fill in these details for oneself than it would be to go through a detailed argument written by us.

This completes our discussion of the proof of Lemma 4.43. We are left with the task of proving Lemma 4.41. Let us first reformulate it in terms of cubes.

Let Δ be a family of cubes on E, as in §I.3.1. For each $\epsilon, \mu > 0$ set

(4.44) $\mathscr{B}'_s(\epsilon, \mu) = \{Q \in \Delta:$ there is a special ϵ-box M' contained in
$\mathbf{R}^n \setminus E$ such that $\operatorname{dist}(M', Q) \leq \operatorname{diam} Q$,
$\epsilon \operatorname{diam} Q \leq \operatorname{diam} M' \leq \operatorname{diam} Q$, and there exist
$u^j \in E$, $1 \leq j \leq n+2$, which lie in the inside I',
of M' and which satisfy $\pi_n(u^j) < \pi_n(u^{j+1})$
$- \mu \operatorname{diam} M'$ for $j = 1, \ldots, n+1\}$.

LEMMA 4.45. *If E satisfies the hypotheses of Proposition 4.35, then $\mathscr{B}'_s(\epsilon, \mu)$ satisfies a Carleson packing condition for all $\epsilon, \mu > 0$.*

It is easy to see that Lemma 4.45 implies Lemma 4.41.

In preparation for the proof of Lemma 4.45 we give a couple of lemmas which will aid us in organizing the cubes efficiently.

LEMMA 4.46. *Define $\widetilde{\Delta}$ to be the set of $Q \in \Delta$ such that*

$$\operatorname{dist}(Q, 0) > 10 \operatorname{diam} Q.$$

Then $\Delta \setminus \widetilde{\Delta}$ satisfies a Carleson packing condition, and every element of $\widetilde{\Delta}$ is contained in a maximal element of $\widetilde{\Delta}$.

The proof of this is straightforward, and we omit it. We care about this lemma because the iterative constructions which we shall perform need to start somewhere, so it is convenient to work with $\widetilde{\Delta}$, with all of its maximal cubes, rather than Δ, which has none. Of course, we have defined $\widetilde{\Delta}$ in such a way that $\Delta \setminus \widetilde{\Delta}$ is insignificant.

We also want to be able to avoid unpleasant interactions between these maximal cubes. The next three lemmas will be used to take care of that.

LEMMA 4.47. *For each maximal cube Q_0 in $\widetilde{\Delta}$ there is only a bounded number of other maximal cubes Q in $\widetilde{\Delta}$ such that $9Q$ intersects $9Q_0$.*

[Recall that λQ is defined in (I.3.7).]

The main point of the proof of this is the following observation: if $Q \in \widetilde{\Delta}$ is maximal, then

$$10 \operatorname{diam} Q < \operatorname{dist}(0, Q) \le C \operatorname{diam} Q,$$

so

(4.48) $$\operatorname{diam} Q < \operatorname{dist}(x, 0) < C \operatorname{diam} Q$$

for all $x \in 9Q$. Hence $\operatorname{diam} Q_0 \approx \operatorname{diam} Q$ if Q and Q_0 are maximal elements of $\widetilde{\Delta}$ such that $9Q_0$ intersects $9Q$, and the lemma is a simple consequence of this.

Let $\mathscr{M}$ denote the set of maximal elements of $\widetilde{\Delta}$. Note that $\widetilde{\Delta} = \{Q \in \Delta : Q \subseteq Q_0 \text{ for some } Q_0 \in \mathscr{M}\}$, by Lemma 4.46 and the definition of $\widetilde{\Delta}$.

LEMMA 4.49. *There are subsets $\mathscr{M}_i$, $1 \le i \le N_1$, of $\mathscr{M}$ such that $\mathscr{M} = \bigcup_{i=1}^{N_1} \mathscr{M}_i$ and $9Q_1 \cap 9Q_2 = \varnothing$ whenever $Q_1, Q_2 \in \mathscr{M}_i$, $Q_1 \ne Q_2$. Here N_1 depends on innocuous constants like n, d, and the regularity constant for E.*

This is a consequence of Lemma 4.47 and a simple recursive "coding" argument. [You simply assign each element of $\mathscr{M}$, one by one, to an $\mathscr{M}_i$, in such a way that you never destroy the above disjointness property. You can do this as long as N_1 is larger than the bounded number mentioned in Lemma 4.47.]

LEMMA 4.50. *For each $k_0 > 1$ there is an $N_2 > 1$ and subsets $\mathscr{A}_j$ of Δ, $1 \le j \le N_2$, such that:*

(4.51) $$\Delta = \bigcup_{j=1}^{N_2} \mathscr{A}_j;$$

(4.52) *if some $\mathscr{A}_j$ has an element in each of Δ_l and Δ_m, l, $m \in \mathbf{Z}$, then either $l = m$ or $|l - m| > k_0$;*

(4.53) *if $Q_1, Q_2 \in \mathscr{A}_j \cap \Delta_l$ for some j and l, then either $Q_1 = Q_2$ or $\operatorname{dist}(Q_1, Q_2) > k_0(\operatorname{diam} Q_1 + \operatorname{diam} Q_2)$.*

[Remember that Δ_l is as in §I.3.1.]

The proof of this is an easy exercise, and we omit it.

4.2. THE CONVERSE WHEN $d = 1$

We shall employ Lemma 4.50 in the following manner. Suppose that k_0 has been chosen, and let the $\mathscr{A}_j$'s be as above. Given $i \in \{1, \ldots, N_1\}$ and $j \in \{1, \ldots, N_2\}$ set

(4.54) $\tilde{\Delta}(i, j) = \{Q \in \tilde{\Delta} : Q$ is contained in an element of $\mathscr{M}_i$,

and Q is itself an element of $\mathscr{A}_j\}$.

Then $\tilde{\Delta} = \bigcup_{i=1}^{N_1} \bigcup_{j=1}^{N_2} \tilde{\Delta}(i, j)$, and the elements of any particular $\tilde{\Delta}(i, j)$ are well separated from each other.

This concludes the organizational preliminaries. Let us now come back to Lemma 4.45.

LEMMA 4.55. *Let ϵ, $\mu > 0$ and $q > 0$ be given. Then we can find $k_0 > 1$ (large) and $\theta > 0$ (small) so that for each $i \in \{1, \ldots, N_1\}$ and $j \in \{1, \ldots, N_2\}$ there is a Lipschitz mapping $f : E \to \mathbf{R}$ such that*

(4.56) $$\inf_{x \in Q} \gamma_q(x, 5 \operatorname{diam} Q) \geq \theta$$

whenever $Q \in \tilde{\Delta}(i, j) \cap \mathscr{B}'_s(\epsilon, \mu)$. [Here N_2 is as in Lemma 4.50, and $\gamma_q(x, t)$ is associated to f as in the statement of Proposition 4.35 (i.e., $\gamma_q(x, t) = \gamma_q^{(\infty)}(x, t)$; see (4.1)).]

Of course, k_0 and θ are allowed to depend on innocuous constants like n, d, and the regularity constant for E, in addition to ϵ, μ, and q.

Notice that Lemma 4.55 implies Lemma 4.45, since $\Delta \setminus \tilde{\Delta}$ satisfies a Carleson packing condition (Lemma 4.46) and $\tilde{\Delta}$ is the union of the $\tilde{\Delta}(i, j)$'s.

There is one last reduction that we want to make.

LEMMA 4.57. *Let ϵ, $\mu > 0$ and $q > 0$ be given. Then we can find $k_0 > 1$ (large) and $\theta > 0$ (small) so that for each $Q_0 \in \mathscr{M}$ and $j \in \{1, \ldots, N_2\}$ there is a Lipschitz mapping $f : E \to \mathbf{R}$ with norm ≤ 1 such that for each $Q \in \mathscr{B}'_s(\epsilon, \mu) \cap \mathscr{A}_j$ with $Q \subseteq Q_0$ we have that*

(4.58) $$\inf_{x \in Q} \gamma_q(x, 5 \operatorname{diam} Q) \geq \theta.$$

Let us assume Lemma 4.57 for the moment and see why it implies Lemma 4.55. Let ϵ, μ, and q be given, and let k_0 and θ be as in Lemma 4.57. Fix $i \in \{1, \ldots, N_1\}$ and $j \in \{1, \ldots, N_2\}$. By definitions we have

$$\tilde{\Delta}(i, j) = \{Q \in \mathscr{A}_j : Q \subseteq Q_0 \text{ for some } Q_0 \in \mathscr{M}_i\}.$$

For each $Q_0 \in \mathscr{M}_i$ let $f_{Q_0} : E \to \mathbf{R}$ be the mapping promised by Lemma 4.57. We may as well assume that f_{Q_0} vanishes somewhere on Q_0; otherwise, we can subtract a constant from f_{Q_0} to force this to happen. Because f_{Q_0} is Lipschitz with norm ≤ 1, we get that

(4.59) $$\sup_{6Q_0} |f_{Q_0}| \leq 6 \operatorname{diam} Q_0.$$

Define $g : \bigcup_{Q_0 \in \mathcal{M}_i}(6Q_0) \to \mathbf{R}$ by $g = f_{Q_0}$ on $6Q_0$ for each $Q_0 \in \mathcal{M}_i$. It is not hard to check that g is Lipschitz with norm ≤ 6. [When you estimate $|g(x) - g(y)|$, there are two cases. If x and y both lie in $6Q_0$ for the same $Q_0 \in \mathcal{M}_i$, then you can use the fact that f_{Q_0} is Lipschitz with norm ≤ 1. Now suppose that $x \in 6Q_0$ and $y \in 6Q_0'$, where Q_0, $Q_0' \in \mathcal{M}_i$ but $Q_0 \neq Q_0'$. Then $9Q_0 \cap 9Q_0' = \varnothing$, so

$$\text{dist}(x, Q_0') > 8 \operatorname{diam} Q_0' \quad \text{and} \quad \text{dist}(y, Q_0) > 8 \operatorname{diam} Q_0.$$

Since $x \in 6Q_0$ and $y \in 6Q_0'$, we get that $|x-y|$ is larger than both $3 \operatorname{diam} Q_0'$ and $3 \operatorname{diam} Q_0$, and hence

$$|x - y| \geq \operatorname{diam} Q_0' + \operatorname{diam} Q_0 \geq \tfrac{1}{6}|g(x)| + \tfrac{1}{6}|g(y)| \geq \tfrac{1}{6}|g(x) - g(y)|.]$$

Hence g can be extended to a Lipschitz mapping f from $\mathbf{R}^n$ to $\mathbf{R}$. (This extension can be produced via a formula similar to the one given in the paragraph after (4.7).)

It is not hard to check that f satisfies the conclusions of Lemma 4.55 using the corresponding properties for the f_{Q_0}'s given by Lemma 4.57. [This uses in particular the fact that if $Q \subseteq Q_0$, $Q_0 \in \mathcal{M}_i$, and $x \in Q$, then $B(x, 5 \operatorname{diam} Q) \cap E \subseteq 6Q_0$. This fact is needed to make sure that we did not mess up (4.58) when building g.] This proves that Lemma 4.57 does imply Lemma 4.55.

It remains to prove Lemma 4.57. This is, at long last, the real crux of the matter.

Let ϵ, μ and q be given. Let $k_0 > 1$ be large, to be chosen later. Fix $Q_0 \in \mathcal{M}$ and $j \in \{1, \ldots, N_2\}$.

We are going to build our Lipschitz function f by an iterative procedure. We shall initially get a Lipschitz function with large norm, but then we shall renormalize.

Choose $l \in \mathbf{Z}$ so that $Q_0 \in \Delta_l$. We are going to construct a sequence of functions $\{f_m\}_{m=-1}^{\infty}$ on E with the following properties:

(4.60) $\qquad$ if $m \geq 0$, then $f_m = f_{m-1}$ unless $\mathcal{A}_j \cap \Delta_{l-m} \neq \varnothing$;

(4.61) $\qquad$ for each $m \geq 0$ and $z \in E$ there is a Lipschitz function
$\phi : \mathbf{R} \to \mathbf{R}$ with norm ≤ 1 such that $f_m(x) = \phi(\pi_n(x))$
for all $x \in E \cap B(z, C_0^{-1}\epsilon^2 2^{l-m})$.

This constant C_0 will depend only on innocuous quantities like n, d, and the regularity constant for E.

In view of (4.52), (4.60) says that $f_m = f_{m-1}$ for most m's. When reading (4.61), one should keep in mind that 2^{l-m} is approximately the diameter of the cubes in Δ_{l-m}. Thus, (4.61) says that f_m looks like a function of x_n alone on scales that are small compared to the diameters of the cubes in Δ_{l-m}.

The reason that (4.61) is desirable is that it is easier to mess around with functions of 1 variable without losing control over their Lipschitz norms than it is to do the same thing with functions of several variables. The reason that we are able to impose (4.61) is that E can only touch the tops of the relevant boxes and not their sides or bottoms. If this were not the case, then the global compatibility conditions stemming from (4.61) would be intractable.

Let us now explain how the f_m's are constructed. Set $f_{-1} \equiv 0$. Let $m \geq 0$ be given, and assume that f_p has been constructed for $p < m$, in accordance with (4.60) and (4.61). We want to build f_m.

We may as well assume that $\mathscr{A}_j \cap \Delta_{l-m} \neq \varnothing$, since otherwise we simply take $f_m = f_{m-1}$. Define $\mathscr{R}_m \subseteq \Delta$ by

$$\mathscr{R}_m = \{Q : Q \in \Delta_{l-m} \cap \mathscr{A}_j \cap \mathscr{B}'_s(\epsilon, \mu) \text{ and } Q \subseteq Q_0\}.$$

(Here "$\mathscr{R}$" stands for "relevant".) Set

(4.62) $$f_m = f_{m-1} \quad \text{on } E \setminus \left(\bigcup_{Q \in \mathscr{R}_m} 3Q\right).$$

Let $Q \in \mathscr{R}_m$ be given, and let us define f_m on $3Q$. Notice that $4Q \cap 4Q' = \varnothing$ if $Q' \in \mathscr{R}_m$, $Q' \neq Q$, by (4.53) (assuming that k_0 is large enough), so the definition of f_m on $3Q$ will be independent of our choices for the other cubes in $\mathscr{R}_m$.

Let $M' = M'(Q)$ be a special ϵ-box associated to Q as in the definition (4.44) of $\mathscr{B}'_s(\epsilon, \mu)$. Let $I' = I'(Q)$ denote its inside. By definitions we have

(4.63) $$\operatorname{dist}(x, Q) \leq 2 \operatorname{diam} Q \quad \text{for all } x \in M' \cup I'.$$

In particular, $I' \cap E \subseteq 3Q$. Set

$$f_m = f_{m-1} \quad \text{on } 3Q \setminus I'.$$

Thus it remains to specify f_m on $I' \cap E$.

Let $u^1, \ldots, u^{n+2}$ be associated to M' as in the definition of $\mathscr{B}'_s(\epsilon, \mu)$. Let $\phi : \mathbf{R} \to \mathbf{R}$ be a 1-Lipschitz function such that $f_{m-1}(x) = \phi(\pi_n(x))$ whenever $x \in E \cap B(u^1, 10 \operatorname{diam} Q)$. The existence of such a ϕ follows from (4.60) and (4.61) if k_0 is large enough. Indeed, we are assuming that $\mathscr{A}_j \cap \Delta_{l-m} \neq \varnothing$, and hence $\mathscr{A}_j \cap \Delta_{l-p} = \varnothing$ if $0 < |p-m| \leq k_0$, by (4.52). From (4.60) we conclude that $f_{m-1} = f_a$, where $a = m - k_0$ if $m - k_0 \geq -1$, and $a = -1$ if $m - k_0 \leq -1$. The desired representation for f_{m-1} now follows from (4.61) if k_0 is large enough.

Let $a, b \in \mathbf{R}$ be such that

$$\pi_n(I') = [a, b].$$

Thus $a \leq \pi_n(u^1) < \pi_n(u^2) < \cdots < \pi_n(u^{n+2}) \leq b$. We are going to take

(4.64) $$f_m(x) = \psi_Q(\pi_n(x)) \quad \text{on } I' \cap E,$$

where $\psi_Q : [a, b] \to \mathbf{R}$ is a Lipschitz function with norm ≤ 1 such that

(4.65) $$\psi_Q = \phi \quad \text{on} \quad \left[\pi_n(u^{n+2}), b\right].$$

We will specify ψ_Q later; for now we shall use only the conditions on ψ_Q that were just stated to derive some important properties of f_m.

There is a useful fact that we want to record before proceeding. To state it we need to introduce a little more notation.

Let $\widehat{M}'$ denote the convex hull of M'. To understand $\widehat{M}'$ in concrete terms notice that if $M'(\epsilon)$ denotes the standard ϵ-box, then

(4.66) $$\widehat{M}'(\epsilon) = \left\{(x_1, \ldots, x_n) \in \mathbf{R}^n : 0 \leq x_n \leq 10\epsilon \text{ and } \sup_{1 \leq i \leq n-1} |x_i| \leq 1\right\}.$$

Define $\tilde{\phi} : \mathbf{R} \to \mathbf{R}$ by

$$\begin{aligned}\tilde{\phi} &= \phi(t) & \text{when } t \geq b, \\ &= \psi_Q(t) & \text{when } a \leq t \leq b, \\ &= \psi_Q(a) & \text{when } t \leq a.\end{aligned}$$

Because of (4.65), $\tilde{\phi}$ is continuous, and it is easy to see that $\tilde{\phi}$ is Lipschitz with norm 1. The fact that we need is

(4.67) $$f_m(x) = \tilde{\phi}(\pi_n(x)) \quad \text{on} \quad \widehat{M}' \cap E.$$

This is not hard to check, by chasing definitions. [Do not forget that $M' \cap E = \varnothing$.]

Let us check that f_m satisfies (4.61). Let $z \in E$ be given, and set $B = B(z, C_0^{-1} \epsilon^2 2^{l-m})$. [Of course, we have not specified C_0 yet, but it will be chosen very soon.] There are three possibilities:

(i) $B \cap I'(Q) = \varnothing$ for all $Q \in \mathscr{R}_m$;
(ii) $B \subseteq I'(Q)$ for some $Q \in \mathscr{R}_m$;
(iii) B intersects $I'(Q)$ for some $Q \in \mathscr{R}_m$, but B is not contained in $I'(Q)$.

In the first case we have $f_m = f_{m-1}$ on $B \cap E$, and we are in business because of the induction hypothesis. The second case is taken care of by (4.64). The third case is a little more interesting.

Suppose that B satisfies (iii) and that $Q \in \mathscr{R}_m$ is as in (iii). It suffices to show that

(4.68) $$B \subseteq \widehat{M}'(Q),$$

because of (4.67). [Here again we let $\widehat{M}'(Q)$ denote the convex hull of $M'(Q)$.] To show that (4.68) is true, we observe first that

(4.69) $$\operatorname{dist}(I'(Q), \mathbf{R}^n \setminus \widehat{M}'(Q)) \geq C^{-1} \epsilon \operatorname{diam} M'(Q),$$

where C is a very computable constant. This inequality is easy to derive from the definitions, by reducing to the case of the standard ϵ-box. Next, notice that

(4.70) $$\epsilon^2 2^{l-m} \leq C\epsilon^2 \operatorname{diam} Q \leq C\epsilon \operatorname{diam} M'(Q).$$

[The second inequality comes from the fact that the ϵ-box $M'(Q)$ is associated to Q as in the definition (4.44) of $\mathscr{B}'_s(\epsilon, \mu)$.] If C_0 is large enough, then (4.68) follows from (4.69), (4.70), and the requirement in (iii) that B intersect $I'(Q)$. This is the only place where we need to impose a condition on the size of C_0; any C_0 which is large enough for this to work is okay for (4.61).

That finishes the proof that f_m satisfies (4.61).

Next we want to show that

(4.71) $$\|f_m - f_{m-1}\|_\infty \leq C 2^{l-m}$$

for all m. If $\mathscr{A}_j \cap \Delta_{l-m} = \varnothing$ then $f_m \equiv f_{m-1}$ and there is nothing to prove, so we may as well assume that $\mathscr{A}_j \cap \Delta_{l-m} \neq \varnothing$. By construction we also have that $f_m = f_{m-1}$ on $E \setminus \bigcup_{Q \in \mathscr{R}_m} I'(Q)$, so we need only look at $f_m - f_{m-1}$ on $I'(Q)$ for each Q in $\mathscr{R}_m$.

Fix such a Q. Let ϕ, ψ_Q, and $[a, b]$ be as before (around (4.64) and (4.65)). Thus if $x \in I'(Q) \cap E$ then we have $f_m(x) = \psi_Q(\pi_n(x))$ and $f_{m-1}(x) = \phi(\pi_n(x))$. We also have $\phi(b) = \psi_Q(b)$, by (4.65), and we are assuming that ϕ and ψ_Q are both 1-Lipschitz. Hence

$$\sup_{I(Q) \cap E} |f_m - f_{m-1}| \leq 2(b-a) \leq 2 \operatorname{diam} I'(Q).$$

Of course, $\operatorname{diam} I'(Q) \leq \operatorname{diam} M'(Q)$ by definitions, and $\operatorname{diam} M'(Q) \leq \operatorname{diam} Q$ since $M'(Q)$ corresponds to Q as in the definition (4.44) of $\mathscr{B}'_s(\epsilon, \mu)$. Thus $|f_m - f_{m-1}| \leq C 2^{l-m}$ on $I'(Q) \cap E$. This proves (4.71).

From (4.71) we conclude that the sequence $\{f_m\}$ converges to a function $f : E \to \mathbf{R}$ and that

(4.72) $$\|f - f_m\|_\infty \leq C 2^{l-m}$$

for all m.

LEMMA 4.73. *f is Lipschitz with norm $\leq C\epsilon^{-2}$.*

Here C depends only on innocuous constants, but the main point is that it does not depend on Q_0.

Let z, z' in E be given, and choose m so that

$$C_0^{-1} \epsilon^2 2^{l-m-1} \leq |z - z'| < C_0^{-1} \epsilon^2 C 2^{l-m},$$

where C_0 is as in (4.61). Then

$$|f_m(z) - f_m(z')| \leq |z - z'|$$

by (4.61), while

$$|f_m(z) - f(z)| + |f_m(z') - f(z')| \leq C2^{l-m}$$

by (4.72). Lemma 4.73 now follows easily.

We are almost finished now. All that remains is for us to choose the ψ_Q's in such a way that we have the right sort of lower bounds on the γ_q's (as in (4.58)).

Let $\gamma_q(x, t)$ be associated to f in the usual way, and let $\gamma_{q,m}(x, t)$ be the corresponding quantity for f_m. Fix $Q \in \mathscr{B}'_s(\epsilon, \mu) \cap \mathscr{A}_j$ such that $Q \subseteq Q_0$, and choose $m \geq 0$ so that $Q \in \Delta_{l-m}$ (and thus $Q \in \mathscr{R}_m$). For each $x \in Q$ we have that

(4.74) $\quad \gamma_{q,m}(x, 5 \operatorname{diam} Q) \leq C \|f - f_m\|_\infty (\operatorname{diam} Q)^{-1} + \gamma_q(x, 5 \operatorname{diam} Q),$

as one can see by chasing definitions. On the other hand, we have that $\mathscr{A}_j \cap \Delta_{l-m}$ is nonempty, because it contains Q, so $\mathscr{A}_j \cap \Delta_{l-a} = \emptyset$ for $a = m+1, m+2, \ldots, m+k_0$ by (4.52). Thus (4.60) implies that $f_m = f_{m+k_0}$, so

$$\|f - f_m\|_\infty = \|f - f_{m+k_0}\|_\infty \leq C2^{l-m-k_0}$$

by (4.72). Inserting this into (4.74) we get

(4.75) $\quad\quad \gamma_{q,m}(x, 5 \operatorname{diam} Q) \leq C2^{-k_0} + \gamma_q(x, 5 \operatorname{diam} Q).$

Keep in mind that we get to take k_0 to be as large as we want, so the first term on the right will be very small.

We need to show that we can choose ψ_Q in such a way as to get a good lower bound for $\gamma_{q,m}(x, 5 \operatorname{diam} Q)$. This lower bound will not depend on any of the other specific choices that are made in the construction of f_m (only on the general properties discussed above).

Let $M'(Q)$ denote the ϵ-box associated to Q as in the construction of f_m, and let $u^1, \ldots, u^{n+2} \in I'(Q)$ be the points associated to $M'(Q)$ as described in the definition (4.44) of $\mathscr{B}'_s(\epsilon, \mu)$. We get to choose ψ_Q subject only to the constraints that it be a Lipschitz function with norm ≤ 1 which satisfies (4.65). If we are given real numbers $\alpha_1, \ldots, \alpha_{n+2}$, then we can find such a ψ_Q that also satisfies

(4.76) $\quad\quad \psi_Q(\pi_n(u^p)) = \alpha_p, \quad 1 \leq p \leq n+2,$

so long as

(4.77) $\quad \alpha_{n+2} = \phi(\pi_n(u^{n+2}))$ and $|\alpha_p - \alpha_{p+1}| \leq \mu \operatorname{diam} M'(Q)$
$$\text{for } p = 1, 2, \ldots, n+1.$$

Here ϕ is the function which arises in the construction of f_m and which is discussed in the paragraph just before the one which contains (4.64). [The sufficiency of (4.77) stems from the fact that

$$\pi_n(u^p) < \pi_n(u^{p+1}) - \mu \operatorname{diam} M'(Q)$$

for $p = 1, \ldots, n+1$. See (4.44).]

The next lemma will help us to select the α_p's. [This task would be substantially simpler if q were infinite.]

LEMMA 4.78. *There is a $C_1 > 0$, which depends only on n, and a set $F \subseteq \mathbf{R}^{n+2}$ with $n+1$ elements, such that each $\alpha \in F$, $\alpha = \{\alpha_p\}_{p=1}^{n+2}$, satisfies (4.77), and also*

$$(4.79) \qquad \max_{\alpha \in F} \inf_{a \in A(\mathbf{R}^n)} \max_{1 \leq p \leq n+2} |a(v^p) - \alpha_p| \geq C_1^{-1} \mu \operatorname{diam} M'(Q)$$

for any family of $n+2$ points $v^1, \ldots, v^{n+2}$ in $\mathbf{R}^n$.

Here $A(\mathbf{R}^n)$ denotes the set of real-valued affine functions on $\mathbf{R}^n$. We shall derive Lemma 4.78 from the following.

LEMMA 4.80. *There is a set $F_1 \subseteq \mathbf{R}^{n+1}$ with $n+1$ elements such that each $\beta = \{\beta_p\}_{p=1}^{n+1}$ in F_1 satisfies $\max_p |\beta_p| \leq \mu \operatorname{diam} M'(Q)$, and also*

$$(4.81) \qquad \max_{\beta \in F_1} \inf_{w \in H} \max_{1 \leq p \leq n+1} |w_p - \beta_p| \geq C_1^{-1} \mu \operatorname{diam} M'(Q)$$

for all hyperplanes H in $\mathbf{R}^{n+1}$ which pass through the origin.

Lemma 4.80 is trivial. Lemma 4.78 is really just a dressed-up version of Lemma 4.80. This is not hard to verify, and we omit the details, except to supply two observations. The first is that

$$\{w \in \mathbf{R}^{n+1} : \text{there is an } a \in A(\mathbf{R}^n) \text{ such that } w_p = a(v^p) - a(v^{p+1})$$
$$\text{for each } p, \ 1 \leq p \leq n+1\}$$

is a linear subspace of $\mathbf{R}^{n+1}$ with dimension $\leq n$, no matter how the v^p's are chosen. The second observation is that

$$2 \max_{1 \leq p \leq n+2} |a(v^p) - \alpha_p| \geq \max_{1 \leq p \leq n+1} |a(v^p) - a(v^{p+1}) - (\alpha_p - \alpha_{p+1})|.$$

We could also mention that the α's in Lemma 4.78 correspond to the β's in Lemma 4.80 via $\beta_p = \alpha_p - \alpha_{p+1}$.

We are going to choose ψ_Q after we select an α from F. To decide how α should be chosen, we need to make some more computations.

For each $t > 0$ let $\Omega(t)$ denote the cartesian product of $B(u^p, t) \cap E$, $p = 1, \ldots, n+2$, and let $\pi_p : \Omega(t) \to B(u^p, t) \cap E$ denote the projection onto the pth factor. From (4.79) we have that

$$\left(\sum_{\alpha \in F} \inf_{a \in A(\mathbf{R}^n)} \sum_{p=1}^{n+2} |a(\pi_p(\omega)) - \alpha_p|^q \right)^{1/q}$$
$$\geq C_1^{-1} \mu \operatorname{diam} M'(Q) \geq C_1^{-1} \mu \epsilon \operatorname{diam} Q$$

for all $\omega \in \Omega(t)$ (see also (4.44)), and hence
$$\left(\sum_{\alpha \in F} t^{-d(n+2)} \int_{\Omega(t)} \inf_{a \in A(\mathbf{R}^n)} \sum_{p=1}^{n+2} |a(\pi_p(\omega)) - \alpha_p|^q \, d\omega \right)^{1/q} \geq C^{-1} \mu \epsilon \operatorname{diam} Q. \tag{4.82}$$

Here "$d\omega$" denotes the $(n+2)$-fold product of d-dimensional Hausdorff measure (restricted to E).

If we interchange the order of the integral and the infimum in (4.82) then we increase the left-hand side. We can then interchange the sum in p with the integral and simplify to obtain

$$\left(\sum_{\alpha \in F} \inf_{a \in A(\mathbf{R}^n)} \sum_{p=1}^{n+2} t^{-d} \int_{B(u^p, t) \cap E} |a(y) - \alpha_p|^q \, dy \right)^{1/q} \geq C^{-1} \mu \epsilon \operatorname{diam} Q.$$

This implies that

$$\max_{\alpha \in F} \inf_{a \in A(\mathbf{R}^n)} \left(\sum_{p=1}^{n+2} t^{-d} \int_{B(u^p, t) \cap E} |a(y) - \alpha_p|^q \, dy \right)^{1/q} \geq C^{-1} \mu \epsilon \operatorname{diam} Q. \tag{4.83}$$

This is almost what we want but not quite. Define $h_t : \mathbf{R} \to \mathbf{R}$ by $h_t(s) = 0$ when $|s| \leq t$, and $h_t(s) = |s| - t$ otherwise. Then

$$\max_{\alpha \in F} \inf_{a \in A(\mathbf{R}^n)} \left(\sum_{p=1}^{n+2} t^{-d} \int_{B(u^p, t) \cap E} h_t(a(y) - \alpha_p)^q \, dy \right)^{1/q} \tag{4.84}$$

is bounded from below by $C^{-1} \mu \epsilon \operatorname{diam} Q - Ct$.

Fix $t > 0$, $t < \epsilon \operatorname{diam} M'(Q)/10$, which is small enough so that (4.84) is $\geq C^{-1} \mu \epsilon \operatorname{diam} Q$, but which is not too small, so that

$$t > C(\epsilon, \mu)^{-1} \operatorname{diam} Q. \tag{4.85}$$

Now we choose $\alpha \in F$ so that

$$\inf_{a \in A(\mathbf{R}^n)} \left(\sum_{p=1}^{n+2} t^{-d} \int_{B(u^p, t) \cap E} h_t(a(y) - \alpha_p)^q \, dy \right)^{1/q} \geq C^{-1} \mu \epsilon \operatorname{diam} Q. \tag{4.86}$$

Finally we take ψ_Q to be any function which satisfies (4.76) (with this choice of α) as well as the conditions that were specified in the construction of f_m. We are left with the task of producing a suitable lower bound for $\gamma_{q,m}(x, 5 \operatorname{diam} Q)$.

We claim that

$$\inf_{a \in A(\mathbf{R}^n)} \left(\sum_{p=1}^{n+2} t^{-d} \int_{B(u^p, t) \cap E} |f(y) - a(y)|^q \, dy \right)^{1/q} \geq C^{-1} \mu \epsilon \operatorname{diam} Q. \tag{4.87}$$

To see this recall that f_m is Lipschitz with norm ≤ 1 on $\widehat{M'}(Q) \cap E$, where $\widehat{M'}(Q)$ denotes the convex hull of $M'(Q)$. [See (4.67).] Hence

$$|f_m(y) - \alpha_p| = |f_m(y) - f_m(u^p)| \leq t \quad \text{when } y \in B(u^p, t) \cap E,$$

so $|f_m(y) - a(y)| \geq h_t(a(y) - \alpha_p)$ on $B(u^p, t) \cap E$. Thus (4.87) follows from (4.86).

Combining (4.87) with (4.85) we get that

(4.88) $$\gamma_{q,m}(x, 5 \operatorname{diam} Q) \geq C(\epsilon, \mu)^{-1}.$$

The right side does not depend on x, Q, or k_0, so we conclude from (4.75) that we can choose k_0 large enough (depending on ϵ and μ) so that

(4.89) $$\gamma_q(x, 5 \operatorname{diam} Q) \geq \theta_1,$$

where $\theta_1 > 0$ depends on ϵ and μ (and various innocuous constants) but not on x or Q.

Thus we have succeeded in constructing a Lipschitz function $f : E \to \mathbf{R}$ that satisfies all of the requirements of Lemma 4.57 except that its Lipschitz norm may be larger than 1. Since we do know (from Lemma 4.73) that the Lipschitz norm of f is $\leq C\epsilon^{-2}$, this problem can be solved by renormalizing.

This completes the proof of Lemma 4.57. Et voilà.

REMARK 4.90. It is a good exercise to think about why we do not know how to prove that the WALA implies uniform rectifiability when $d > 1$. In our construction of the f_m's we needed to be able to build Lipschitz functions (the ψ_Q's) which satisfied certain boundary conditions while also exhibiting substantially "bad" behavior in some places. We also had to do this in such a way that the Lipschitz norm of the f_m's remained bounded as m increases. It is much harder to do all these things when $d > 1$.

Although we can imagine more complicated methods that might avoid some of these difficulties (using, for instance, the results of [DS3]), we have not managed to do much more than simply imagine such an argument.

4.3. A more abstract version of the WALA.

The starting point for this section is the following question: Which features of affine functions played a significant role in the proof of Proposition 4.35? The slightly surprising answer is that we really only used the fact that the collection of all affine functions is a finite-dimensional vector space. This leads to the following generalization of the WALA.

Definition 4.91. Let E be a d-dimensional regular set in $\mathbf{R}^n$, and let Δ be a family of cubes in E, as in §I.3.1. We say that E satisfies the GWALA (generalized WALA) if for each $Q \in \Delta$ there is a vector space A_Q of real-valued functions on E such that

(4.92) $$\sup_{Q \in \Delta} \dim A_Q < \infty$$

and such that there is a $q > 0$ so that for each Lipschitz function $f : E \to \mathbf{R}$ we have that

(4.93) $$\{Q \in \Delta : (\operatorname{diam} Q)^{-1} \inf_{a \in A_Q} \left(\frac{1}{|Q|} \int_{2Q} |f - a|^q \right)^{1/q} > \epsilon \}$$

satisfies a Carleson packing condition for every $\epsilon > 0$.

There is a minor technical problem with this definition, which is that we have not demanded that the elements of the A_Q's be measurable, and yet we are integrating them in (4.93). We resolve this problem by interpreting the integral in (4.93) as an outer integral. [Recall that the outer integral of a nonnegative function g on E is defined to be the infimum of $\int_E h$ over all measurable functions h such that $h \geq g$.] Of course, this problem is particularly irrelevant when $q = \infty$, which is probably the most interesting case anyway.

We have defined the GWALA in terms of cubes instead of the upper half space $E \times \mathbf{R}_+$ to avoid other minor technical problems of a similar nature. It is easy to check that the definition does not depend on the specific choice of the family of cubes Δ in a substantial way.

This generalization of the WALA is not as ridiculous as it might seem. The chief advantage of the GWALA over the WALA is that it is much more intrinsic. Because the WALA is defined in terms of affine functions, it is inextricably tied to the ambient space $\mathbf{R}^n$, while one could formulate a version of the GWALA for an abstract set E which is equipped with a measure and a distance function.

This issue manifests itself well when one considers the behavior of these conditions under a mapping. Suppose that E is a d-dimensional regular set in $\mathbf{R}^n$, and let $\phi : E \to \mathbf{R}^n$ be another embedding of E into $\mathbf{R}^n$. If ϕ is bilipschitz and E satisfies the GWALA, then $\phi(E)$ does too, and this follows from definition-chasing. It is not so obvious that the WALA is preserved by a bilipschitz mapping, and in fact we know this to be true only when $d = 1$, in which case we know that the WALA is equivalent to uniform rectifiability.

In connection with the question of bilipschitz invariance it is natural to look at some other generalizations of the WALA besides the GWALA. Consider, for instance, the version of the GWALA that arises by adding the restriction that all the A_Q's are the same. This condition still has the feature that it is bilipschitz-invariant by definition, but it is also significantly less general than the GWALA, at least in appearance. Alternatively we could allow the A_Q's to depend on Q but impose some restrictions on how often A_Q can change. It would seem natural, for example, to consider the condition obtained by requiring that there be a coronization $(\mathscr{B}, \mathscr{G}, \mathscr{F})$ of E such that A_Q does not change as Q varies within a single $S \in \mathscr{F}$. This variant of the GWALA is also bilipschitz-invariant.

Let us now proceed to the main result of this section.

THEOREM 4.94. *Let E be a d-dimensional regular set in $\mathbf{R}^n$, $0 < d < n$, and suppose that E satisfies the GWALA. Then E also satisfies the WNB (see Definition II.3.44), and in particular E must be uniformly rectifiable when $d = 1$ (see Proposition II.3.45).*

Let E be given. The proof that E satisfies the WNB is almost exactly the same as it was in the case of the WALA, because the affine funtions played almost no role in the argument. The few modifications that are needed are as follows.

Set $N = \sup\{\dim A_Q : Q \in \Delta\}$. The definitions of $\mathscr{B}_s(\epsilon, \mu)$ and $\mathscr{B}'_s(\epsilon, \mu)$ should be changed so that there are $N + 2$ points u^j in (4.40) and (4.44) instead of $n + 2$. This necessitates some typographical changes throughout §4.2, but one does not really need to do any thinking until one gets to Lemmas 4.78 and 4.80. The modifications which are needed there are rather straightforward, and we leave the details as an exercise, except for one minor remark.

Suppose that we are given $N + 2$ points $v^1, \ldots, v^{N+2}$ in $\mathbf{R}^n$. For each $Q \in \Delta$ consider the set

$$L_Q = \{w \in \mathbf{R}^{N+1} : \text{there is an } a \in A_Q \text{ such that}$$
$$w_p = a(v^p) - a(v^{p+1}) \text{ for each } p, \ 1 \le p \le N + 1\},$$

which is a linear subspace of $\mathbf{R}^{N+1}$ with dimension $\le \dim A_Q \le N$. When A_Q is the space of affine functions, we actually have $\dim L_Q \le n - 1 = \dim A_Q - 1$, but in general we only know that $\dim L_Q \le \dim A_Q$. That is why we only needed $n + 2$ points u^j in §4.2, while in the present circumstances we need $N + 2$, which corresponds to $n + 3$ when A_Q is simply the space of affine functions for each $Q \in \Delta$.

CHAPTER 5

The Weak Constant Density Condition

This chapter is primarily devoted to the proof of Theorem I.2.56. We give a preliminary soft result in the first section, from which the $d = 1, 2$ cases can be derived easily. The second, third, and fourth sections deal with the codimension 1 case, which is more difficult. In the last section we consider a variant of the WCD.

5.1. Compactness will only get you so far.
Fix d and n, $0 < d < n$.
Let CD (for "constant densities") denote the set of nonempty closed sets A in $\mathbf{R}^n$ such that $A = \operatorname{supp} \alpha$ for some measure α which satisfies

(5.1) $\qquad \alpha(B(a, r)) = r^d \quad$ for all $a \in \operatorname{supp} \alpha$ and all $r > 0$.

Clearly CD contains the set of all d-planes in $\mathbf{R}^n$, but there are also some nonlinear examples known.

EXAMPLE 5.2. Consider $A \subseteq \mathbf{R}^4$ defined by

$$A = \{x \in \mathbf{R}^4 : x_1^2 = x_2^2 + x_3^2 + x_4^2\}.$$

Then $A \in$ CD (with $d = 3$ and $n = 4$). (See Proposition 3.8 in [KoP].)

PROPOSITION 5.3. *Let E be a d-dimensional regular set in $\mathbf{R}^n$. If E satisfies the WCD, then E lies in* Approx (*CD*).

Recall that the WCD and Approx(·) are defined in Definitions I.2.55 and I.2.21, respectively. Roughly speaking, the WCD requires that, at most location and scales, E can be realized as the support of a measure which almost satisfies (5.1), while $E \in \operatorname{Approx}(CCD)$ means that, at most locations and scales, E can be approximated by a set which actually does lie in CD.

We shall prove Proposition 5.3 (later in this section) using a compactness argument. This seems rather natural, although undesirably mystical. Before giving the proof we want to discuss the consequences of Proposition 5.3.

COROLLARY 5.4. *The WCD implies the BWGL (and hence uniform rectifiability) when $d = 1$ or 2.*

[Recall that the BWGL is defined in Definition I.2.2 and the fact that it implies uniform rectifiability was stated as Theorem I.2.4.]

Corollary 5.4 follows from Proposition 5.3 once we know that CD consists only of lines or 2-planes when $d = 1$ or 2, respectively. This is true and was proved in [**P**], although, as was pointed out in [**P**], the $d = 1$ case is elementary.

Of course, when $d = 1$ we also have the alternate proof given in §1.3 of Part II, which has the advantage of being more direct by dint of avoiding the compactness argument.

When $d > 2$ the story is more complicated because of the presence of nonlinear sets in CD. In general there is no known characterization of the elements of CD, but there is one in the codimension 1 case, due to Kowalski and Preiss.

THEOREM 5.5 [**KoP**]. *Suppose that $d = n - 1$, $n \geq 4$. Then CD consists precisely of the hyperplanes in $\mathbf{R}^n$ and rotations and translations of the set*

(5.6) $$\{x \in \mathbf{R}^n : x_1^2 = x_2^2 + x_3^2 + x_4^2\}.$$

Theorem 5.5 and Proposition 5.3 combine to tell us a lot about sets which satisfy the WCD when $d = n - 1$. Unfortunately they do not tell us enough for us to be able simply to invoke one of our results from Part II to conclude that the given set is uniformly rectifiable. Additional arguments are required, and these will be given in the next sections.

In a word, the problem is that the set (5.6) is singular. Although the results in Chapter II.3 allow certain types of singularities, the set (5.6) does not have the right kind of geometry for us to be able to use those results. Instead we are going to show that these singularities generally have to stay away from the places that matter. We shall be able to do this in large measure because we know exactly what the singularities look like and, in particular, that they cannot appear or disappear too suddenly.

These issues are illustrated well by the following model problem. For this we consider one-dimensional sets in $\mathbf{R}^2$. Set

$$\mathscr{A}_1 = \{A \subseteq \mathbf{R}^2 : A \text{ is a line or a union of two lines}\},$$

and denote by $\mathscr{A}_2$ the subset of $\mathscr{A}_1$ whose elements are either lines or unions of perpendicular lines. Consider $\text{Approx}(\mathscr{A}_i)$, $i = 1, 2$, which is defined in accordance with Definition I.2.21. The model problem is to show that the elements of $\text{Approx}(\mathscr{A}_2)$ are all uniformly rectifiable. This is in fact true for $\text{Approx}(\mathscr{A}_1)$, by the results of Chapter II.3, but that is much harder. $\text{Approx}(\mathscr{A}_2)$ can be treated by the techniques of this chapter, because the singularities cannot appear or disappear too suddenly.

It is easy to imagine how similar issues could arise when attempting to prove that the WCD implies uniform rectifiability in other dimensions and codimensions. Although there is no general characterization of the elements of CD, Kirchheim and Preiss [**KiP**] have proved that the elements of CD are always analytic varieties. In fact, they show that each element of CD can be realized as the zero set of a globally defined real-analytic function, and this function is given explicitly in terms of the associated measure α (from the definition of CD). If the possible behavior of the singularities of the elements of CD were better understood, one might be able to extend Theorem I.2.56 to the general case with methods which resemble those employed in the next sections.

We should mention a related result of Preiss [**P**], which was instrumental in his proof of the fact that an H^d-measurable set in $\mathbf{R}^n$ with finite measure is rectifiable if its densities exist almost everywhere (i.e., if the upper and lower densities are equal a.e.). Preiss showed that if a given element of CD is not a d-plane, then it has to be far from being flat at infinity. This result comes with an estimate, but for our purposes it has a serious drawback, which is that you do not know how far out (to infinity) you have to go before this phenomena manifests itself. It is impossible to know this in advance because of Example 5.2; you can start off in a very smooth part of the cone, far away from the singularity, and it could be a long time before the truth emerges.

Let us now prove Proposition 5.3. This will be rather similar to the proof of Lemma 2.36 in §2.4 except that we shall organize the arguments a little differently.

Fix d and n, $0 < d < n$. Let $\mathrm{RM}(C_0)$ denote the set of (Ahlfors) regular measures of dimension d on $\mathbf{R}^n$ with constant $\leq C_0$. That is, $\mathrm{RM}(C_0)$ consists of the nonnegative Borel measures μ on $\mathbf{R}^n$ which are not identically zero and which satisfy

$$(5.7) \qquad C_0^{-1} R^d \leq \mu(B(x, R)) \leq C_0 R^d$$

for all $x \in \operatorname{supp} \mu$ and all $R > 0$.

We say that a sequence $\{\mu_j\}$ of nonnegative Borel measures on $\mathbf{R}^n$ converges weakly to the Borel measure μ if

$$\lim_{j \to \infty} \int_{\mathbf{R}^n} \phi \, d\mu_j = \int_{\mathbf{R}^n} \phi \, d\mu$$

for all continuous functions ϕ on $\mathbf{R}^n$ which have compact support.

LEMMA 5.8. *Let $C_0 > 0$ be given.*

  (a) *Every sequence in $\mathrm{RM}(C_0)$ has a subsequence which converges weakly to some Borel measure on $\mathbf{R}^n$.*
  (b) *If a sequence in $\mathrm{RM}(C_0)$ converges weakly to a Borel measure μ on $\mathbf{R}^n$, then either $\mu \equiv 0$ or $\mu \in \mathrm{RM}(C_0)$.*

Part (a) follows from standard results in functional analysis. Part (b) is a minor variant of Lemma 2.42 and is fairly straightforward in any case.

LEMMA 5.9. *Suppose that $\{\mu_j\} \subseteq \mathrm{RM}(C_0)$ and $\mu_j \to \mu$ weakly. Then for every ball B in $\mathbf{R}^n$ we have that*

$$(5.10) \qquad \lim_{j \to \infty} \left(\sup_{p \in B \cap \mathrm{supp}\,\mu} \mathrm{dist}(p, \mathrm{supp}\,\mu_j) \right) = 0$$

and

$$(5.11) \qquad \lim_{j \to \infty} \left(\sup_{p \in B \cap \mathrm{supp}\,\mu_j} \mathrm{dist}(p, \mathrm{supp}\,\mu) \right) = 0.$$

For the purposes of this lemma we are using the convention that the supremum of anything over the empty set is 0. Also, we take $\mathrm{dist}(p, \mathrm{supp}\,\mu)$ to be ∞ if $\mathrm{supp}\,\mu = \varnothing$.

Lemma 5.9 is nothing but the obvious generalization of Lemmas 2.41 and 2.43 to arbitrary dimensions, and we shall not repeat the (easy) proof.

Lemmas 5.8 and 5.9 contain the basic information about compactness that we shall need. Now we want to focus our attention on some special classes of measures that are pertinent to Proposition 5.3.

Let $\mathscr{U}$ (for "uniform") denote the set of Borel measures μ on $\mathbf{R}^n$ which are not identically zero and which satisfy

$$\mu(B(x, R)) = R^d$$

for all $x \in \mathrm{supp}\,\mu$ and $R > 0$. Given $C_0 > 0$, $k > 0$, and $\epsilon > 0$, let $\mathscr{U}(C_0, k, \epsilon)$ denote the set of $\mu \in RM(C_0)$ such that

$$|\mu(B(x, R)) - R^d| \leq \epsilon$$

whenever $x \in \mathrm{supp}\,\mu$, $|x| \leq k$, and $0 < R \leq k$. Let $\mathscr{U}_0$ and $\mathscr{U}_0(C_0, k, \epsilon)$ denote the corresponding classes of measures which satisfy the additional constraint that $0 \in \mathrm{supp}\,\mu$.

LEMMA 5.12. *Let $C_0 > 0$ be given in addition to sequences $\{\epsilon_j\}$ and $\{k_j\}$ of positive numbers such that $\epsilon_j \to 0$ and $k_j \to \infty$. Let $\{\mu_j\}$ be a sequence of Borel measures on $\mathbf{R}^n$ with $\mu_j \in \mathscr{U}_0(C_0, k_j, \epsilon_j)$, and assume that $\mu_j \to \mu$ weakly. Then $\mu \in \mathscr{U}_0$.*

The proof of this is straightforward, and we omit it. Combining this with compactness we can get the following, which is the crux of the matter.

LEMMA 5.13. *Let $C_0 > 0$ and $\eta > 0$ be given. There exist $k > 0$ (large) and $\epsilon > 0$ (small) so that if $\mu \in \mathscr{U}_0(C_0, k, \epsilon)$, then there is an $A \in CD$ such that*

$$(5.14) \qquad \sup_{p \in (\mathrm{supp}\,\mu) \cap B(0, 1)} \mathrm{dist}(p, A) \leq \eta$$

and

$$(5.15) \qquad \sup_{a \in A \cap B(0, 1)} \mathrm{dist}(a, \mathrm{supp}\,\mu) \leq \eta.$$

Suppose not. Then we can find $C_0 > 0$, $\eta > 0$, $\{k_j\}$, $\{\epsilon_j\} \subseteq \mathbf{R}_+$, and $\{\mu_j\} \subseteq RM(C_0)$ such that $k_j \to \infty$, $\epsilon_j \to 0$, $\mu_j \in \mathscr{U}_0(C_0, k_j, \epsilon_j)$ for all j, and the conclusion of Lemma 5.13 fails for each μ_j. In view of Lemma 5.8(a) we may assume that $\{\mu_j\}$ converges weakly to some measure μ (by passing to a subsequence if necessary). Lemma 5.12 tells us that $\mu \in \mathscr{U}_0$, so $A = \operatorname{supp} \mu$ lies in CD, by definition. From Lemma 5.9 it follows that (5.14) and (5.15) are indeed true if μ is replaced by μ_j and j is large enough. This contradicts our hypothesis about the μ_j's, and Lemma 5.13 is proved.

All that remains of the proof of Proposition 5.3 is some definition-chasing.

Given $x \in \mathbf{R}^n$ and $t > 0$, define $\tau_{x,t} : \mathbf{R}^n \to \mathbf{R}^n$ by $\tau_{x,t}(y) = t^{-1}(y - x)$. Thus $\tau_{x,t}$ maps $B(x, t)$ onto $B(0, 1)$.

If μ is a measure on $\mathbf{R}^n$, let $\tau_{x,t}(\mu)$ be the measure defined by

$$T_{x,t}(\mu)(V) = t^d \mu(\tau_{x,t}(V)).$$

Observe that $T_{x,t}$ preserves the classes $RM(C_0)$ and $\mathscr{U}$.

LEMMA 5.16. *Let E be a d-dimensional regular set in $\mathbf{R}^n$, and suppose that E satisfies the WCD. Then there is a $C_0 > 0$ so that for every $k \geq 1$ and $\epsilon > 0$ the complement of the following set in $E \times \mathbf{R}_+$ is a Carleson set:*

$$\{(x, t) \in E \times \mathbf{R}_+ : \text{ there is a } \mu \in \mathscr{U}_0(C_0, k, \epsilon)$$
$$\text{ such that } E = \operatorname{supp}(T_{x,t}(\mu))\}.$$

The converse to this is also true, but we do not need it.

When $k = 1$ this is nothing but a recasting of Definition I.2.55 in new notation. It is not hard to extend this to all $k \geq 1$, using in particular the simple observation that the mapping

$$(x, t) \mapsto (x, kt)$$

pulls Carleson sets in $E \times \mathbf{R}_+$ back to Carleson sets.

With Lemma 5.16 in hand, Proposition 5.3 follows directly from Lemma 5.13 and the definitions.

5.2. The codimension 1 case, part 1.

Fix $n \geq 4$, and set $d = n - 1$.

Let E be a d-dimensional regular set in $\mathbf{R}^n$, and let Δ be a family of cubes in E, as in §I.3.1. For technical reasons it will be convenient for us to impose an additional condition on our cubes:

(5.17) if Q, $Q' \in \Delta$ and $Q \subseteq Q'$, then either $Q = Q'$

or $\operatorname{diam} Q \leq \frac{1}{2} \operatorname{diam} Q'$.

Let us explain why we can do this. Remember that Δ is the union of Δ_l, $l \in \mathbf{Z}$, as in §I.3.1. However, there is no reason why a cube Q cannot lie in both Δ_l and Δ_m, even though $l \neq m$, as long as $|l - m|$ is not too large. We can get (5.17) by simply redefining the Δ_l's in such a way that

$$\Delta_l = \Delta_{jN} \quad \text{when } jN \leq l < (j+1)N.$$

If N is large enough, then (5.17) will be true. This change in the Δ_l's will not destroy the other important properties of the cubes (listed in §I.3.1), although it may make the constants a little worse.

Given $\epsilon > 0$ (small) let $\mathcal{N}(\epsilon)$ ("$\mathcal{N}$" for "nice") denote the set of cubes $Q \in \Delta$ for which there exists an $A \in $ CD such that

(5.18) $\qquad \operatorname{dist}(x, A) \leq \epsilon \operatorname{diam} Q \quad$ for all $\ x \in 10Q$

and

(5.19) $\quad \operatorname{dist}(a, E) \leq \epsilon \operatorname{diam} Q$
$\qquad\qquad$ whenever $\ a \in A \ $ satisfies $\ \operatorname{dist}(a, Q) \leq 10 \operatorname{diam} Q$.

[Remember that λQ was defined in (I.3.7).]

LEMMA 5.20. *If E satisfies the WCD, then $\Delta \backslash \mathcal{N}(\epsilon)$ satisfies a Carleson packing condition for all ϵ, i.e.,*

$$\sum_{\substack{Q \in \Delta \backslash \mathcal{N}(\epsilon) \\ Q \subseteq Q_0}} |Q| \leq C(\epsilon) |Q_0|$$

for all $Q_0 \in \Delta$.

This follows easily from Proposition 5.3.

Given $\epsilon > 0$ define $\mathcal{G}(\epsilon)$ ("$\mathcal{G}$" for "good") to be the set of $Q \in \Delta$ for which there exists $A \in $ CD such that (5.18) and (5.19) are true, and also

(5.21) $\qquad \operatorname{dist}(\operatorname{sing}(A), Q) \geq 7 \operatorname{diam} Q$.

Here "sing(A)" refers simply to the set of nonsmooth points on A. To be explicit, $\operatorname{sing}(A) = \varnothing$ when A is a hyperplane (in which case we interpret the left side of (5.21) to be $+\infty$), and $\operatorname{sing}(A) = \{x \in \mathbf{R}^n : x_1 = x_2 = x_3 = x_4 = 0\}$ when A is given by (5.6).

We want to show that if E satisfies the WCD, then $\Delta \backslash \mathcal{G}(\epsilon)$ satisfies a Carleson packing condition for all $\epsilon > 0$. Once we do that we will be in good shape. The idea behind the proof of this is that we can track the movement of sing(A) as we go from cube to cube, and because the singularity set is always rather small, it will never be the case that very many cubes will be affected. This is admittedly rather vague, but we hope that it will give the reader some idea of why we care about the next lemmas.

LEMMA 5.22. *Let Q and $\widehat{Q}$ be cubes in E, with $\widehat{Q}$ the parent of Q. Suppose that $\widehat{Q} \in \mathcal{G}(\epsilon)$ and $Q \in \mathcal{N}(\epsilon)$. If ϵ is small enough, then Q lies in $\mathcal{G}(\epsilon)$.*

This gives an example of a sense in which the singularities cannot appear too suddenly.

This is basically a trivial consequence of the definitions, but let us be careful. Observe first that we could have $Q = \widehat{Q}$, in which case there is nothing to prove. So suppose that $Q \neq \widehat{Q}$.

5.2. THE CODIMENSION 1 CASE, PART 1

Let A, $\widehat{A} \in$ CD correspond to Q and $\widehat{Q}$ as in the definitions of $\mathscr{G}(\epsilon)$ and $\mathscr{N}(\epsilon)$. Thus

(5.23) $\qquad \text{dist}(x, A), \leq \epsilon \operatorname{diam} Q \quad \text{for all } x \in 10Q,$

(5.24) $\quad \text{dist}(a, E) \leq \epsilon \operatorname{diam} Q \quad \text{whenever } a \in A$
$\qquad\qquad\qquad\qquad \text{and dist}(a, Q) \leq 10 \operatorname{diam} Q,$

(5.25) $\qquad \text{dist}(x, \widehat{A}) \leq \epsilon \operatorname{diam} \widehat{Q} \quad \text{for all } x \in 10\widehat{Q}, \text{ and}$

(5.26) $\quad \text{dist}(b, E) \leq \epsilon \operatorname{diam} \widehat{Q} \quad \text{whenever } b \in \widehat{A}$
$\qquad\qquad\qquad\qquad \text{and dist}(b, \widehat{Q}) \leq 10 \operatorname{diam} \widehat{Q},$

and we also have

(5.27) $\qquad \text{dist}(\operatorname{sing}(\widehat{A}), \widehat{Q}) \geq 7 \operatorname{diam} \widehat{Q}.$

Assume that $Q \notin \mathscr{G}(\epsilon)$, so that there is a $p \in \operatorname{sing}(A)$ which satisfies

$$\text{dist}(p, Q) < 7 \operatorname{diam} Q.$$

Because of (5.24) we can find a point $z \in E$ such that

$$|p - z| \leq \epsilon \operatorname{diam} Q.$$

In particular, $\text{dist}(z, Q) \leq (7 + \epsilon) \operatorname{diam} Q$, so $z \in 9Q$ if $\epsilon \leq 1$.

From (5.23)–(5.26) we have that E is very well approximated (bilaterally) by both A and $\widehat{A}$ inside $B(z, \operatorname{diam} Q)$, so they have to be close to each other inside $B(z, \operatorname{diam} Q)$ we well. The bottom line is that

$$\text{dist}(p, \operatorname{sing}(\widehat{A})) \leq C\epsilon \operatorname{diam} \widehat{Q}$$

if ϵ is sufficiently small, because $p \in \operatorname{sing}(A)$. Since

$$\begin{aligned} \text{dist}(p, \widehat{Q}) &\leq \text{dist}(p, Q) \\ &\leq 7 \operatorname{diam} Q \\ &\leq 4 \operatorname{diam} \widehat{Q} \quad [\text{by (5.17)}], \end{aligned}$$

this contradicts (5.27) as soon as ϵ is small enough.

The next lemma says that cubes which are nice but not good normally have a lot of good descendents.

LEMMA 5.28. *There exists $M \in \mathbf{Z}_+$ (large) so that the following is true. Let $l \in \mathbf{Z}$ be given, and suppose that $Q_0 \in \Delta_l \cap \mathscr{N}(\epsilon)$. Set $\mathscr{T} = \{Q \in \Delta_{l-M} : Q \subseteq Q_0, Q \in \mathscr{N}(\epsilon), \text{ but } Q \notin \mathscr{G}(\epsilon)\}$. Then*

(5.29) $$\left| \bigcup_{\mathscr{T}} Q \right| \leq \frac{1}{10} |Q_0|$$

if ϵ is small enough.

Let l, Q_0 be as above, and let M be large and ϵ be small, to be specified soon. Let A_0 be such that (5.18) and (5.19) are satisfied (with Q replaced by Q_0). We claim that

(5.30) $$\operatorname{dist}(Q, \operatorname{sing}(A_0)) \leq 8 \operatorname{diam} Q$$

for all $Q \in \mathcal{T}$, at least if $\epsilon 2^M$ is sufficiently small.

The proof of (5.30) is very similar to the proof of Lemma 5.22. Let $Q \in \mathcal{T}$ be given, and let $A \in \mathrm{CD}$ be such that (5.18) and (5.19) are satisfied, but

$$\operatorname{dist}(\operatorname{sing}(A), Q) < 7 \operatorname{diam} Q.$$

Choose $p \in \operatorname{sing}(A)$ such that

(5.31) $$\operatorname{dist}(p, Q) < 7 \operatorname{diam} Q.$$

We also have that $\operatorname{dist}(p, E) \leq \epsilon \operatorname{diam} Q$ (since Q and A satisfy (5.19)). As in the proof of Lemma 5.22, we now observe that E is well-approximated (bilaterally) by both A and A_0 inside $B(p, \operatorname{diam} Q)$, so A and A_0 have to approximate each other well inside that same ball. More precisely we have that

$$\sup_{a \in A \cap B(p, \operatorname{diam} Q)} \operatorname{dist}(a, A_0) \leq C 2^M \epsilon \operatorname{diam} Q$$

and

$$\sup_{b \in A_0 \cap B(p, \operatorname{diam} Q)} \operatorname{dist}(b, A) \leq C 2^M \epsilon \operatorname{diam} Q.$$

The 2^M appears because the estimates for the extent to which A_0 approximates E near Q_0 are in terms of $\operatorname{diam} Q_0$, and $\operatorname{diam} Q_0 \leq C 2^M \operatorname{diam} Q$. If $2^M \epsilon$ is small enough, then we have that

$$\operatorname{dist}(p, \operatorname{sing}(A_0)) \leq C 2^M \epsilon \operatorname{diam} Q.$$

If $C 2^M \epsilon \leq 1$, we get (5.30) from (5.31).

If $\operatorname{sing}(A_0) = \varnothing$, then we have that $\mathcal{T}$ is empty, and there is nothing to prove. Otherwise, $\operatorname{sing}(A_0)$ is an $(n-4)$-plane, which we shall denote by P. From (5.30) we have that

(5.32) $$\bigcup_{\mathcal{T}} Q \subseteq \{x \in Q_0 : \operatorname{dist}(x, P) \leq C 2^{-M} \operatorname{diam} Q_0\}$$

(since $\operatorname{diam} Q \leq C 2^{-M} \operatorname{diam} Q_0$ for all $Q \in \Delta_{l-M}$). This clearly implies (5.29) if M is large enough, since E is regular and $\dim P < n-1$.

Let us say a few words about the quantifiers here. We only need to assume that M is large in order to derive (5.29) from (5.32). Thus we can really choose M at the beginning, in a way that depends only on innocuous constants like d, n, and the regularity constant for E. In particular, M does not depend on ϵ, and it is fine for ϵ to depend on M.

As we said before, we want to show that $\Delta \backslash \mathcal{G}(\epsilon)$ satisfies a Carleson packing condition for all $\epsilon > 0$. Lemmas 5.20, 5.22, and 5.28 encode all the information that comes from the geometry that we need to do this. The remaining ingredient that we need is more general, so we state and prove it in a separate section.

5.3. A general lemma about Carleson packing conditions.

Let E be a d-dimensional regular set in $\mathbf{R}^n$. For the present purposes it does not matter at all what d and n are. Let Δ be a family of cubes on E, as in §I.3.1.

LEMMA 5.33. *Suppose that $\mathcal{N} \subseteq \Delta$ and that $\Delta \backslash \mathcal{N}$ satisfies a Carleson packing condition. Let $\mathcal{G}$ be a subset of $\mathcal{N}$ with the following two properties*:

(i) *if $Q \in \mathcal{N}$ and the parent of Q lies in $\mathcal{G}$, then $Q \in \mathcal{G}$*;
(ii) *if $Q_0 \in \mathcal{N} \cap \Delta_l$, then $\mathcal{T} = \{Q \in \Delta_{l-M} : Q \subseteq Q_0, \ Q \in \mathcal{N} \backslash \mathcal{G}\}$ satisfies $|\bigcup_{\mathcal{T}} Q| \leq \frac{1}{10}|Q_0|$.*

Here M is some positive integer which is not allowed to depend on Q_0 or l. Then $\Delta \backslash \mathcal{G}$ satisfies a Carleson packing condition.

The proof of this uses only standard techniques, but unfortunately it requires more of them than usual. We begin with the following.

LEMMA 5.34. *Same assumptions as in Lemma 5.33. Suppose that $S \subseteq \mathcal{N}$ has the property that if $Q_1, Q_2, Q_3 \in \Delta$, $Q_1, Q_3 \in S$, and $Q_1 \subseteq Q_2 \subseteq Q_3$, then $Q_2 \in S$. Then $S \backslash \mathcal{G}$ satisfies a Carleson packing condition.*

Let $\mathcal{N}$, $\mathcal{G}$, and S be as above, and set $\mathcal{A} = S \backslash \mathcal{G}$. For $j = 1, 2, \ldots, M$, set

$$\mathcal{A}_j = \mathcal{A} \cap \left(\bigcup_{l \equiv j \bmod M} \Delta_l \right).$$

We need only show that each $\mathcal{A}_j$ satisfies a Carleson packing condition. According to Lemma 1.31 it suffices to check that

(5.35) $\qquad |\cup \{Q \in \mathcal{A}_j : Q \subseteq Q_0, \ Q \neq Q_0\}| \leq \frac{1}{2}|Q_0|$

for all $Q_0 \in \mathcal{A}_j$ (and each j).

Fix $Q_0 \in \mathcal{A}_j$. By definitions $Q_0 \in \mathcal{N}$. Fix $l \in \mathbf{Z}$, $l \equiv j \bmod M$, such that $Q_0 \in \Delta_l$, and let $\mathcal{T}$ be as in the statement of Lemma 5.33. To prove (5.35) it is enough to show that

(5.36) $\qquad \bigcup \{Q \in \mathcal{A}_j : Q \subseteq Q_0, \ Q \neq Q_0\} \subseteq \bigcup_{\mathcal{T}} Q.$

Fix a cube $T \in \mathcal{A}_j$ such that $T \subseteq Q_0$ but $T \neq Q_0$. Let Q_1 be the element of Δ_{l-M} which contains T so that $Q_1 \subseteq Q_0$. If $Q_1 \in \mathcal{T}$, then T is contained in the right side of (5.36), and we are in business.

To show that $Q_1 \in \mathcal{T}$, we need only show that $Q_1 \in \mathcal{N} \backslash \mathcal{G}$. We certainly have $Q_1 \in \mathcal{N}$, because T, $Q_0 \in \mathcal{A}_j \subseteq S$, so $Q_1 \in S$, while $S \subseteq \mathcal{N}$ by assumption. If Q_1 were an element of $\mathcal{G}$, then T would be too, because assumption (i) above (in the statement of Lemma 5.33) could be applied to show that all the descendents of Q_1 which are elements of S (and T in particular) must lie in $\mathscr{G}$. This would contradict the assumption that $T \notin \mathscr{G}$, so we must have $Q_1 \notin \mathscr{G}$. Thus $Q_1 \in \mathscr{N}\backslash\mathscr{G}$, as desired. This proves (5.36), and Lemma 5.34 as well.

Let us now use Lemma 5.34 to prove Lemma 5.33.

Fix $R \in \Delta$. We want to show that

(5.37) $$\sum_{\substack{Q \in \mathscr{N}\backslash\mathscr{G} \\ Q \subseteq R}} |Q| \leq C|R|.$$

Set $\Delta(R) = \{Q \in \Delta : Q \subseteq R\}$. Recall from §I.3.2 (see the remark just after Definition 3.13) that a subset S of Δ is said to be semicoherent if it has a unique maximal element which contains all the other elements of S, and if $Q_1, Q_2, Q_3 \in \Delta$, $Q_1 \subseteq Q_2 \subseteq Q_3$, and $Q_1, Q_3 \in S$ imply that $Q_2 \in S$. [This is not quite literally the same as the definition given in §I.3.2, but it is easily seen to be equivalent.] Let $\mathscr{F}$ denote the collection of subsets S of $\mathscr{N} \cap \Delta(R)$ which are semicoherent and maximal as subsets of $\mathscr{N} \cap \Delta(R)$. It is easy to generate the elements of $\mathscr{F}$ by simple processes. Observe, for instance, that the collection of maximal cubes of elements of $\mathscr{F}$ is the same as the collection of cubes in $\Delta(R) \cap \mathscr{N}$ whose parents do not lie in $\Delta(R) \cap \mathscr{N}$. Once you know the maximal cube of an S in $\mathscr{F}$, you can generate the rest of S by simply going down until you have to stop.

Given $S \in \mathscr{F}$, let $Q(S)$ denote its maximal element. Then

(5.38) $$\sum_{S \in \mathscr{F}} |Q(S)| \leq C|R|.$$

This follows from the observation above that the parent of each $Q(S)$ does not lie in $\Delta(R) \cap \mathscr{N}$ and our assumption that $\Delta\backslash\mathscr{N}$ satisfies a Carleson packing condition.

On the other hand we have that

$$\sum_{Q \in S\backslash\mathscr{G}} |Q| \leq C|Q(S)|$$

for every $S \in \mathscr{F}$, by Lemma 5.34 (since $Q \subseteq Q(S)$ for all $Q \in S$). This and (5.38) imply (5.37), because $\Delta(R) \cap \mathscr{N} = \bigcup_{\mathscr{F}} S$. This proves Lemma 5.53.

5.4. The codimension 1 case, part 2.

We return now to the same notation and assumptions as in §5.2.

LEMMA 5.39. *If E satisfies the WCD, then $\Delta\backslash\mathscr{G}(\epsilon)$ satisfies a Carleson packing condition for all $\epsilon > 0$.*

Clearly $\mathscr{G}(\epsilon)$ becomes smaller as ϵ decreases, so we may as well assume that ϵ is as small as we want.

We want to apply Lemma 5.33, with $\mathscr{G} = \mathscr{G}(\epsilon)$ and $\mathscr{N} = \mathscr{N}(\epsilon)$. We can do this, as long as ϵ is small enough, because Lemmas 5.20, 5.22, and

5.28 ensure that the hypotheses of Lemma 5.33 are satisfied. This proves Lemma 5.39.

We are almost finished now, but there is one last observation to make. Given $\delta > 0$ (small) and $k > 1$ (large), let $\mathscr{G}(\delta, k)$ denote the set of cubes $Q \in \Delta$ for which there exists an $A \in \text{CD}$ such that

(5.40) $$\text{dist}(x, A) \leq \delta \operatorname{diam} Q \quad \text{for all } x \in kQ,$$

(5.41) $$\text{dist}(a, E) \leq \delta \operatorname{diam} Q \quad \text{whenever } a \in A \text{ satisfies}$$
$$\text{dist}(a, Q) \leq k \operatorname{diam} Q, \text{ and}$$

(5.42) $$\text{dist}(\operatorname{sing}(A), Q) \geq k \operatorname{diam} Q.$$

LEMMA 5.43. *If E satisfies the WCD, then $\Delta \backslash \mathscr{G}(\delta, k)$ satisfies a Carleson packing condition for every $\delta > 0$ and $k > 1$.*

This follows easily from Lemma 5.39. The point is that for every δ and k there is a $N > 1$ and an $\epsilon > 0$ so that $Q \in \mathscr{G}(\delta, k)$ whenever the Nth ancestor of Q lies in $\mathscr{G}(\epsilon)$.

Once we have Lemma 5.43 it is easy to prove that the WCD implies uniform rectifiability (in the codimension 1 case) using the results of Part II. For instance, if k is not too small, then for each $Q \in \mathscr{G}(\delta, k)$ we have that E is bilaterally approximated by a Lipschitz graph near Q, so we can use the results of §II.4.2. However, if you take k to be really large, then for each $Q \in \mathscr{G}(\delta, k)$ we have that E is well approximated by a hyperplane near Q, so we can use the BWGL. In both cases we are using Theorem 5.5 to conclude that if $A \in \text{CD}$, $d = n - 1$, and $a \in A$ is far away from $\operatorname{sing}(A)$, then A is very nice near a.

5.5. The weak dyadic density condition.

The weak dyadic density condition (WDD) is defined in much the same way as the WCD was (in Definition I.2.55). The main changes are that only dyadic radii are considered, and the "density" is allowed to vary from point to point. [This last is something of a red herring, because it can be prevented by relatively mild conditions. If, for instance, the definition of the WCD were altered to allow for this possibility, then the resulting condition would still be equivalent to the WCD.]

Let E be a d-dimensional regular set in $\mathbf{R}^n$. Given $\epsilon > 0$ and $C_0 > 0$, let $\mathscr{G}_{dd}(C_0, \epsilon)$ denote the (good) set of $(x, t) \in E \times \mathbf{R}_+$ for which there is a measure $\mu = \mu_{x,t}$ such that $\operatorname{supp} \mu = E$, μ is regular with constant $\leq C_0$ (i.e, μ satisfies Definition I.2.50, or, equivalently, $\mu \in RM(C_0)$), and

(5.44) $$|\mu(B(y, 2^{m+1})) - 2^d \mu(B(y, 2^m))| \leq \epsilon t^d$$

for all $y \in E \cap B(x, t)$ and all integers m, $2^m \leq t$.

DEFINITION 5.45. *E satisfies the weak dyadic density condition (WDD) if there is a $C_0 > 0$ so that the complement of $\mathscr{G}_{dd}(C_0, \epsilon)$ in $E \times \mathbf{R}_+$ is a Carleson set for every $\epsilon > 0$.*

Clearly the WCD implies the WDD trivially. We do not know whether the converse is true in any dimension or codimension. Our custom for treating conditions that we do not understand is to look at what happens when we toss in the WGL.

THEOREM 5.46. *The WDD together with the WGL imply uniform rectifiability (without restrictions on the dimensions).*

The converse to this is of course true, since uniform rectifiability implies the WCD (by Theorem I.2.52) and also the geometric lemma I.1.59.

We are going to prove Theorem 5.46 using a compactness argument much like the one given in §1. If we were to do this proof from scratch, we might do it a little differently, but under the circumstances it makes sense to employ the same basic approach and some of the same lemmas as before.

Fix n and d, $0 < d < n$. Let $\mathrm{RM}(C_0)$ denote the set of (d-dimensional) regular measures on $\mathbf{R}^n$ with constant $\leq C_0$, as in §1. For each $C_0 > 0$, $k > 0$, and $\epsilon > 0$ let $\mathscr{D}_0(C_0, k, \epsilon)$ denote the set of measures $\mu \in \mathrm{RM}(C_0)$ such that:

(i) $0 \in \mathrm{supp}\,\mu$;
(ii) $|\mu(B(y, 2^{m+1})) - 2^d \mu(B(y, 2^m))| \leq \epsilon$ whenever $y \in \mathrm{supp}\,\mu$ and $m \in \mathbf{Z}$ satisfy $|y| \leq k$ and $2^m \leq k$;
(iii) there exists a d-plane P such that $\mathrm{dist}(y, P) \leq \epsilon$ whenever $y \in \mathrm{supp}\,\mu$ and $|y| \leq k$.

Also let $\mathscr{D}$ denote the set of (d-dimensional) regular measures μ on $\mathbf{R}^n$ which satisfy

$$\mu(B(y, 2^{m+1})) = 2^d \mu(B(y, 2^m))$$

for all $y \in \mathrm{supp}\,\mu$ and $m \in \mathbf{Z}$ and for which there is a d-plane P such that $\mathrm{supp}\,\mu \subseteq P$. (Of course, there are not so many elements of $\mathscr{D}$, as we shall soon see.)

In the next lemma we give a more convenient formulation of the hypotheses of Theorem 5.46.

LEMMA 5.47. *Let E be a d-dimensional regular set in $\mathbf{R}^n$, and suppose that E satisfies the WDD and the WGL. Then there is a $C_0 > 0$ so that for every $\epsilon > 0$ and every $k > 0$ the complement of the following subset of $E \times \mathbf{R}_+$ is a Carleson set:*

$$\{(x, t): \text{if } l \in \mathbf{Z} \text{ satisfies } 2^l \leq t < 2^{l+1}, \text{ then there is a}$$
$$\mu \in \mathscr{D}_0(C_0, k, \epsilon) \text{ such that } E = \mathrm{supp}(T_{x, 2^l}(\mu))\}.$$

Recall that "$T_{x,t}$" was defined just before Lemma 5.16.

Like Lemma 5.16, Lemma 5.47 is essentially a recasting of definitions, and we leave its proof as an exercise.

To prove Theorem 5.46 it suffices to show that the WDD together with the WGL imply the BWGL (see Theorem I.2.4). This will be an immediate consequence of the next lemma.

5.5. THE WEAK DYADIC DENSITY CONDITION

LEMMA 5.48. *Let $C_0 > 0$ and $\eta > 0$ be given. Then there exists $k > 0$ (large) and $\epsilon > 0$ (small) so that if $\mu \in \mathscr{D}_0(C_0, k, \epsilon)$, then there is a d-plane P such that*

$$\sup_{z \in (\operatorname{supp}\mu) \cap B(0,1)} \operatorname{dist}(z, P) \leq \eta \tag{5.49}$$

and

$$\sup_{p \in P \cap B(0,1)} \operatorname{dist}(p, \operatorname{supp}\mu) \leq \eta. \tag{5.50}$$

Suppose not. Then we can find $C_0 > 0$, $\eta > 0$, $\{k_j\}$, $\{\epsilon_j\} \subseteq \mathbf{R}_+$, and $\{\mu_j\} \subseteq \operatorname{RM}(C_0)$ such that $k_j \to \infty$, $\epsilon_j \to 0$, $\mu_j \in \mathscr{D}_0(C_0, k_j, \epsilon_j)$ for each j, but the conclusion of Lemma 5.48 fails for each j. In view of Lemma 5.8(a) we may assume that $\{\mu_j\}$ converges weakly to some measure μ (by passing to a subsequence if necessary). From Lemmas 5.9 and 5.8 it follows that $0 \in \operatorname{supp}\mu$ and $\mu \in \operatorname{RM}(C_0)$. If we can prove that the support of μ is a d-plane, then we shall have a contradiction, because Lemma 5.9 would then imply that the μ_j's do satisfy the conclusion of Lemma 5.48 when j is large enough.

Let P_j be a d-plane which satisfies

$$\operatorname{dist}(y, P_j) \leq \epsilon_j \quad \text{whenever } y \in \operatorname{supp}\mu_j \text{ and } |y| \leq k_j. \tag{5.51}$$

Such a d-plane exists because $\mu_j \in \mathscr{D}_0(C_0, k_j, \epsilon_j)$. Since $0 \in \operatorname{supp}\mu_j$, we have in particular that

$$\operatorname{dist}(0, P_j) \leq \epsilon_j. \tag{5.52}$$

It follows from Lemma 5.9 that there is a d-plane P which passes through the origin and which satisfies

$$\operatorname{supp}\mu \subseteq P. \tag{5.53}$$

Notice that P is unique, because a d-dimensional regular measure cannot be supported in a $(d-1)$-plane. It remains to show that $\operatorname{supp}\mu = P$. We begin with a technical observation.

LEMMA 5.54. *Let $x, y \in \mathbf{R}^n$ and $s, t > 0$ be given, and suppose that*

$$|x - y| + |s - t| < t/2$$

and also that $x \in \operatorname{supp}\mu$. Then

$$|\mu(B(x, s)) - \mu(B(y, t))| \leq C(|x-y| + |s-t|)t^{d-1}.$$

Let A denote the symmetric difference of $B(x, s)$ and $B(y, t)$. Set $\rho = |x-y|+|s-t|$. Then the intersection of A with the d-plane P can be covered by $\leq Ct^{d-1}\rho^{-(d-1)}$ balls of radius ρ. Because μ is a (d-dimensional) regular measure, we conclude that $\mu(A) \leq C\rho t^{d-1}$, which proves the lemma.

The main point now is that $\mu \in \mathscr{D}$. This follows from the definition of $\mathscr{D}(C_0, k, \epsilon)$ and the obvious limiting argument, with Lemmas 5.9 and 5.54 providing assistance with the technicalities.

To finish the proof of Lemma 5.48, it remains to show the following.

LEMMA 5.55. *If $\nu \in \mathscr{D}$, then* $\operatorname{supp} \nu$ *is a d-plane.*

Since the support of every element of $\mathscr{D}$ is contained in a d-plane, by definition, this lemma can be reformulated as follows. Suppose that ν is a d-dimensional regular measure on $\mathbf{R}^d$ and that

(5.56) $$\nu(B(x, 2^{m+1})) = 2^d \nu(B(x, 2^m))$$

for all $x \in \operatorname{supp} \nu$ and $m \in \mathbf{Z}$. Then $\operatorname{supp} \nu = \mathbf{R}^d$.

Notice that

$$\lim_{m \to \infty} 2^{-md} |\nu(B(x, 2^m)) - \nu(B(y, 2^m))| = 0$$

for all $x, y \in \mathbf{R}^d$. This is really just a special case of Lemma 5.54. Combining this with (5.56) we get that

$$\nu(B(x, 1)) = \nu(B(y, 1))$$

for all $x, y \in \operatorname{supp} \nu$. Hence there is a constant $a > 0$ so that

(5.57) $$\nu(B(x, 2^m)) = a 2^{md}$$

for all $x \in \operatorname{supp} \nu$ and $m \in \mathbf{Z}$. We may as well assume that a is simply the volume of the unit ball in $\mathbf{R}^d$, since we can always multiply ν by a positive constant.

Let λ denote Lebesgue measure on $\mathbf{R}^d$. By the Radon-Nikodym theorem there is an $f \in L^\infty(\mathbf{R}^d)$ such that $\nu = f \lambda$. From the Lebesgue differentiation theorem we have that $f = 1$ a.e. on the support of ν. Of course, $f = 0$ a.e. on $\mathbf{R}^d \setminus \operatorname{supp} \nu$. That is to say, $f = \chi_V$ a.e., where $V = \operatorname{supp} \nu$. Since we also have

$$\nu(B(x, 2^m)) = |B(x, 2^m)|$$

whenever $x \in V$, we conclude that $\mathbf{R}^d \setminus V$ has measure zero. This proves Lemma 5.55, and the proof of Theorem 5.46 is now complete.

PART IV

Direct Arguments for Some Stability Results

CHAPTER 1

Stability of Various Versions of the Geometric Lemma

1.1. The statements.
Let E be a d-dimensional regular set in $\mathbf{R}^n$. Given $q \in (0, \infty]$, define $\beta_q(x, t)$ for $x \in E$ and $t > 0$ as in (I.1.45), i.e.,

$$(1.1) \qquad \beta_q(x, t) = \inf_P \left(t^{-d} \int_{E \cap B(x,t)} (t^{-1} \operatorname{dist}(y, P))^q \, dy \right)^{1/q},$$

where the infimum is taken over all d-planes P in $\mathbf{R}^n$. When $q = \infty$ the L^q-average over $E \cap B(x, t)$ should be replaced by a supremum, as in (I.1.46).

DEFINITION 1.2. E satisfies the (p, q)-geometric lemma, $0 < q \leq \infty$, $0 < p < \infty$, if

$$\beta_q(x, t)^p \, dx \frac{dt}{t}$$

defines a Carleson measure on $E \times \mathbf{R}_+$.

When $p = 2$ we encountered this condition before in §§1.3 and 1.4 in Part I. In particular we know that E is uniformly rectifiable if and only if it satisfies the $(2, q)$-geometric lemma, where $1 \leq q < 2d/d - 2$ when $d \geq 2$ and $1 \leq q \leq \infty$ when $d = 1$. (See the discussion between (I.1.59) and (I.1.60) in §I.1.4.) We also know that uniform rectifiability is stable under "the big pieces functor", so the same must be true of the $(2, q)$-geometric lemma when q lies in the range described above. The proof of this stability result (obtained via the equivalence with uniform rectifiability) is rather laborious, and we shall give in this chapter a proof which is more direct and which works in greater generality.

THEOREM 1.3. *Let E be a d-dimensional set in $\mathbf{R}^n$, and let p and q be given, $0 < p < \infty$, $0 < q \leq \infty$. Assume that*

$$(1.4) \qquad \frac{1}{q} - \frac{1}{p} + \frac{1}{d} > 0.$$

Suppose that there exist constants $C_0 > 0$ and $\theta > 0$ so that the following holds: for each $x \in E$ and $t > 0$ there is a d-dimensional regular set $\tilde{E}$, with regularity constant $\leq C_0$, which satisfies the (p, q)-geometric lemma with Carleson constant $\leq C_0$, and also

$$(1.5) \qquad |E \cap \tilde{E} \cap B(x, t)| \geq \theta t^d.$$

Then E also satisfies the (p, q)-geometric lemma.

An interesting special case of this result (which is already nontrivial) is the fact that the union of two regular sets which satisfy the (p, q)-geometric lemma also satisfies the (p, q)-geometric lemma. There is a simpler proof of this fact, which does not require the assumption (1.4), but we shall not give it here.

The natural version of the (p, q)-geometric lemma when $p = \infty$ is the weak geometric lemma (WGL). Recall from Definition I.1.71 that E satisfies the WGL if

$$(1.6) \qquad \{(x, t) \in E \times \mathbf{R}_+ : \beta_\infty(x, t) > \epsilon\}$$

is a Carleson set for every $\epsilon > 0$. As was pointed out in §I.1.5, this condition is equivalent to the one you get when you replace $\beta_\infty(x, t)$ by $\beta_q(x, t)$ for any $q > 0$, because of (I.1.73); thus the choice of q no longer matters when $p = \infty$.

There is an analogue of Theorem 1.3 for the WGL, but before we state it we need to record a definition.

DEFINITION 1.7. For each $C_0 > 0$ and $\gamma : (0, 1] \to (0, \infty)$ we let WGL(C_0, γ) denote the collection of d-dimensional regular sets E in $\mathbf{R}^n$ with regularity constant $\leq C_0$ such that the Carleson constant for (1.6) is $\leq \gamma(\epsilon)$ for every $\epsilon \in (0, 1)$.

We should perhaps say that for the purposes of this definition d and n should be fixed, although we do not want to burden our notation by making this explicit.

THEOREM 1.8. *Let E be a d-dimensional regular set in $\mathbf{R}^n$. Suppose that there exists $\theta > 0$, $C_0 > 0$, and $\gamma : (0, 1] \to (0, \infty)$ such that for each $x \in E$ and $t > 0$ there is an $\tilde{E} \in WGL(C_0, \gamma)$ which satisfies*

$$|E \cap \tilde{E} \cap B(x, t)| \geq \theta t^d.$$

Then E also satisfies the WGL.

This was also observed by Peter Jones.

Theorems 1.3 and 1.8 are proved in the next three sections. We first reduce to an estimate that is easier to verify (in §2), and then we give (in §3) the main step in the argument, where we show how to estimate the β's for one set E in terms of the β's for a nearby set $\widetilde{E}$. In the fourth section we simply combine these ingredients to finish the proof.

We should point out that if we did not care about Theorem 1.3 then we could give a much shorter proof of Theorem 1.8. The devoted reader will notice this when we get to the technical lemmas. We should also point out that Theorem 1.8 will be used in the next chapter, but we shall not need Theorem 1.3 (although it would do just as well).

1.2. A John-Nirenberg-Strömberg lemma for Carleson packing conditions.

Let E be a d-dimensional regular set in $\mathbf{R}^n$, as before. Let Δ be a family of cubes on E as in §I.3.1. Although we shall be dealing with other regular sets, we shall only be working with cubes on E.

Given $Q \in \Delta$ and $q \in (0, \infty]$ set

$$(1.9) \qquad \beta_q(Q) = \inf_P \left(\frac{1}{|Q|} \int_{2Q} \left(\frac{\operatorname{dist}(x, P)}{\operatorname{diam} Q} \right)^q dx \right)^{1/q},$$

where the infimum is taken over all d-planes. (As usual, the L^q-average should be replaced by a supremum when $q = \infty$.) It is easy to see that E satisfies the (p, q)-geometric lemma if and only if there is a $C > 0$ so that

$$(1.10) \qquad \sum_{Q \subseteq R} \beta_q(Q)^p |Q| \leq C|R| \quad \text{for all } R \in \Delta,$$

while E satisfies the WGL if and only if for each $\epsilon > 0$ there is a $C(\epsilon) > 0$ so that

$$(1.11) \qquad \sum_{\substack{Q \subseteq R \\ \beta_\infty(Q) > \epsilon}} |Q| \leq C(\epsilon) |R| \quad \text{for all } R \in \Delta.$$

For the purposes of proving Theorems 1.3 and 1.8 the following criterion will be helpful.

LEMMA 1.12. *Let* $\alpha : \Delta \to [0, \infty)$ *be given, and suppose that there exist* $N > 0$ *and* $\eta > 0$ *such that*

$$(1.13) \qquad \left| \left\{ x \in R : \sum_{\substack{Q \ni x \\ Q \subseteq R}} \alpha(Q) \leq N \right\} \right| \geq \eta |R|$$

for all $R \in \Delta$. *Then there is a* $C > 0$ *such that*

(1.14) $$\sum_{Q \subseteq R} \alpha(Q)|Q| \leq C|R|$$

for all $R \in \Delta$.

Lemma 1.12 is basically well known, and it is proved using a standard John-Nirenberg-Strömberg argument. We include a sketch of the proof for the reader's convenience. We shall not be doing anything else in this section, so the reader may find it convenient to skip it.

Incidentally, Lemma 1.12 becomes a little simpler (and more geometric) if we restrict ourselves to the case where α takes values in $\{0, 1\}$. For Theorem 1.8 this special case is all that matters.

Let α, N, and η be given, as above. Fix $R \in \Delta$. Set

$$F_1 = \left\{ x \in R : \sum_{\substack{Q \ni x \\ Q \subseteq R}} \alpha(Q) \leq N \right\}$$

and $G_1 = R \setminus F_1$.

If $x \in G_1$, then there is a cube $Q \in \Delta$ such that $x \in Q$, $Q \subseteq R$, and $Q \subseteq G_1$. This follows from the definition of G_1. Let $\{R_{1,j}\}$ be an enumeration of the maximal cubes contained in G_1, so that

$$G_1 = \bigcup_j R_{1,j}.$$

For each j set

$$F_{2,j} = \left\{ x \in R_{1,j} : \sum_{\substack{Q \ni x \\ Q \subseteq R_{1,j}}} \alpha(Q) \leq N \right\},$$

$G_{2,j} = R_{1,j} \setminus F_{2,j}$, $F_2 = \bigcup_j F_{2,j}$, and $G_2 = \bigcup_j G_{2,j}$. Notice that

(1.15) $$|F_{2,j}| \geq \eta |R_{1,j}|,$$

by hypothesis.

As before it is true that if $x \in G_{2,j}$ for some j, then there is a cube Q such that $x \in Q$, $Q \subseteq R_{1,j}$, and $Q \subseteq G_{2,j}$. Let $\{R_{2,k}\}$ be an enumeration of all the maximal cubes of all the $G_{2,j}$'s, so that

$$G_2 = \bigcup_k R_{2,k}$$

and each $R_{2,k}$ is contained in some $R_{1,j}$. Actually, the $R_{2,k}$'s could also be described as the maximal cubes contained in G_2; the maximality of the $R_{2,k}$'s in G_2 follows from the fact that each $F_{2,j}$ is nonempty, by (1.15).

1.2. A LEMMA FOR CARLESON PACKING CONDITIONS

By repeating this process we get two sequences $\{F_l\}_{l\geq 1}$ and $\{G_l\}_{l\geq 1}$ of subsets of R such that

$$G_l = F_{l+1} \cup G_{l+1} \text{ and } G_l \cap F_l = \emptyset \text{ when } l \geq 1,$$

$R = F_1 \cup G_1$, and

$$|F_{l+1}| \geq \eta |G_l|.$$

Using this and $|F_1| \geq \eta |R|$ we get that

(1.16) $$|G_l| \leq (1-\eta)^l |R|.$$

Let us explain why we also have that

(1.17) $$\sum_{\substack{Q \ni x \\ Q \subseteq R}} \alpha(Q) \leq lN \text{ when } x \in F_l.$$

When $l = 1$ this is true by definition. Suppose that $l = 2$, and let $x \in F_2$ be given. Let j be such that $x \in F_{2,j}$. Then

(1.18) $$\sum_{\substack{Q \ni x \\ Q \subseteq R_{1,j}}} \alpha(Q) \leq N$$

by definition. Because $R_{1,j}$ is a maximal cube in G_1, its parent intersects F_1, and so

(1.19) $$\sum_{\substack{Q \subseteq R \\ Q \supseteq R_{1,j} \\ Q \neq R_{1,j}}} \alpha(Q) \leq N.$$

Combining this with (1.18) gives (1.17) when $l = 2$. The larger values of l can be dealt with using induction and the same argument as we just used to go from $l = 1$ to $l = 2$. This requires no new ideas but plenty of notation, and we omit the details.

From (1.17) we obtain that

$$\left\{ x \in R : \sum_{\substack{Q \ni x \\ Q \subseteq R}} \alpha(Q) > lN \right\} \subseteq G_l,$$

and hence

$$\left| \left\{ x \in R : \sum_{\substack{Q \ni x \\ Q \subseteq R}} \alpha(Q) > lN \right\} \right| \leq (1-\eta)^l |R|,$$

by (1.16). This implies that

$$\int_R \left(\sum_{\substack{Q \ni x \\ Q \subseteq R}} \alpha(Q) \right) dx = \sum_{Q \subseteq R} \alpha(Q)|Q| \leq C(N, \eta)|R|,$$

i.e., (1.14) is true. This proves Lemma 1.12.

1.3. Two lemmas on approximations of regular sets by d-planes.

Let E be a d-dimensional regular set in $\mathbf{R}^n$, with an associated family of cubes Δ. Let $\widetilde{E}$ be some other d-dimensional regular set, which we think of as intersecting E in a substantial way. We are going to estimate the β's for E in terms of the β's for $\widetilde{E}$. In the next section we shall use these estimates to derive Theorems 1.3 and 1.8 from Lemma 1.12.

In what follows we permit our constants C to depend on the regularity constants for E and $\widetilde{E}$ without further mention. In the applications we have a bound on the regularity constant for $\widetilde{E}$ anyway.

Let $\tilde{\beta}_q(x, t) : \widetilde{E} \times \mathbf{R}_+ \to \mathbf{R}$ be defined exactly as in (1.1) but with E replaced by $\widetilde{E}$. Define $\tilde{\delta} : \mathbf{R}^n \to \mathbf{R}$ by

$$\tilde{\delta}(y) = \mathrm{dist}(y, \widetilde{E}),$$

and set $E_1 = E \setminus \widetilde{E}$.

LEMMA 1.20. *Let $x \in E \cap \widetilde{E}$ and $Q \in \Delta$ be given, with $x \in Q$. Then for each q, $0 < q \leq \infty$, we have*

(1.21) $\qquad \beta_q(Q) \leq C(q)\tilde{\beta}_q(x, 6\,\mathrm{diam}\,Q) + C(q)I_q(Q),$

where

(1.22) $\qquad I_q(Q) = \left\{ |Q|^{-1} \int_{2Q \cap E_1} \left[\tilde{\delta}(y)(\mathrm{diam}\,Q)^{-1} \right]^q dy \right\}^{1/q}.$

When $q = \infty$ we interpret the definition of $I_q(Q)$ to mean

(1.23) $\qquad I_\infty(Q) = \sup_{y \in 2Q \cap E_1} \left[\tilde{\delta}(y)(\mathrm{diam}\,Q)^{-1} \right].$

Fix x and Q as above, and let P be a d-plane which is optimal for $\tilde{\beta}_q(x, 6\,\mathrm{diam}\,Q)$. We certainly have

$$\beta_q(Q) \leq \left(\frac{1}{|Q|} \int_{2Q} \left(\frac{\mathrm{dist}(y, P)}{\mathrm{diam}\,Q} \right)^q dy \right)^{1/q}.$$

1.3. APPROXIMATING REGULAR SETS BY d-PLANES

It is very easy to prove (1.21) when $q = \infty$, so we leave that to the reader. [Of course, that is the only case which is needed for the proof of Theorem 1.8.] Thus we restrict ourselves to $q \in (0, \infty)$. This is a little tricky because we have to compare integrals on E to integrals on $\widetilde{E}$.

Define $\rho : \mathbf{R}^n \to \mathbf{R}$ by

$$\rho(y) = \mathrm{dist}(y, P).$$

Clearly we have

(1.24)
$$\begin{aligned}\beta_q(Q) &\leq C \left(\frac{1}{|Q|} \int_{2Q \cap \widetilde{E}} \left(\frac{\rho(y)}{\mathrm{diam}\, Q} \right)^q dy \right)^{1/q} \\ &\quad + C \left(\frac{1}{|Q|} \int_{2Q \cap E_1} \left(\frac{\rho(y)}{\mathrm{diam}\, Q} \right)^q dy \right)^{1/q} \\ &\leq C \tilde{\beta}_q(x, 2\,\mathrm{diam}\, Q) + C \left(\frac{1}{|Q|} \int_{2Q \cap E_1} \left(\frac{\rho(y)}{\mathrm{diam}\, Q} \right)^q dy \right)^{1/q}.\end{aligned}$$

[These and future constants may depend on q, but we will not bother to make that explicit.] When $y \in 2Q \cap E_1$ we want to control $\rho(y)^q$ in terms of $\tilde{\delta}(y)^q$ and a suitable average of $\rho(z)^q$ for certain points $z \in \widetilde{E}$. We have to choose these points with some care in order to ensure that not too many z's correspond to one y. To account for this, we define and estimate an appropriate multiplicity function.

Define $M(\cdot, \cdot)$ on $\widetilde{E} \times \mathbf{R}_+$ by

(1.25)
$$M(z, s) = \int_{\substack{w \in E_1 \\ |w - z| \leq 2\tilde{\delta}(w) \\ \tilde{\delta}(w) \leq s}} \tilde{\delta}(w)^{-d} dw.$$

Roughly speaking, this is trying to measure the number of times that a point $z \in \widetilde{E}$ might be "chosen" by some $w \in E_1$. The parameter s is needed for localization.

LEMMA 1.26. *For every $u \in \mathbf{R}^n$ and $s > 0$ we have that*

(1.27)
$$\int_{B(u, s) \cap \widetilde{E}} M(z, s)\, dz \leq Cs^d.$$

Simply apply Fubini's theorem:

$$\int_{B(u,s)\cap\widetilde{E}} M(z,s)\,dz = \int_{B(u,s)\cap\widetilde{E}} \int_{\substack{w\in E_1 \\ |w-z|\leq 2\tilde{\delta}(w) \\ \tilde{\delta}(w)\leq s}} \tilde{\delta}(w)^{-d}\,dw\,dz$$

$$\leq \int_{w\in E_1\cap B(u,3s)} \tilde{\delta}(w)^{-d} \left\{ \int_{\widetilde{E}\cap B(w,2\tilde{\delta}(w))} dz \right\} dw$$

$$\leq C \int_{E_1\cap B(u,3s)} dw \leq Cs^d.$$

Now that we have estimated the multiplicity function, we want to look at the sets where it is not too large. For each $y \in E_1$, define $A(y) \subseteq \widetilde{E}$ by

(1.28) $\quad A(y) = \{z \in \widetilde{E} : z \in B(y, 2\tilde{\delta}(y)) \text{ and } M(z, 2\tilde{\delta}(y)) \leq K\},$

where K is chosen to be sufficiently large so that

(1.29) $\quad\quad\quad\quad\quad\quad |A(y)| \geq C^{-1}\tilde{\delta}(y)^d.$

Lemma 1.26 ensures that we can choose a K which does not depend on Q or y and which is large enough so that (1.29) is true. [Notice that $|\widetilde{E} \cap B(y, 2\tilde{\delta}(y))| \geq C^{-1}\tilde{\delta}(y)^d$, because $\widetilde{E}$ is regular.]

Let us use $A(y)$ to analyze the last term in (1.24). For each $y \in E_1 \cap 2Q$ we have that

$$\rho(y)^q \leq C\tilde{\delta}(y)^q + C\tilde{\delta}(y)^{-d}\int_{A(y)} \rho(z)^q\,dz,$$

and hence

(1.30) $\quad \left(\dfrac{1}{|Q|} \displaystyle\int_{2Q\cap E_1} \left(\dfrac{\rho(y)}{\operatorname{diam} Q}\right)^q dy \right)^{1/q} \leq CI_q(Q) + CJ_q(Q),$

where I_q is as in (1.22) and

$$J_q(Q) = \left\{ |Q|^{-1} \int_{2Q\cap E_1} \tilde{\delta}(y)^{-d} \int_{A(y)} \left(\frac{\rho(z)}{\operatorname{diam} Q}\right)^q dz\,dy \right\}^{1/q}.$$

We need to control $J_q(Q)$.

1.3. APPROXIMATING REGULAR SETS BY d-PLANES

Notice that $\tilde{\delta}(y) \leq 2 \operatorname{diam} Q$ when $y \in 2Q$, since $x \in Q \cap \widetilde{E}$. In particular, we have that
$$A(y) \subseteq B(x, 6 \operatorname{diam} Q) \cap \widetilde{E}.$$
From Fubini's theorem we get that
(1.31)
$$J_q(Q)^q \leq |Q|^{-1} \int_{z \in B(x, 6 \operatorname{diam} Q) \cap \widetilde{E}} \left(\frac{\rho(z)}{\operatorname{diam} Q}\right)^q \left\{ \int_{\substack{y \in E_1 \cap 2Q \\ A(y) \ni z}} \tilde{\delta}(y)^{-d} \, dy \right\} dz.$$

LEMMA 1.32. *For each $z \in \widetilde{E}$ we have that*

(1.33)
$$\int_{\substack{y \in E_1 \cap 2Q \\ A(y) \ni z}} \tilde{\delta}(y)^{-d} \, dy \leq K.$$

Fix $z \in \widetilde{E}$. Choose $y_0 \in E_1 \cap 2Q$ such that $z \in A(y_0)$ and $\tilde{\delta}(y_0)$ is approximately maximal:

(1.34) $\quad \tilde{\delta}(y_0) \geq \frac{1}{2} \sup\{\tilde{\delta}(y) : y \in E_1 \cap 2Q \text{ and } z \in A(y)\}.$

(If no such y exists, then there is nothing to prove.) Since $z \in A(y_0)$, we have that $M(z, 2\tilde{\delta}(y_0)) \leq K$, i.e.,

(1.35)
$$\int_{\substack{y \in E_1 \\ |y-z| \leq 2\tilde{\delta}(y) \\ \tilde{\delta}(y) \leq 2\tilde{\delta}(y_0)}} \tilde{\delta}(y)^{-d} \, dy \leq K.$$

This implies (1.33): if $y \in E_1 \cap 2Q$ and $z \in A(y)$, then $|z - y| \leq 2\tilde{\delta}(y)$, by definition of $A(y)$, while $\tilde{\delta}(y) \leq 2\tilde{\delta}(y_0)$ comes from (1.34).

Applying Lemma 1.32 to (1.31) we obtain
$$J_q(Q) \leq C \tilde{\beta}_q(x, 6 \operatorname{diam} Q).$$
This completes the proof of Lemma 1.20, because of (1.30) and (1.24).

In order to make use of Lemma 1.20 we need to be able to control $I_q(Q)$. To do this it is convenient to consider a variant $\widehat{I}_q(Q)$ of $I_q(Q)$, which is defined by simply taking the right side of (1.22) and intersecting the domain of integration with $\{y \in E_1 : \tilde{\delta}(y) \leq 2 \operatorname{diam} Q\}$. (When $q = \infty$, we define $\widehat{I}_q(Q)$ by modifying (1.23) in the analogous way. In this definition we consider the supremum over the empty set to be zero.) In the context of Lemma 1.20 this change is irrelevant, because

(1.36) $\quad I_q(Q) = \widehat{I}_q(Q) \quad \text{whenever } Q \cap \widetilde{E} \neq \emptyset.$

IV.1. STABILITY OF THE GEOMETRIC LEMMA

LEMMA 1.37. *If $0 < p < \infty$, $0 < q \leq \infty$, and $\frac{1}{q} - \frac{1}{p} + \frac{1}{d} > 0$, then there is a $C > 0$ such that*

$$\sum_{Q \subseteq T} \hat{I}_q(Q)^p |Q| \leq C|T| \quad \text{for all } T \in \Delta. \tag{1.38}$$

We should perhaps mention that the C in (1.38) depends only on p, q, n, d, and the regularity constant for E.

Assume first that $q = p$, so that $q < \infty$ in particular. Then we can interchange the order of summation and integration to get

$$\sum_{Q \subseteq T} \hat{I}_q(Q)^q |Q| = \sum_{Q \subseteq T} \int_{\substack{y \in 2Q \cap E_1 \\ \tilde{\delta}(y) \leq 2 \operatorname{diam} Q}} \left[\tilde{\delta}(y)(\operatorname{diam} Q)^{-1} \right]^q dy$$

$$= \int_{2T \cap E_1} \left\{ \sum_{\substack{2Q \ni y \\ 2 \operatorname{diam} Q \geq \tilde{\delta}(y)}} \left[\tilde{\delta}(y)(\operatorname{diam} Q)^{-1} \right]^q \right\} dy$$

$$\leq C \int_{2T \cap E_1} dy \leq C|T|.$$

When $q < p$ we can reduce to the $q = p$ case using the simple observation that

$$\hat{I}_q(Q) \leq C.$$

Now suppose that $q > p$. We are going to reduce this to the $q = p$ case too, but this is a little trickier.

Let us first check that

$$\hat{I}_\infty(Q)^{(r+d)/r} \leq C(r) \hat{I}_r(\hat{Q}) \tag{1.39}$$

holds for each $r > 0$, where $\hat{Q}$ is the smallest cube containing Q which satisfies $\operatorname{diam} \hat{Q} > 2 \operatorname{diam} Q$. Indeed, if $y \in 2Q \cap E_1$ and $\tilde{\delta}(y) \leq 2 \operatorname{diam} Q$, then

$$\left[\tilde{\delta}(y)(\operatorname{diam} Q)^{-1} \right]^{r+d} \leq C|Q|^{-1} \int_{B(y, \tilde{\delta}(w)/10) \cap E_1} \left[\tilde{\delta}(w)(\operatorname{diam} Q)^{-1} \right]^r dw$$

$$\leq C \hat{I}_r(\hat{Q})^r.$$

For the first inequality we are using the regularity of E (as well as the definitions of E_1 and $\tilde{\delta}$).

Next we want to use (1.39) to control $\hat{I}_q(Q)$ in terms of $\hat{I}_r(\hat{Q})$ when $r < q$. If $r, q > 0$ and $r \leq q$, then we have that

$$\hat{I}_q(Q) \leq \hat{I}_r(Q)^{r/q} \hat{I}_\infty(Q)^{1-r/q} \leq C \hat{I}_r(\hat{Q})^s,$$

where

$$\text{(1.40)} \qquad s = \left[\frac{r}{q} + \frac{r}{(r+d)}\left(1 - \frac{r}{q}\right)\right],$$

because of (1.39). Hence

$$\text{(1.41)} \qquad \hat{I}_q(Q)^p \leq C\hat{I}_r(\hat{Q})^{sp}.$$

Now we are ready to handle the $p < q$ case of Lemma 1.37. Since we have already proved the lemma when $p \geq q$, we conclude from (1.41) that (1.38) is true when $p < q$ as long as we can find an $r \in (0, q]$ such that $sp \geq r$. This last inequality is equivalent to

$$\frac{1}{q} + \frac{1}{(r+d)}\left(1 - \frac{r}{q}\right) \geq \frac{1}{p},$$

which is itself equivalent to

$$\frac{1}{q} - \frac{1}{p} + \frac{1}{d} \geq \frac{1}{d} - \frac{1}{(r+d)}\left(1 - \frac{r}{q}\right).$$

The right side is always positive, but it can be made arbitrarily small by choosing r small enough. Thus there is an $r \in (0, q]$ such that $sp \geq r$ if and only if

$$\frac{1}{q} - \frac{1}{p} + \frac{1}{d} > 0.$$

This proves Lemma 1.37.

Let us record the analogue of Lemma 1.37 which is relevant for Theorem 1.8.

LEMMA 1.42. *For each $\epsilon > 0$,*

$$\text{(1.43)} \qquad \{Q \in \Delta : \hat{I}_\infty(Q) > \epsilon\}$$

satisfies a Carleson packing condition (with a constant which depends only on ϵ, n, d, and the regularity constant for E).

This can be derived from the fact that Lemma 1.37 holds for $q = \infty$ and merely some $p < \infty$. However, there is a more trivial proof, which we now sketch.

It is not hard to see that (1.43) is the same as

(1.44)
$$\{Q \in \Delta : \text{there is a } y \in 2Q \text{ such that } \epsilon \operatorname{diam} Q < \tilde{\delta}(y) \leq 2 \operatorname{diam} Q\},$$

just by chasing definitions. Consider instead

(1.45)
$$\{T \in \Delta : \text{there is a } y \in T \text{ such that } 2\operatorname{diam} T \leq \tilde{\delta}(y) \leq k \operatorname{diam} T\}.$$

If $k > 1$ is large enough, then for each Q in (1.44) we can find a $T \subseteq 3Q$ such that T lies in (1.45) and $\operatorname{diam} T \geq C^{-1} \operatorname{diam} Q$. It is easy to check that (1.44) must then satisfy a Carleson packing condition whenever (1.45) does.

On the other hand, no point x in E can lie in more than a bounded number of cubes T in (1.45). This follows from the simple observation that

$$\operatorname{diam} T \leq \tilde{\delta}(x) \leq (k+1)\operatorname{diam} T$$

for all $x \in T$ when T lies in (1.45), since $\tilde{\delta}$ is Lipschitz with norm ≤ 1. This proves Lemma 1.42.

1.4. The proof of the theorems.

Let E be a d-dimensional regular set in $\mathbf{R}^n$, with its associated family of cubes Δ. Assume that E satisfies the hypotheses of Theorem 1.3 with some choices of p, q, C_0, and θ. We want to use Lemma 1.12 to show that (1.10) holds.

Let $R \in \Delta$ be given. We know from Lemma I.3.5 that there is a ball B centered on E such that $B \cap E \subseteq R$ and $\operatorname{diam} B \geq C^{-1}\operatorname{diam} R$. Our assumptions on E imply that there is a d-dimensional regular set $\widetilde{E}$ with regularity constant $\leq C_0$ which satisfies

(1.46) $$|R \cap \widetilde{E}| \geq C^{-1}\theta|R|$$

and also the (p,q)-geometric lemma with constant $\leq C_0$. The constant in (1.46), like all the others in this section, does not depend on R.

Let us now apply Lemma 1.20 with this choice of $\widetilde{E}$. For each $x \in R \cap \widetilde{E}$ we have that

$$\sum_{\substack{Q \ni x \\ Q \subseteq R}} \beta_q(Q)^p \leq C \sum_{\substack{Q \ni x \\ Q \subseteq R}} \tilde{\beta}_q(x, 6\operatorname{diam} Q)^p + C \sum_{\substack{Q \ni x \\ Q \subseteq R}} \widehat{I}_q(Q)^p,$$

because of (1.21) and (1.36). Integrating this over $R \cap \widetilde{E}$ we get

(1.47) $$\int_{R \cap \widetilde{E}} \left(\sum_{\substack{Q \ni x \\ Q \subseteq R}} \beta_q(Q)^p \right) dx$$
$$\leq C \int_{R \cap \widetilde{E}} \int_0^{10\operatorname{diam} R} \tilde{\beta}_q(x, t)^p \frac{dt}{t} dx + C \sum_{Q \subseteq R} \widehat{I}_q(Q)^p |Q|.$$

We have used here the fact that

$$\tilde{\beta}_q(x, s) \leq C\tilde{\beta}_q(x, t) \quad \text{when } s \leq t \leq 2s,$$

and we interchanged the order of summation and integration to get the $\widehat{I}_q(Q)$ term in the form above.

We can control the two terms on the right side of (1.47) using Lemma 1.37 and the assumption that $\widetilde{E}$ satisfies the (p,q)-geometric lemma. This gives

(1.48) $$\int_{R \cap \widetilde{E}} \left(\sum_{\substack{Q \ni x \\ Q \subseteq R}} \beta_q(Q)^p \right) dx \leq C(\operatorname{diam} R)^d + C|R| \leq C|R|.$$

This together with (1.46) implies that there exist N, $\eta > 0$ (which do not depend on R) such that

$$\left|\left\{x \in R : \sum_{\substack{Q \ni x \\ Q \subseteq R}} \beta_q(Q)^p \leq N\right\}\right| \geq \eta |R|.$$

We can now conclude from Lemma 1.12 that (1.10) holds, i.e., E satisfies the (p, q)-geometric lemma. This proves Theorem 1.3.

The proof of Theorem 1.8 proceeds along similar lines, and we omit the details.

CHAPTER 2

Stability Properties of the Corona Decomposition

The main purpose of this chapter is to provide a fairly direct proof of the fact that BPLG (see Definition I.1.26) implies the existence of a corona decomposition. This is interesting in its own right, but it is also useful for the proof of Theorem I.3.42 (concerning generalized corona decompositions), as we discussed in §I.3.3.

We shall in fact prove a more general result about the stability of the corona decomposition under the big pieces functor. We shall also provide a direct proof of the fact (I.3.33) that Lipschitz graphs admit corona decompositions.

We shall rely heavily in this chapter on the terminology of Chapter I.3 (cubes, coronizations, and corona decompositions), which the reader may wish to review.

2.1. Corona decompositions revisited.

PROPOSITION 2.1. *Let E be a d-dimensional regular set in $\mathbf{R}^n$, with an associated family of cubes Δ (as in §I.3.1). Then E admits a corona decomposition if and only if for each ϵ, $\delta > 0$ there is a coronization $(\mathcal{B}, \mathcal{G}, \mathcal{F})$ of E and an assignment $Q \mapsto P_Q$ of a d-plane to each cube $Q \in \mathcal{G}$ such that*

(2.2) $\operatorname{dist}(x, P_Q) \leq \epsilon \operatorname{diam} Q$ *when* $x \in 2Q$ *and* $Q \in \mathcal{G}$, *and*
(2.3) $\operatorname{Angle}(P_Q, P_{Q(S)}) \leq \delta$ *whenever* $Q \in S$, $S \in \mathcal{F}$.

We are following here our usual custom of denoting by $Q(S)$ the (unique) maximal element of $S \in \mathcal{F}$.

When we refer to the angle between two d-planes P_1 and P_2 we really mean the largest angle between them. The precise definition is not really important; all that matters is that $\operatorname{Angle}(P_1, P_2)$ should be invariant under translation of P_1 or P_2, that $\operatorname{Angle}(P_1, P_2)$ should define a distance function on the Grassmann manifold of d-planes in $\mathbf{R}^n$ which pass through the origin, and that this distance function should be compatible with the usual topology on the Grassmann manifold. In particular, the angle should be zero exactly when P_1 and P_2 are parallel.

The "only if" portion of Proposition 2.1 is an immediate consequence of the definition of a corona decomposition. The converse is neither completely trivial nor terribly difficult. An argument can be found in §8 of [**DS2**], but the situation there was somewhat more complicated than it is here, because it was necessary in [**DS2**] to pay more attention to the dependence on the parameters. Another argument is given in §4 of [**S4**]. This argument would have to be modified for the present circumstances, because [**S4**] dealt only with the codimension 1 case.

Proposition 2.1 provides a rather convenient criterion for the existence of a corona decomposition. To understand this criterion it is helpful to look at what happens when you remove (2.3). If you do that you get a characterization of the WGL (see Lemma I.3.10); this follows from Lemma I.3.22. In practice, when one tries to verify that a set admits a corona decomposition, one first proves that it satisfies the WGL, and then one worries about (2.3) afterward. One of the nice features of (2.3) is that it fits well with stopping-time arguments.

2.2. Corona constructions and Lipschitz functions.

In this section we shall provide a proof of (I.3.33), i.e., the fact that Lipschitz graphs admit corona decompositions (with bounds). As we pointed out in §I.3.2, this is known, but we do not know such a good reference. At any rate, we consider it worthwhile to include a proof for the reader's convenience. This will also provide us with some opportunities to illustrate some useful techniques from harmonic analysis.

For the record let us emphasize that nothing in this section is due to us.

The first step is to reduce the problem to one about Lipschitz functions. For the rest of this section we are going to be working with functions on $\mathbf{R}^d$, and so we let Δ denote the usual family of dyadic cubes on $\mathbf{R}^d$.

PROPOSITION 2.4. *Let* $f : \mathbf{R}^d \to \mathbf{R}$ *be a Lipschitz function with norm* ≤ 1. *For each* $\epsilon, \delta > 0$ *there is a coronization* $(\mathscr{B}, \mathscr{G}, \mathscr{F})$ *of* $\mathbf{R}^d$ *and an assigment* $Q \to A_Q$ *of an affine function on* $\mathbf{R}^n$ *to each cube* $Q \in \mathscr{G}$ *such that*

(2.5) $\quad\quad \sup_{2Q} |f - A_Q| \leq \epsilon \operatorname{diam} Q \;\; \textit{for all} \;\; Q \in \mathscr{G}, \;\; \textit{and}$

(2.6) $\quad\quad |\nabla A_Q - \nabla A_{Q(S)}| \leq \delta \;\; \textit{whenever} \;\; Q \in S, \;\; S \in \mathscr{F}.$

Furthermore, the constants in the Carleson packing conditions for $\mathscr{B}$ *and the maximal cubes of* $\mathscr{G}$ *(i.e., (I.3.15) and (I.3.18)) can be chosen so that they depend on* d, ϵ, *and* δ, *but not* f.

It is easy to prove that Lipschitz graphs admit corona decompositions once you have Propositions 2.4 and 2.1. To be honest, there is a minor problem when the codimension is larger than 1, because then the analogue of Proposition 2.4 for vector-valued functions is needed. This is not a serious issue, because the same proof still works in that case. Alternatively, one can reduce the vector-valued case to the **R**-valued situation with the aid of the not-so-deep general fact that if $(\mathscr{B}_1, \mathscr{G}_1, \mathscr{F}_1)$ and $(\mathscr{B}_2, \mathscr{G}_2, \mathscr{F}_2)$ are two coronizations of $\mathbf{R}^d$, then a third coronization $(\mathscr{B}, \mathscr{G}, \mathscr{F})$ can be produced by taking $\mathscr{B} = \mathscr{B}_1 \cup \mathscr{B}_2$, $\mathscr{G} = \mathscr{G}_1 \cap \mathscr{G}_2$, and $\mathscr{F} = \{S_1 \cap S_2 : S_1 \in \mathscr{F}_1, S_2 \in \mathscr{F}_2,$ and $S_1 \cap S_2 \neq \varnothing\}$. We leave it to the reader as an exercise to verify that $(\mathscr{B}, \mathscr{G}, \mathscr{F})$ is indeed a coronization of $\mathbf{R}^d$.

It is also easy to prove that bilipschitz images of $\mathbf{R}^d$ in $\mathbf{R}^n$ admit corona decompositions using Propositions 2.1 and 2.4. This is a little messier than the case of Lipschitz graphs but not much more so. We omit the details.

Proposition 2.4 is simply a formulation of Carleson's corona construction which is adapted to Lipschitz functions. We shall derive it from a more traditional version for L^∞ functions. An excellent general reference for Carleson's corona construction and related topics is [G3].

Proposition 2.4 should be viewed as a quantitative version of the fact that Lipschitz functions are differentiable almost everywhere. This should not be confused with the other quantifiable aspect of differentiability, where one looks at the size of the derivatives. Classical analysis has traditionally been more concerned with this second issue than the first, but this second issue does not make sense in the geometrical setting; there is no such thing as the "size" of a tangent plane. Of course, the quantifying of differentiability as given in Proposition 2.4 does have a geometric counterpart, namely, the corona decomposition.

Notice that Proposition 2.4 reduces to the WALA (Definition I.2.47) if (2.6) is removed, in the same way that Proposition 2.1 reduces to the WGL when (2.3) is removed.

The next result is a "no-frills" version of the corona construction for L^∞ functions, which we shall strengthen afterward.

PROPOSITION 2.7. *Given $h \in L^\infty(\mathbf{R}^d)$, $\|h\|_\infty \leq 1$, and $\eta > 0$, there is a coronization $(\mathscr{B}, \mathscr{G}, \mathscr{F})$ of $\mathbf{R}^d$ such that*

$$\left| \frac{1}{|Q|} \int_Q h - \frac{1}{|Q(S)|} \int_{Q(S)} h \right| \leq \eta \quad \forall Q \in S \tag{2.8}$$

for all $S \in \mathscr{F}$. The Carleson constants for this coronization can be chosen so that they depend on d and η but not h.

This result should be viewed as providing a quantitative version of the fact that

$$\lim_{\substack{Q \ni x \\ |Q| \to 0}} \frac{1}{|Q|} \int_Q h = h(x)$$

for almost all $x \in \mathbf{R}^d$.

Proposition 2.7 is very well known, but we include a proof for the sake of completeness. We begin by throwing away a few cubes.

LEMMA 2.9. *Set $\widetilde{\Delta} = \{Q \in \Delta : 0 \notin 2Q\}$. Then $\Delta \setminus \widetilde{\Delta}$ satisfies a Carleson packing condition, and every element of $\widetilde{\Delta}$ is contained in a maximal element of $\widetilde{\Delta}$.*

This is very easy to check, and we omit the details.

Let h and η be given, as in the statement of Proposition 2.7. Set $\mathscr{B} = \Delta \setminus \tilde{\Delta}$ and $\mathscr{G} = \tilde{\Delta}$.

LEMMA 2.10. *There is a partition $\mathscr{F}$ of $\mathscr{G}$ which has the following properties*:

  (i) *each $S \in \mathscr{F}$ satisfies the coherence condition* (I.3.17);
  (ii) *each S satisfies* (2.8); *and*
  (iii) *each minimal cube of each $S \in \mathscr{F}$ has a child which does not satisfy the inequality in* (2.8).

This is not hard to prove, by running the obvious stopping-time argument. Given a maximal cube Q_0 in $\mathscr{G}$, we can generate an $S \in \mathscr{F}$ by putting Q_0 into S and then adding to S the descendents of Q_0, generation by generation, which have the property that they and their siblings all satisfy the inequality in (2.8). (Recall that two cubes are said to be siblings if they have the same parent.) Whenever we run into a cube which has a child that does not satisfy the inequality in (2.8), then we stop, i.e., we do not add into S any of the descendents of that cube. After doing this for all the maximal elements of $\mathscr{G}$, we can simply remove the subsets of $\mathscr{G}$ that have been so generated and repeat the process for the collection $\mathscr{G}_1$ of cubes in $\mathscr{G}$ that remain. By iterating this procedure indefinitely we can produce the desired partition $\mathscr{F}$ of $\mathscr{G}$.

It remains to show that $\{Q(S) : S \in \mathscr{F}\}$ satisfies a Carleson packing condition.

Given $S \in \mathscr{F}$, let $m(S)$ denote the set of minimal cubes in S, and let $cm(S)$ denote the set of children of the minimal cubes in S. Set

$$M(S) = \bigcup_{Q \in m(S)} Q \quad \text{and} \quad Z(S) = Q(S) \setminus M(S).$$

Thus $Z(S)$ consists of the points in $Q(S)$ which lie in arbitrarily small elements of S, and hence

(2.11) $\qquad Z(S) \cap Z(S') = \emptyset \quad \text{when} \quad S, S' \in \mathscr{F}, \, S \neq S'.$

It is easy to check that this implies

(2.12) $\qquad \{Q(S) : S \in \mathscr{F}, \, |Z(S)| \geq \frac{1}{2}|Q(S)|\}$

satisfies a Carleson packing condition. The case where $|M(S)| \geq \frac{1}{2}|Q(S)|$ is more complicated. For this we shall use L^2 estimates.

Given $S \in \mathscr{F}$, define $h_S : \mathbf{R}^d \to \mathbf{R}$ as follows:

$h_S(x) = 0$ when $x \notin Q(S)$;

$h_S(x) = h(x) - \dfrac{1}{|Q(S)|} \displaystyle\int_{Q(S)} h$ when $x \in Z(S)$;

$h_S(x) = \dfrac{1}{|Q|} \displaystyle\int_Q h - \dfrac{1}{|Q(S)|} \displaystyle\int_{Q(S)} h$ when $x \in Q$ for some $Q \in cm(S)$.

LEMMA 2.13. (a) $\int_{\mathbf{R}^d} h_S = 0$ for each $S \in \mathscr{F}$.

(b) $\int_{\mathbf{R}^d} h_S h_{S'} = 0$ whenever $S, S' \in \mathscr{F}$ and $S \neq S'$.

(c) $\int_{\mathbf{R}^d} |h_S|^2 \geq 2^{-d} \eta^2 |M(S)|$.

(d) If $T \in \Delta$, then $\| \sum_{\substack{S \in \mathscr{F} \\ Q(S) \subseteq T}} h_S \|_{L^\infty} \leq 2$.

Part (a) is straightforward, and we leave it as an exercise. To prove (b) we may as well assume that $Q(S)$ is contained in $Q(S')$, since otherwise the reverse is true or they are disjoint. Since $S \neq S'$, we must have that $Q(S) \subseteq R$ for some $R \in cm(S')$. Hence $h_{S'}$ is constant on $Q(S)$, so (b) follows from (a). Part (c) is an immediate consequence of the definition of h_S and (iii) in Lemma 2.10, because they imply that $|h_S| \geq \eta$ on a child of each $Q \in m(S)$.

Let us check (d). Let Q_j, $j \in J$, be an enumeration of the maximal elements of $\{Q \in \Delta : Q \subseteq T$ and $Q = Q(S)$ for some $S \in \mathscr{F}\}$. The definitions unravel to give

$$\sum_{\substack{S \in \mathscr{F} \\ Q(S) \subseteq T}} h_S = \sum_{j \in J} \left(h - \frac{1}{|Q_j|} \int_{Q_j} h \right) \chi_{Q_j}.$$

This implies (d).

Let us use the lemma to prove that

$$\mathscr{F}' = \{Q(S) : S \in \mathscr{F}, \ |M(S)| \geq \tfrac{1}{2}|Q(S)|\}$$

satisfies a Carleson packing condition. Let $T \in \Delta$ be given. We have that

$$\sum_{\substack{S \in \mathscr{F}' \\ Q(S) \subseteq T}} |Q(S)| \leq 2^{d+1} \eta^{-2} \sum_{\substack{S \in \mathscr{F} \\ Q(S) \subseteq T}} \int |h_S|^2$$

$$= 2^{d+1} \eta^{-2} \int \left| \sum_{\substack{Q \in \mathscr{F} \\ Q(S) \subseteq T}} h_S \right|^2 \quad \text{[by orthogonality]}$$

$$\leq 2^{d+1} \eta^{-2} \int_T 4 = 2^{d+3} \eta^{-2} |T|.$$

This completes the proof of Proposition 2.7.

Although Proposition 2.7 is the "no-frills" version of the corona construction, it does incorporate the main points. To strengthen it we shall remove some more "bad" cubes. The next lemma will provide us with the necessary Carleson packing estimates.

Recall from Definition I.3.23 that two cubes Q_2, $Q_2 \in \Delta$ are said to be N-close if

$$N^{-1} \operatorname{diam} Q_1 \le \operatorname{diam} Q_2 \le N \operatorname{diam} Q_1 \text{ and}$$
$$\operatorname{dist}(Q_1, Q_2) \le N(\operatorname{diam} Q_1 + \operatorname{diam} Q_2).$$

LEMMA 2.14. *Let* $\alpha : \Delta \to \Delta$ *be given, such that* $\alpha(Q)$ *is* N*-close to* Q *for some* $N > 0$ *and all* $Q \in \Delta$. *Then for each* $g \in L^2(\mathbf{R}^n)$ *we have that*

$$\sum_{Q \in \Delta} \left| \frac{1}{|Q|} \int_Q g - \frac{1}{|\alpha(Q)|} \int_{\alpha(Q)} g \right|^2 |Q| \le C(N) \int_{\mathbf{R}^d} |g|^2.$$

For each $Q \in \Delta$ define $k_Q \in L^2(\mathbf{R}^d)$ by

$$k_Q = |Q|^{1/2} \left(\frac{1}{|Q|} \chi_Q - \frac{1}{|\alpha(Q)|} \chi_{\alpha(Q)} \right).$$

The lemma states that
$$g \mapsto \{\langle g, k_Q \rangle\}_{Q \in \Delta}$$
defines a bounded linear mapping from $L^2(\mathbf{R}^d)$ to $l^2(\Delta)$. Here $\langle \, , \, \rangle$ denotes the usual inner product on L^2. By duality this is equivalent to the statement that

(2.15)
$$\left\| \sum_{Q \in \Delta} \lambda_Q k_Q \right\|_{L^2}^2 \le C(N) \sum_{Q \in \Delta} |\lambda_Q|^2$$

for all families of real numbers $\{\lambda_Q\}_{Q \in \Delta}$. This is in turn implied by the condition that

(2.16)
$$\sum_{Q' \in \Delta} \left| \sum_{Q \in \Delta} \langle k_{Q'}, k_Q \rangle \lambda_Q \right|^2 \le C(N) \sum_{Q \in \Delta} |\lambda_Q|^2$$

for all $\{\lambda_Q\}_{Q \in \Delta}$. [Actually, (2.15) also implies (2.16), but we will shall not need that.]

To prove (2.16) we use the method of Schur. The first step is to show that

(2.17)
$$\sum_{Q \in \Delta} |\langle k_{Q'}, k_Q \rangle| \, |Q|^s \le C(N, s) |Q'|^s$$

for all $Q' \in \Delta$ when $\frac{1}{2} - \frac{1}{d} < s < \frac{1}{2}$. This is not hard to verify. Fix $Q' \in \Delta$. If $|Q| \ge |Q'|$ one uses the estimate

$$|\langle k_{Q'}, k_Q \rangle| \le C(N) \left(\frac{|Q'|}{|Q|} \right)^{1/2}$$

together with the fact that $\langle k_{Q'}, k_Q \rangle = 0$ unless $\operatorname{dist}(Q, Q') \le C(N) \operatorname{diam} Q$. When $|Q| \le |Q'|$ one uses the estimate

$$|\langle k_{Q'}, k_Q \rangle| \le C(N) \left(\frac{|Q|}{|Q'|} \right)^{1/2}$$

together with the fact that $\langle k_{Q'}, k_Q \rangle = 0$ unless
$$\operatorname{dist}(Q, \partial Q' \cup \partial(\alpha(Q'))) \leq C(N) \operatorname{diam} Q.$$
For this last observation it is helpful to remember that $\int k_Q = 0$.

Once we have (2.17) we can derive (2.16) using Jensen's inequality (or Cauchy-Schwarz), which implies that

$$\left| \sum_{Q \in \Delta} \langle k_{Q'}, k_Q \rangle \lambda_Q \right|^2$$

$$= |Q'|^{2s} \left| \sum_{Q \in \Delta} |Q'|^{-s} \langle k_{Q'}, k_Q \rangle |Q|^{s} (|Q|^{-s} \lambda_Q) \right|^2$$

$$\leq C(N, s) |Q'|^{2s} \sum_{Q \in \Delta} |Q'|^{-s} |\langle k_{Q'}, k_Q \rangle| \, |Q|^s (|Q|^{-s} |\lambda_Q|)^2.$$

From here (2.16) follows easily, by using (2.17) again, but with the roles of Q and Q' reversed. [Notice that we have only needed (2.17) to be true for some $s \in \mathbf{R}$.] This proves Lemma 2.14.

Before proceeding let us record the appropriate version of Lemma 2.14 for L^∞ functions.

LEMMA 2.18. *Let $\alpha : \Delta \to \Delta$ and $N > 0$ be as in Lemma 2.14. Then for each $g \in L^\infty(\mathbf{R}^d)$ and every $R \in \Delta$ we have that*

$$\sum_{Q \subseteq R} \left| \frac{1}{|Q|} \int_Q g - \frac{1}{|\alpha(Q)|} \int_{\alpha(Q)} g \right|^2 |Q| \leq C(N) \|g\|_{L^\infty}^2 |R|.$$

This follows easily from Lemma 2.14: Given $R \in \Delta$, the sum above is not changed if we replace g with $g\chi_{10NR}$, which permits us to reduce to the L^2-estimate in Lemma 2.14.

COROLLARY 2.19. *Let $N, \eta > 0$ be given, and also $h \in L^\infty(\mathbf{R}^d)$, with $\|h\|_\infty \leq 1$. Denote by $\mathscr{G}(\eta, N)$ the collection of $Q \in \Delta$ such that*

$$\left| \frac{1}{|Q|} \int_Q h - \frac{1}{|Q'|} \int_{Q'} h \right| \leq \eta$$

for all $Q' \in \Delta$ that are N-close to Q. Then $\Delta \setminus \mathscr{G}(\eta, N)$ satisfies a Carleson packing condition, with a bound which depends on d, η, and N but not h.

This is an immediate consequence of Lemma 2.18.

COROLLARY 2.20. *Let $N, \eta > 0$ and $h \in L^\infty(\mathbf{R}^d)$ be given, with $\|h\|_\infty \leq 1$, and let $\mathscr{G}(\eta, N)$ be as in Corollary 2.19. Then Proposition 2.7 remains true if we add the requirement that $\mathscr{G} \subseteq \mathscr{G}(\eta, N)$.*

This is easy to derive from Proposition 2.7 and Corollary 2.19, using also the following general fact.

LEMMA 2.21. *If* $(\mathscr{B}, \mathscr{G}, \mathscr{F})$ *is a coronization of* $\mathbf{R}^d$ *and if* $\widetilde{\mathscr{G}}$ *is a subset of* Δ *such that* $\Delta \setminus \widetilde{\mathscr{G}}$ *satisfies a Carleson packing condition, then there is another coronization* $(\mathscr{B}', \mathscr{G}', \mathscr{F}')$ *of* $\mathbf{R}^d$ *such that* $\mathscr{G}' \subseteq \mathscr{G} \cap \widetilde{\mathscr{G}}$ *and each* $S' \in \mathscr{F}'$ *is contained in some* S *in* $\mathscr{F}$.

We have, in fact, already proved this result, in the proof of Lemma I.3.22. Alternatively, Lemma 2.21 can be derived from Lemma I.3.22.

We are going to derive Proposition 2.4 from Corollary 2.20, but we need one more piece of information.

LEMMA 2.22. *For every* $\epsilon > 0$ *there is an* $\eta > 0$ *(small) and an* $N > 0$ *(large) so that if* $f : \mathbf{R}^d \to \mathbf{R}$ *is a Lipschitz function with norm* ≤ 1 *and if* $Q \in \Delta$ *satisfies*

$$\left| \frac{1}{|Q|} \int_Q \nabla f - \frac{1}{|Q'|} \int_{Q'} \nabla f \right| \leq \eta$$

for all $Q' \in \Delta$ *which are* N-*close to* Q, *then*

$$\sup_{2Q} |f - A_Q| \leq \epsilon \operatorname{diam} Q,$$

where A_Q *is the affine function on* $\mathbf{R}^d$ *that agrees with* f *at the center of* Q *and also satisfies*

$$\nabla A_Q = \frac{1}{|Q|} \int_Q \nabla f.$$

Clearly Proposition 2.4 follows from this lemma and Corollary 2.20 (with $h = \nabla f$). Actually, there is a minor technical problem, which is that we only claimed Corollary 2.20 for real-valued functions, while $h = \nabla f$ is $\mathbf{R}^d$-valued. This is, of course, not a serious issue, for the same reasons as discussed in the paragraph just after the statement of Proposition 2.4.

Let us now prove Lemma 2.22. Let ϵ and f be given, and let Q be as above, with η and N to be specified later.

Define $g : \mathbf{R}^d \to \mathbf{R}$ by $g = f - A_Q$. Thus g is Lipschitz with norm ≤ 2, and

(2.23) $\qquad \left| \dfrac{1}{|Q'|} \displaystyle\int_{Q'} \nabla g \right| \leq \eta \quad \text{when } Q' \in \Delta \text{ is } N\text{-close to } Q.$

The idea is that we can use (2.23) to control g, because g can be represented in terms of an integral of ∇g against a reasonably smooth function. There are many ways to implement this idea, and we are going to use one that is rather crude but simple.

Let ϕ be a smooth nonnegative function on $\mathbf{R}^d$ such that $\operatorname{supp} \phi \subseteq B(0, 1)$ and $\int \phi = 1$. Set $\phi_t(x) = t^{-d} \phi(t^{-1} x)$ for $t > 0$, so that $\int \phi_t = 1$ too. It is easy to check that

(2.24) $\qquad \|g - \phi_t * g\|_\infty \leq 2t.$

2.2. CORONA CONSTRUCTIONS AND LIPSCHITZ FUNCTIONS

We want to estimate $\phi_t * g$ for t small by controlling its gradient, which we write as

$$\nabla(\phi_t * g)(x) = \int \phi_t(x-y) \nabla g(y) \, dy.$$

Let k be an integer such that $2^k \leq t$, and let Δ_k denote the set of dyadic cubes in $\mathbf{R}^d$ with sidelength 2^k. Clearly

$$\left| \phi_t(x-y) - \sum_{R \in \Delta_k} \phi_t(x-z_R) \chi_R(y) \right| \leq C 2^k t^{-d-1},$$

where z_R denotes the center of R. Hence

(2.25)
$$\left| \nabla(\phi_t * g)(x) - \sum_{R \in \Delta_k} \phi_t(x-z_R) \int_R \nabla g(y) \, dy \right|$$
$$\leq C 2^k t^{-d-1} \int_{B(x,dt)} |\nabla g| \leq C 2^k t^{-1}.$$

Now let us restrict our attention to $x \in 10Q$ and $t \leq \text{diam}\, Q$. [At the end we shall in fact take t to be $\epsilon \, \text{diam}\, Q/8$.] If we also require that

(2.26)
$$2^{-k} \text{diam}\, Q + 100d \leq N,$$

then we have that

$$\left| \int_R \nabla g \right| \leq \eta |R|$$

whenever $R \in \Delta_k$ satisfies $\phi_t(x - z_R) \neq 0$, because of (2.23). Under these conditions we get from (2.25) that

(2.27) $\displaystyle |\nabla(\phi_t * g)(x)| \leq \sum_{\substack{R \in \Delta_k \\ \phi_t(x-z_R) \neq 0}} C t^{-d} \eta |R| + C 2^k t^{-1} \leq C\eta + C 2^k t^{-1}.$

Let us now estimate $\phi_t * g$ itself using this information about its gradient. Since g vanishes at the center of Q, by definition, $|\phi_t * g| \leq 2t$ there, because of (2.24). This together with (2.27) gives

$$\sup_{2Q} |\phi_t * g| \leq 2t + C(\eta + 2^k t^{-1}) \text{diam}\, Q,$$

and hence

$$\sup_{2Q} |g| \leq 4t + C(\eta + 2^k t^{-1}) \text{diam}\, Q.$$

The right side of this is $< \epsilon \, \text{diam}\, Q$ if we choose our parameters correctly. We first take $t = \epsilon \, \text{diam}\, Q/8$. If η is small enough and N is large enough, then we can choose k so that (2.26) is satisfied and also

$$C(\eta + 2^k t^{-1}) < \epsilon/2.$$

Thus we can choose η and N in such a way as to ensure that $|g| < \epsilon$ on $2Q$. This proves Lemma 2.22.

REMARK 2.28. Notice that the fact that $\mathbf{R}^d$ satisfies the WALA (see Definition I.2.47) follows from Lemma 2.22 and Corollary 2.19. Theorem I.1.40 can be proved with similar methods, but the argument is somewhat more complicated.

2.3. The statement of the main result.

Given a collection $\mathscr{E}$ of d-dimensional regular subsets of $\mathbf{R}^n$, let $\mathrm{BP}(\mathscr{E})$ denote the class of all d-dimensional regular sets E in $\mathbf{R}^n$ which have big pieces of elements of $\mathscr{E}$. That is, $E \in \mathrm{BP}(\mathscr{E})$ if there exists $\theta > 0$ so that for each $x \in E$ and $r > 0$ there is an $F \in \mathscr{E}$ such that

$$|E \cap B(x,r) \cap F| \geq \theta r^d.$$

For instance, if $\mathrm{LG}(k)$ denotes the set of d-dimensional Lipschitz graphs in $\mathbf{R}^n$ with constant $\leq k$, then

$$\bigcup_{k>0} \mathrm{BP}(\mathrm{LG}(k))$$

is the class of d-dimensional regular sets in $\mathbf{R}^n$ which have BPLG (see Definition I.1.26). Notice that this is different from

$$\mathrm{BP}\left(\bigcup_k \mathrm{LG}(k)\right),$$

because the quantifiers are different.

THEOREM 2.29. *Let $\mathscr{E}$ be a collection of d-dimensional regular sets in $\mathbf{R}^n$ whose regularity constants (in the sense of Definition I.1.13) are uniformly bounded. Suppose also that every element of $\mathscr{E}$ admits a corona decomposition, with uniform bounds. Then every d-dimensional regular set in $\mathrm{BP}(\mathscr{E})$ also admits a corona decomposition.*

When we say that the elements of $\mathscr{E}$ admit corona decompositions with uniform bounds we mean that the constants for the Carleson packing conditions which enter into Definition I.3.19 through Definition I.3.13 can be taken to be independent of the particular element of $\mathscr{E}$ under consideration. These Carleson constants, though, will depend on the parameters η and θ in Definition I.3.19.

Theorem 2.29 follows from the results of [DS2], but we want to provide a more direct proof for the reasons that were discussed in §I.3.3.

In view of the fact that Lipschitz graphs admit corona decompositions, with suitable bounds, a special case of Theorem 2.29 is the result that E admits a corona decomposition if E has BPLG. This was used in §I.3.3 to prove Theorem I.3.42 (concerning generalized corona decompositions).

The proof of Theorem 2.29 is given in the next two sections.

2.4. Preliminaries.

Fix d and n, $0 < d < n$. For the rest of this chapter we assume that E and $\mathscr{E}$ are given to us as in Theorem 2.29.

2.4. PRELIMINARIES

LEMMA 2.30. *E satisfies the WGL.*

This follows from Theorem 1.8 and the simple observation that regular sets which admit corona decompositions must satisfy the WGL in particular.

Let $\epsilon > 0$ and $\delta > 0$ be given and fixed for the rest of the chapter. We are going to show that E satisfies the conditions described in Proposition 2.1. We may as well assume that ϵ is much smaller than δ (since that simply strengthens the conditions that we must prove), and we shall impose more specific restrictions of this type later on (in Lemmas 2.35 and 2.53).

Let Δ be a family of cubes on E as in §I.3.1. Set

$$(2.31) \quad \mathscr{G}(\epsilon) = \left\{ Q \in \Delta : \text{ there is a } d\text{-plane } P_Q \text{ such that } \sup_{x \in 2Q} \text{dist}(x, P_Q) \leq \epsilon \operatorname{diam} Q \right\},$$

so that $\Delta \setminus \mathscr{G}(\epsilon)$ satisfies a Carleson packing condition, by Lemma 2.30. To each $Q \in \mathscr{G}(\epsilon)$ we assign a d-plane P_Q as in (2.31). This d-plane is not unique, but it almost is, as the next result shows.

LEMMA 2.32. *Let F be a d-dimensional regular set in $\mathbf{R}^n$ with regularity constant $\leq C_0$. Let A be a subset of F such that*

$$(2.33) \quad 0 < (\operatorname{diam} A)^d \leq C_0 |A|.$$

Suppose that P_1 and P_2 are d-planes in $\mathbf{R}^n$ which satisfy

$$\sup_{a \in A} \text{dist}(a, P_i) \leq \alpha \operatorname{diam} A, \quad i = 1, 2.$$

Then $\text{Angle}(P_1, P_2) \leq C_1 \alpha$, *and*

$$(2.34) \quad \text{dist}(z, P_2) \leq C_1 \alpha \operatorname{diam} A + C_1 \alpha \, \text{dist}(z, A) \quad \text{for all } z \in P_1$$

(and similarly with the roles of P_1 and P_2 reversed). Here C_1 depends only on C_0, n, and d.

The proof of this is more tedious than difficult and we shall merely mention the main points. The idea is that A cannot live in too small a neighborhood of a $(d-1)$-plane, because then regularity of F would force A to have too little mass.

Let us be more precise. If you cover A by a collection of balls of radius t, $t < \operatorname{diam} A$, then there cannot be fewer than $C^{-1}(\operatorname{diam} A/t)^d$ of these balls, because of (2.33). This implies that for each k-plane M in $\mathbf{R}^n$, $k < d$, there must be an $x \in A$ such that $\text{dist}(x, M) \geq C^{-1} \operatorname{diam} A$.

Using this observation it is easy to choose points $a_0, a_1, \ldots, a_d$ in A which are affinely independent, with estimates. That is, the distance from a_j to the (affine) $(j-1)$-plane generated by $a_0, a_1, \ldots, a_{j-1}$ is at least $C^{-1} \operatorname{diam} A$ for each j, $j = 1, 2, \ldots, d$. By hypothesis both P_1 and P_2 must be close to the (affine) d-plane generated by $a_0, a_1, \ldots, a_d$, and from here the conclusions of Lemma 2.32 can be derived with the aid of some elementary linear algebra.

The next lemma will provide us with our candidate for the coronization $(\mathcal{B}, \mathcal{G}, \mathcal{F})$ of E required by Proposition 2.1.

LEMMA 2.35. *There is a subset $\mathcal{G}$ of $\mathcal{G}(\epsilon)$ such that $\mathcal{B} = \Delta \setminus \mathcal{G}$ satisfies a Carleson packing condition and such that there is a partition $\mathcal{F}$ of $\mathcal{G}$ with the following properties*:

(i) *each $S \in \mathcal{F}$ satisfies the coherence condition (I.3.17) and, in particular, has a unique maximal element $Q(S)$;*
(ii) $\mathrm{Angle}(P_Q, P_{Q(S)}) \leq \delta$ *whenever $Q \in S$, $S \in \mathcal{F}$; and*
(iii) *if Q is a minimal cube in S, then either $\mathrm{Angle}(P_Q, P_{Q(S)}) \geq \delta/2$ or one of the children of Q lies in $\mathcal{B}$.*

This is the same as Lemma 7.1 in [DS2], but we shall sketch a proof for the sake of completeness. (It is also quite similar to Lemma 2.10.) Fix some point x_0 in E, and set

$$\mathcal{G} = \{Q \in \mathcal{G}(\epsilon) : x_0 \notin 2Q\}.$$

It is easy to see that $\mathcal{G}(\epsilon) \setminus \mathcal{G}$ satisfies a Carleson packing condition and that every element of $\mathcal{G}$ is contained in a maximal element of $\mathcal{G}$. Set $\mathcal{B} = \Delta \setminus \mathcal{G}$.

Now let us define $\mathcal{F}$. Given a maximal cube Q_0 in $\mathcal{G}$, we generate an $S \in \mathcal{F}$ as follows. We put Q_0 into S and then add to S the descendents of Q_0, generation by generation, such that they and their siblings lie in $\mathcal{G}$ and satisfy the inequality in (ii). Whenever we run into a cube that has at least one child which does not lie in $\mathcal{G}$ or does not satisfy the inequality in (ii), then we stop, i.e., we do not put any of the descendents of that cube into S. We do this for all the maximal cubes in $\mathcal{G}$, and we generate in this way a family $\mathcal{F}_1$ of disjoint coherent subsets of $\mathcal{G}$. Let $\mathcal{G}_1$ be the subset of $\mathcal{G}$ that remains after removing all the $S \in \mathcal{F}_1$, and repeat the process for $\mathcal{G}_1$.

After iterating this procedure indefinitely we get a family $\mathcal{F}$ of disjoint coherent subsets of $\mathcal{G}$. It is not hard to see that $\mathcal{G}$ is exhausted by elements of $\mathcal{F}$. The elements of $\mathcal{F}$ satisfy (i) and (ii) by construction, so we need only check (iii). Let $S \in \mathcal{F}$ and a minimal cube Q in S be given. Because Q is minimal, there must be a child Q' of Q such that $Q' \in \mathcal{B}$ or $\mathrm{Angle}(P_{Q'}, P_{Q(S)}) \geq \delta$. If $Q' \in \mathcal{B}$ we are finished; otherwise $Q' \in \mathcal{G}$, and we have $\mathrm{Angle}(P_Q, P_{Q'}) \leq C\epsilon$, by Lemma 2.32, and hence $\mathrm{Angle}(P_Q, P_{Q(S)}) \geq \delta/2$, if ϵ is small enough compared to δ. This proves Lemma 2.35.

Let $\mathcal{B}$, $\mathcal{G}$, and $\mathcal{F}$ be as in Lemma 2.35. To show that E satisfies the conditions of Proposition 2.1 it remains to prove that

(2.36) $\{Q(S) : S \in \mathcal{F}\}$ satisfies a Carleson packing condition.

This task can be reduced somewhat further, as follows.

2.4. PRELIMINARIES

Given $S \in \mathscr{F}$, let $m(S)$ denote the collection of minimal cubes in S. Let $m_0(S)$ denote the set of $Q \in m(S)$ which have at least one child in $\mathscr{B}$, and set $m_1(S) = m(S) \setminus m_0(S)$. Thus if $Q \in m_1(S)$, then each child of Q lies in $\mathscr{G}$, but $\text{Angle}(P_Q, P_{Q(S)}) \geq \delta/2$. For each $\rho > 0$ set

$$\mathscr{F}_0(\rho) = \left\{ S \in \mathscr{F} : \left| \bigcup_{Q \in m_0(S)} Q \right| \geq \rho |Q(S)|/2 \right\},$$

$$\mathscr{F}_1(\rho) = \left\{ S \in \mathscr{F} : \left| \bigcup_{Q \in m_1(S)} Q \right| \geq (1-\rho)|Q(S)| \right\},$$

$$\mathscr{F}_2(\rho) = \left\{ S \in \mathscr{F} : \left| Q(S) \setminus \bigcup_{Q \in m(S)} Q \right| \geq \rho |Q(S)|/2 \right\},$$

so that $\mathscr{F} = \mathscr{F}_0(\rho) \cup \mathscr{F}_1(\rho) \cup \mathscr{F}_2(\rho)$ for all $\rho > 0$.

LEMMA 2.37. $\{Q(S) : S \in \mathscr{F}_i(\rho)\}$ *satisfies a Carleson packing condition for* $i = 0, 2$ *and every* $\rho > 0$.

When $i = 0$ this can be derived from the Carleson packing condition on $\mathscr{B}$. When $i = 2$ this is a consequence of the observation that

$$\left\{ Q(S) \setminus \left(\bigcup_{Q \in m(S)} Q \right) : S \in \mathscr{F} \right\}$$

is a family of pairwise disjoint subsets of E. This disjointness can be derived from the fact that the S's in $\mathscr{F}$ are coherent and disjoint as subsets of Δ. For both $i = 0$ and $i = 2$ we leave the details as an exercise.

To prove (2.36) we are now left with showing that $\{Q(S) : S \in \mathscr{F}_1(\rho)\}$ satisfies a Carleson packing condition for some $\rho > 0$. To this end we shall prove the following in the next section.

LEMMA 2.38. *There exist* $\rho > 0$, $\eta > 0$, *and* $N > 0$ *so that*

$$(2.39) \quad \left| \left\{ x \in R : \sum_{\substack{S \in \mathscr{F}_1(\rho) \\ Q(S) \subseteq R}} \chi_{Q(S)}(x) \leq N \right\} \right| \geq \eta |R|$$

for all $R \in \Delta$.

Once we have proved Lemma 2.38 we shall be finished, because Lemma 1.12 then implies that $\{Q(S) : S \in \mathscr{F}_1(\rho)\}$ does indeed satisfy a Carleson packing condition for this prescribed value of ρ. (Here Lemma 1.12 should be applied with $\alpha : \Delta \to \mathbf{R}_+$ defined by $\alpha(Q) = 1$ if $Q = Q(S)$ for some $S \in \mathscr{F}_1(\rho)$, $\alpha(Q) = 0$ otherwise.)

2.5. The proof of Lemma 2.38.

We continue with the same assumptions and notation as in the preceding section.

Let $R \in \Delta$ be given. We want to show that (2.39) holds for suitable choices of ρ, η, and N (which do not depend on R).

According to Lemma I.3.5 we can find a ball B centered on E such that $B \cap E \subseteq R$ and $\operatorname{diam} B \geq C^{-1} \operatorname{diam} R$. Hence there is an $\widetilde{E} \in \mathscr{E}$ such that

$$(2.40) \qquad |R \cap \widetilde{E}| \geq C^{-1}\theta|R|.$$

Here $\mathscr{E}$ is as in Thereom 2.29, and θ is a positive number that depends on E but not on R or $\widetilde{E}$.

Given $\gamma > 0$, set

$$R_\gamma = \{x \in R : |Q \cap \widetilde{E}| \geq \gamma|Q| \text{ whenever } Q \in \Delta \text{ and } x \in Q \subseteq R\},$$

$$\mathscr{F}_1(\rho, \gamma) = \{S \in \mathscr{F}_1(\rho) : Q(S) \subseteq R \text{ and } |Q(S) \cap \widetilde{E}| \geq \gamma|Q(S)|\}.$$

We shall derive (2.39) from the next two lemmas.

LEMMA 2.41. *If $\gamma > 0$ is small enough, then*

$$|R_\gamma| \geq \gamma|R|.$$

LEMMA 2.42. *For each $\gamma > 0$ there exists $\rho > 0$ and $K > 1$ so that*

$$(2.43) \qquad \sum_{S \in \mathscr{F}_1(\rho, \gamma)} |Q(S)| \leq K|R|.$$

None of the constants in these lemmas depend on R.

Let us assume these two lemmas for the moment and derive Lemma 2.38 from them. Let γ be as in Lemma 2.41, and then choose ρ and K according to Lemma 2.42. Observe that

$$\{S \in \mathscr{F}_1(\rho) : Q(S) \subseteq R \text{ and } Q(S) \cap R_\gamma \neq \varnothing\} \subseteq \mathscr{F}_1(\rho, \gamma),$$

and hence

$$\int_{R_\gamma} \left(\sum_{\substack{S \in \mathscr{F}_1(\rho) \\ Q(S) \subseteq R}} \chi_{Q(S)}(x) \right) dx \leq \sum_{S \in \mathscr{F}_1(\rho, \gamma)} |Q(S)|,$$

by Fubini's theorem. This last is $\leq K|R|$, by Lemma 2.42. This implies (2.39) with $\eta = \gamma/2$ and $N = 2K/\gamma$, by Tchebychev's inequality.

Now let us prove Lemma 2.41. This is quite easy. By definition $R \setminus R_\gamma$ is the union of the cubes $Q \subseteq R$ such that $|Q \cap \widetilde{E}| < \gamma|Q|$. Let $\mathscr{M}$ denote the collection of maximal cubes of this type, so that $R \setminus R_\gamma$ is the disjoint union of the elements of $\mathscr{M}$. This implies that

$$|(R \setminus R_\gamma) \cap \widetilde{E}| \leq \gamma|R \setminus R_\gamma| \leq \gamma|R|.$$

Combining this with (2.40) we get that $|R_\gamma| \geq |R_\gamma \cap \widetilde{E}| \geq \gamma|R|$ if γ is small enough.

2.5. THE PROOF OF LEMMA 2.38

It remains to prove Lemma 2.42. Let $\widetilde{\Delta}$ be a family of cubes on $\widetilde{E}$, as in §I.3.1, so that the corona decomposition for $\widetilde{E}$ (which is promised to us by the hypotheses of Theorem 2.29) is given in terms of $\widetilde{\Delta}$.

Let $\gamma > 0$ be given. For each $S \in \mathscr{F}_1(\rho, \gamma)$ choose a cube $T(Q(S))$ in $\widetilde{\Delta}$ such that

(2.44) $\qquad \operatorname{diam} Q(S) \leq \operatorname{diam} T(Q(S)) \leq C \operatorname{diam} Q(S)$ and

(2.45) $\qquad |Q(S) \cap T(Q(S))| \geq C^{-1}\gamma |Q(S)|.$

Here C depends only on n, d, and the regularity constant for $\widetilde{E}$. We can do this because of the definition of $\mathscr{F}_1(\rho, \gamma)$. This cube $T(Q(S))$ in $\widetilde{E}$ is not necessarily unique, but that does not matter.

Given $\gamma_1 > 0$ and $S \in \mathscr{F}_1(\rho, \gamma)$ set

$$n(S) = \{Q \in m_1(S) : |Q \cap T(Q(S))| \geq \gamma_1 |Q|\}.$$

(Recall that $m_1(S)$ was defined shortly after (2.36).)

LEMMA 2.46. *If γ_1 and ρ are small enough, then*

$$\left| \bigcup_{Q \in n(S)} Q \right| \geq \gamma_1 |Q(S)|$$

for all $S \in \mathscr{F}_1(\rho, \gamma)$.

This is not hard to prove by chasing definitions. [Do not forget that the elements of $m_1(S)$ must be pairwise disjoint.] Of course, the required smallness of γ_1 and ρ depends on γ. Let γ_1 and ρ be chosen as in Lemma 2.46 and fixed from now on.

If $S \in \mathscr{F}_1(\rho, \gamma)$ and $Q \in n(S)$ then we can find a cube $T(Q) \in \widetilde{\Delta}$ such that

(2.47) $\qquad T(Q) \subseteq T(Q(S)),$

(2.48) $\qquad \operatorname{diam} Q \leq \operatorname{diam} T(Q) \leq C \operatorname{diam} Q,$ and

(2.49) $\qquad |Q \cap T(Q)| \geq C^{-1}\gamma_1 |Q|.$

As before we can do this with a constant which depends only on n, d, and the regularity constant for $\widetilde{E}$.

Now we want to bring in the corona decomposition for $\widetilde{E}$. The point is that

(2.50) $\qquad \operatorname{Angle}(P_Q, P_{Q(S)}) \geq \delta/2$ when $Q \in n(S),$

since $n(S) \subseteq m_1(S)$, and we want to show that something similar happens for $\widetilde{E}$. The existence of a corona decomposition for $\widetilde{E}$ will then imply that this does not happen too often.

Let $\widetilde{\varepsilon}$ and $\widetilde{\delta}$ be small, to be chosen later. We can take them to be as small as we wish, so long as they do not depend on R or $\widetilde{E}$.

Because $\widetilde{E}$ admits a corona decomposition we can find a coronization $(\widetilde{\mathscr{B}}, \widetilde{\mathscr{G}}, \widetilde{\mathscr{F}})$ of $\widetilde{E}$ which satisfies the conclusions of Proposition 2.1 with E, Δ, ϵ, and δ replaced by $\widetilde{E}$, $\widetilde{\Delta}$, $\tilde{\epsilon}$, and $\tilde{\delta}$. We put tildes on everything having to do with $\widetilde{E}$; thus each $\widetilde{S} \in \widetilde{\mathscr{F}}$ has a top cube $\widetilde{Q}(\widetilde{S})$, and to each $\widetilde{Q} \in \widetilde{\mathscr{G}}$ we associate a d-plane $\widetilde{P}_{\widetilde{Q}}$ such that

(2.51) $\qquad \text{dist}(x, \widetilde{P}_{\widetilde{Q}}) \leq \tilde{\epsilon} \operatorname{diam} \widetilde{Q} \quad$ for all $x \in 2\widetilde{Q}$.

The analogue of (2.3) in this case is

(2.52) $\qquad \text{Angle}(\widetilde{P}_{\widetilde{Q}}, \widetilde{P}_{\widetilde{Q}(\widetilde{S})}) \leq \tilde{\delta} \quad$ whenever $\widetilde{Q} \in \widetilde{S}$, $\widetilde{S} \in \widetilde{\mathscr{F}}$.

Define $\widetilde{\mathscr{A}_0}$, $\widetilde{\mathscr{A}} \subseteq \widetilde{\Delta}$ by

$$\widetilde{\mathscr{A}_0} = \widetilde{\mathscr{B}} \cup \{\widetilde{Q}(\widetilde{S}) : \widetilde{S} \in \widetilde{\mathscr{F}}\},$$
$$\widetilde{\mathscr{A}} = \widetilde{\mathscr{A}_0} \cup \{\widetilde{Q} \in \widetilde{\Delta} : \widetilde{Q} \text{ is a minimal cube of some } \widetilde{S} \in \widetilde{\mathscr{F}}\}.$$

Then $\widetilde{\mathscr{A}_0}$ satisfies a Carleson packing condition with a constant that depends on $\tilde{\epsilon}$ and $\tilde{\delta}$ but not on $\widetilde{E}$ itself, because of the hypothesis of Theorem 2.29. It is easy to see that the same must be true of $\widetilde{\mathscr{A}}$. [Notice that if $\widetilde{Q}$ is the minimal cube of some $\widetilde{S} \in \widetilde{\mathscr{F}}$, then each of the children of $\widetilde{Q}$ lies in $\widetilde{\mathscr{A}_0}$.]

LEMMA 2.53. *If $\epsilon\delta^{-1}$ is small enough and if $\tilde{\epsilon}$ and $\tilde{\delta}$ are small enough (depending on δ), then for each $S \in \mathscr{F}_1(\rho, \gamma)$ and each $Q \in n(S)$ there is $\widetilde{Q} \in \widetilde{\mathscr{A}}$ such that*

(2.54) $\qquad T(Q) \subseteq \widetilde{Q} \subseteq T(Q(S))$.

The requirements on the size of $\epsilon\delta^{-1}$, $\tilde{\epsilon}$, and $\tilde{\delta}$ also depend on γ_1 and, hence, on γ, but that is fine. [The reader should remember our remark (in the second paragraph after Lemma 2.30) that we may require ϵ to be much smaller than δ. Of course, we are free to take $\tilde{\epsilon}$ and $\tilde{\delta}$ as small as we wish.]

Lemma 2.53 provides the mechanism by which we can transfer information from $\widetilde{E}$ to E.

Let $S \in \mathscr{F}_1(\rho, \gamma)$ and $Q \in n(S)$ be given, so that

(2.55) $\qquad \sup_{x \in Q} \text{dist}(x, P_Q) \leq \epsilon \operatorname{diam} Q \quad$ and

and

(2.56) $\qquad \sup_{x \in Q(S)} \text{dist}(x, P_{Q(S)}) \leq \epsilon \operatorname{diam} Q(S)$.

We may as well assume that neither $T(Q)$ nor $T(Q(S))$ belong to $\widetilde{\mathscr{A}}$, since otherwise the conclusion of the lemma is automatic. Thus we have

(2.57) $\qquad \sup_{x \in T(Q)} \text{dist}(x, \widetilde{P}_{T(Q)}) \leq \tilde{\epsilon} \operatorname{diam} T(Q)$

and

(2.58) $$\sup_{x \in T(Q(S))} \operatorname{dist}(x, \widetilde{P}_{T(Q(S))}) \leq \tilde{\epsilon} \operatorname{diam} T(Q(S)).$$

From Lemma 2.32 we conclude that

(2.59) $$\operatorname{Angle}(P_Q, \widetilde{P}_{T(Q)}) \leq C(\epsilon + \tilde{\epsilon})$$

and

(2.60) $$\operatorname{Angle}(P_{Q(S)}, \widetilde{P}_{T(Q(S))}) \leq C(\epsilon + \tilde{\epsilon}).$$

This uses also (2.48), (2.49), (2.44), and (2.45). These constants C depend on γ and γ_1 as well as innocuous quantities like the regularity constants for E and $\widetilde{E}$ but not on ϵ, $\tilde{\epsilon}$, δ, or $\tilde{\delta}$.

Because of (2.50), we now get that

(2.61) $$\operatorname{Angle}(\widetilde{P}_{T(Q)}, \widetilde{P}_{T(Q(S))}) \geq \delta/10$$

if $\epsilon\delta^{-1}$ and $\tilde{\epsilon}\delta^{-1}$ are small enough. Therefore, $T(Q)$ and $T(Q(S))$ cannot belong to the same $\widetilde{S} \in \widetilde{\mathscr{F}}$ if $\tilde{\delta} < \delta/10$, because of (2.52). This implies that there is indeed a $\widetilde{Q} \in \widetilde{\mathscr{A}}$ which satisfies (2.54), so Lemma 2.53 is established.

Let us now prove (2.43), assuming that ϵ, $\tilde{\epsilon}$, and $\tilde{\delta}$ satisfy the requirements above.

Given $S \in \mathscr{F}_1(\rho, \gamma)$, set

$$\widetilde{\mathscr{A}}(S) = \{\widetilde{Q} \in \widetilde{\mathscr{A}} : \widetilde{Q} \subseteq T(Q(S)) \text{ and } \widetilde{Q} \supseteq T(Q) \text{ for some } Q \in n(S)\}.$$

LEMMA 2.62. $\sum_{\widetilde{Q} \in \widetilde{\mathscr{A}}(S)} |\widetilde{Q}| \geq C^{-1}|Q(S)|$ for each $S \in \mathscr{F}_1(\rho, \gamma)$.

Indeed,

$$\sum_{\widetilde{Q} \in \widetilde{\mathscr{A}}(S)} |\widetilde{Q}| \geq \left| \bigcup_{\widetilde{Q} \in \widetilde{\mathscr{A}}(S)} \widetilde{Q} \right| \geq \left| \bigcup_{Q \in n(S)} T(Q) \right|$$

$$\geq \left| \bigcup_{Q \in n(S)} (Q \cap T(Q)) \right| = \sum_{Q \in n(S)} |Q \cap T(Q)|,$$

since the Q's in $n(S)$ are disjoint. Lemma 2.62 now follows from (2.49) and Lemma 2.46.

LEMMA 2.63. *Each $\widetilde{Q} \in \widetilde{\mathscr{A}}$ can lie in $\widetilde{\mathscr{A}}(S)$ for at most a bounded number of $S \in \mathscr{F}_1(\rho, \gamma)$.*

This is rather easy. If $S \in \mathscr{F}_1(\rho, \gamma)$ and $\widetilde{Q} \in \widetilde{\mathscr{A}}(S)$, then we can find a cube $U(\widetilde{Q})$ in S such that

(2.64) $$\operatorname{diam} U(\widetilde{Q}) \leq \operatorname{diam} \widetilde{Q} \leq C \operatorname{diam} U(\widetilde{Q}) \text{ and}$$
(2.65) $$U(\widetilde{Q}) \cap \widetilde{Q} \neq \varnothing.$$

Indeed, if $\widetilde{Q} \in \widetilde{\mathscr{A}}(S)$, then $\widetilde{Q} \subseteq T(Q(S))$ and $\widetilde{Q} \supseteq T(Q)$ for some $Q \in n(S)$. We can take $U(\widetilde{Q})$ to be the largest ancestor of Q which lies in S and whose diameter is no greater than $\operatorname{diam} \widetilde{Q}$.

On the other hand, there can be only a bounded number of cubes in Δ which satisfy (2.64) and (2.65) for a fixed $\widetilde{Q}$. Since the $S \in \mathscr{F}_1(\rho, \gamma)$ are pairwise disjoint subsets of Δ, we conclude that there is at most a bounded number of S's such that $\widetilde{Q} \in \widetilde{\mathscr{A}}(S)$, as desired.

We are almost finished now with the proof of Lemma 2.42. Set

$$\widetilde{\mathscr{A}}_1 = \bigcup_{S \in \mathscr{F}_1(\rho, \gamma)} \widetilde{\mathscr{A}}(S).$$

Clearly $\widetilde{\mathscr{A}}_1 \subseteq \widetilde{\mathscr{A}}$, but we also have that

$$\widetilde{\mathscr{A}}_1 \subseteq \{\widetilde{Q} \in \widetilde{\mathscr{A}} : \operatorname{diam} \widetilde{Q} \leq C \operatorname{diam} R \text{ and } \operatorname{dist}(\widetilde{Q}, R) \leq C \operatorname{diam} R\},$$

because of the definition of $\mathscr{F}_1(\rho, \gamma)$ (just before Lemma 2.41). Therefore,

(2.66) $$\sum_{\widetilde{Q} \in \widetilde{\mathscr{A}}_1} |\widetilde{Q}| \leq C|R|,$$

since $\widetilde{\mathscr{A}}$ satisfies a Carleson packing condition. From here (2.43) follows easily, with the help of Lemmas 2.62 and 2.63.

This completes the proof of Lemma 2.42 and of Theorem 2.29 as well.

References

[AA] W. K. Allard and F. J. Almgren, Jr., *The structure of stationary one dimensional varifolds with positive density*, Invent. Math. **34** (1976), 83–97.

[B] C. J. Bishop, *Harmonic measures supported on curves*, Dissertation, Univ. of Chicago, 1987.

[BCGJ] C. J. Bishop, L. Carleson, J. B. Garnett, and P. W. Jones, *Harmonic measures supported on curves*, Pacific J. Math. **138** (1989), 233–236.

[BJ] C. J. Bishop and P. W. Jones, *Harmonic measure, L^2 estimates, and the Schwarzian derivative*, J. Analyse Math. (to appear).

[BDS] F. Brackx, R. Delanghe, and F. Sommen, *Clifford Analysis*, Pitman, New York, 1982.

[Ca] A. P. Calderón, *Cauchy integrals on Lipschitz curves and related operators*, Proc. Nat. Acad. Sci. U.S.A. **74** (1977), 1324–1327.

[Ch] F. M. Christ, *Lectures on Singular Integral Operators*, CBMS Regional Conf. Ser. in Math., vol. 77 (1990), Amer. Math. Soc., Providence, RI.

[CDM] R. R. Coifman, G. David, and Y. Meyer, *La solution des conjectures de Calderón*, Adv. in Math. **48** (1983), 144–148.

[CJS] R. R. Coifman, P. W. Jones, and S. Semmes, *Two elementary proofs of the L^2 boundedness of Cauchy integrals on Lipschitz curves*, J. Amer. Math. Soc. **2** (1989), 553–564.

[CMM] R. R. Coifman, A. McIntosh, and Y. Meyer, *L'intégrale de Cauchy définit un opérateur borné sur L^2 pour les courbes lipschitziennes*, Ann. of Math. (2) **116** (1982), 361–388.

[CS] R. R. Coifman and S. Semmes, *L^2 estimates in nonlinear Fourier analysis*, Proceedings of the ICM-90 satellite conference on harmonic analysis at Sendai (S. Igari, ed.), Springer-Verlag, Tokyo, 1991.

[CW] R. R. Coifman and G. Weiss, *Extensions of Hardy spaces and their uses in analysis*, Bull. Amer. Math. Soc. (N.S.) **83** (1977), 569–645.

[D1] G. David, *Opérateurs intégraux singuliers sur certaines courbes du plan complexe*, Ann. Sci. École Norm. Sup. (4) **17** (1984), 157–189.

[D2] _____, *Opérateurs d'intégrale singuliere sur les surfaces régulières*, Ann. Sci. École Norm. Sup. (4) **21** (1988), 225–258.

[D3] _____, *Morceaux de graphes lipschitziens et intégrales singulières sur un surface*, Rev. Mat. Iberoamericana **4** (1988), 73–114.

[D4] _____, *Wavelets and Singular Integrals on Curves and Surfaces*, Lecture Notes in Math., vol. 1465, Springer-Verlag, New York, 1991.

[DJ] G. David and D. Jerison, *Lipschitz approximations to hypersurfaces, harmonic measure, and singular integrals*, Indiana Math. J **39** (1990), 831–845.

[DS1] G. David and S. Semmes, *Harmonic analysis and the geometry of subsets of $\mathbf{R}^n$*, Publ. Mat. **35** (1991), 237–249.

[DS2] _____, *Singular integrals and rectifiable sets in $\mathbf{R}^n$: au-delà des graphes lipschitziens*, Astérisque **193** (1991).

[DS3] _____, *Quantitative rectifiability and Lipschitz mappings*, Trans. Amer. Math. Soc **337** (1993), 885–889.

[Do] J. R. Dorronsoro, *A characterization of potential spaces*, Proc. Amer. Math. Soc. **95** (1985), 21–31.

REFERENCES

[Fl1] K. J. Falconer, *Geometry of Fractal Sets*, Cambridge Univ. Press, Cambridge and New York, 1984.

[Fl2] ———, *Fractal geometry: mathematical foundations and applications*, Wiley, New York, 1990.

[Fg] X. Fang, *The Cauchy integral of Calderón and analytic capacity*, Thesis, Yale Univ., 1990.

[Fe] H. Federer, *Geometric measure theory*, Springer-Verlag, New York, 1969.

[G1] J. Garnett, *Positive length but zero analytic capacity*, Proc. Amer. Math. Soc. **21** (1970), 696–699.

[G2] ———, *Analytic capacity and measure*, Lecture Notes in Math., vol. 297, Springer-Verlag, New York, 1972.

[G3] ———, *Bounded analytic functions*, Academic Press, 1981.

[GJ] J. Garnett and P. W. Jones, *The corona theorem for Denjoy domains*, Acta. Math. **155** (1985), 29–40.

[H] J. E. Hutchinson, $C^{1,\alpha}$ *multiple function regularity and tangent cone behavior for varifolds with second fundamental form in* L^p, Proc. Sympos. Pure Math., vol. 44, Amer. Math. Soc., Providence, RI, 1986, pp. 281–306.

[JK] D. Jerison and C. Kenig, *Boundary behavior of harmonic functions in nontangentially accessible domains*, Adv. Math. **46** (1982), 80–147.

[J1] P. W. Jones, *Square functions, Cauchy integrals, analytic capacity, and harmonic measure*, Harmonic Analysis and Partial Differential Equations (J. Garcia-Cuerva, ed.), Lecture Notes in Math., vol. 1384, Springer-Verlag, New York, 1989.

[J2] ———, *Lipschitz and bi-Lipschitz functions*, Rev. Mat. Iberoamericana **4** (1988), 115–122.

[J3] ———, *Rectifiable sets and the travelling salesman problem*, Invent. Math. **102** (1990), 1–15.

[Jé] J. L. Journé, *Calderón-Zygmund operators, pseudo-differential operators and the Cauchy integral of Calderón*, Lecture Notes in Math., vol. 994, Springer-Verlag, New York, 1983.

[K] C. Kenig, *Weighted* H^p *spaces on Lipschitz domains*, Amer. J. Math. **102** (1980), 129–163.

[KiP] B. Kircheim and D. Preiss, *Uniformly distributed measures in Euclidean spaces*, preprint.

[KoP] O. Kowalski and D. Preiss, *Besicovitch-type properties of measures and submanifolds*, J. Reine Angew. Math. **379** (1987), 115–151.

[Ma1] P. Mattila, *Lecture notes on geometric measure theory*, Departmento de Matemáticas, Universidad de Extremadura, 1986.

[Ma2] ———, *Cauchy integrals and rectifiability of measures in the plane*, Adv. Math. (to appear).

[Ma3] ———, *Geometry of sets and measures in Euclidean spaces*, book manuscript, 1993.

[MP] P. Mattila and D. Preiss (to appear).

[Mu] T. Murai, *A real-variable method for the Cauchy transform, and analytic capacity*, Lecture Notes in Math., vol. 1307, Springer-Verlag, New York, 1988.

[N] L. Nirenberg, *Topics in nonlinear functional analysis*, Courant Institute of Mathematical Sciences, New York, 1974.

[O] K. Okikiolu, *Characterization of subsets of rectifiable curves in* $\mathbf{R}^n$, J. London Math. Soc. **46** (1992), 336–348.

[P] D. Preiss, *Geometry of measures in* $\mathbf{R}^n$: *Distribution, rectifiability, and densities*, Ann. of Math. (2) **125** (1987), 537–643.

[R] W. Rudin, *Functional Analysis*, McGraw-Hill, New York, 1973.

[S1] S. Semmes, *A criterion for the boundedness of singular integrals on hypersurfaces*, Trans. Amer. Math. Soc. **311** (1989), 501–513.

[S2] ———, *Square function estimates and the* $T(b)$ *theorem*, Proc. Amer. Math. Soc. **110** (1990), 721–726.

[S3] ———, *Differentiable function theory on hypersurfaces in* $\mathbf{R}^n$ (*without bounds on their smoothness*), Indiana Math. J. **39** (1990), 985–1004.

[S4] ———, *Analysis vs geometry on a class of rectifiable hypersurfaces in* $\mathbf{R}^n$, Indiana Math. J. **39** (1990), 1005–1035.

[S5] ———, *Chord-arc surfaces with small constant II: good parameterizations*, Adv. Math. **88** (1991), 170–199.
[Si] L. Simon, *Lectures on geometric measure theory*, Proc. Centre Math. Anal. **3** (1983), Australian National University.
[St] E. M. Stein, *Singular integrals and differentiability properties of functions*, Princeton Univ. Press, Princeton, NJ, 1970.
[V] J. Väisälä, *Invariants for quasisymmetric, quasimöbius, and bilipschitz maps*, J. Analyse Math. **50** (1988), 201–233.

Table of Selected Notation

Angle (P_1, P_2)	Section IV.2.1
$\beta(Q)$	usually as in (I.3.8) [but not always; see also (II.3.60)]
$\beta_q(Q)$	(IV.1.9)
$\beta_q(x, t)$, $\beta(x, t)$	(I.1.45), (I.1.46)
$b\beta(x, t)$	(I.2.1)
$c(Q)$	usually as in Lemma I.3.5
$cv(x, t)$	(I.2.6)
$\gamma_q(x, t)$, $\gamma_\infty(x, t)$	(I.1.38), (I.1.39)
$\gamma^{(K)}(x, t)$	(I.2.46)
$\gamma_q^{(K)}(x, t)$	(III.4.1)
$d(x)$	usually as in (II.2.6)
$D(A_1, A_2)$	(I.2.19)
Δ, Δ_j	Section I.3.1
$H^d(\cdot)$	(I.1.1)
$\mathcal{K}_d(\mathbf{R}^n)$	Definition I.1.20
$\widetilde{\mathcal{K}}_d(\mathbf{R}^n)$	(I.2.57) and (I.2.58)
$\mathcal{K}'_d(\mathbf{R}^n)$	just before Theorem I.2.59
λQ	(I.3.7)
$Q(S)$	(I.3.17)
$sy(x, t)$, $sy_q(x, t)$	(I.1.77), (I.1.78)
T_ϵ	(I.1.24)
Z	sometimes as in (II.2.7)

Table of Acronyms

Acronym	Words	Location of Explanation
BABI	Bilateral Approximation by Bilipschitz Images of R^d	Section II.4.1
BAFUP	Bilateral Approximation by Finite Unions of d-Planes	Section I.2.2
BAUP	Bilateral Approximation by Unions of d-Planes	Definition II.3.14
BALG	Bilateral Approximation by Lipschitz Graphs	Definition II.4.6
BPBI	Big Pieces of Bilipschitz Images of R^d	Definition I.1.33
BPLG	Big Pieces of Lipschitz Graphs	Definition I.1.26
BPLI	Big Pieces of Lipschitz Images of R^d	(I.1.60)
BWGL	Bilateral Weak Geometric Lemma	Definition I.2.2
CSL	Central StarLikeness	Definition II.2.99
GWALA	Generalized WALA	Definition III.4.91
GWEC	Generalized Weak Exterior Convexity Condition	Definition II.3.27
LCV	Local Convexity	Definition I.2.7
LS	Local Symmetry	Definition I.1.79
OUWGL	Other Unilateral Weak Geometric Lemma	Definition II.3.13
USFE	Usual Square Function Estimates for the Cauchy kernel	Definitions I.2.35 and I.2.38
VBPBI	Very Big Pieces of Bilipschitz Images of R^d	(I.1.61)
WALA	Weak Approximation of Lipschitz Functions by Affine Functions	Definition I.2.47

WCC	Weak Connectedness Condition	Definition I.2.12
WCD	Weak Constant Density	Definition I.2.55
WCSL	Weaker Central StarLikeness	Definition II.2.101
WDD	Weak Dyadic Density	Definition III.5.45
WEC	Weak Exterior Convexity	Definition I.2.17
WGL	Weak Geometric Lemma	Definition I.1.71
WHIP	Weak Holes Imply Posts	Definition II.3.4
WLCV	Weaker Local Convexity	Definition II.2.90
WLS	Weaker Local Symmetry	Definition II.1.20
WNB	Weak No Boxes	Definition II.3.44
WNM	Weak No Mugs	Definition II.3.40
WNR	Weak No Reels	Definition II.3.49
WTN	Weakly Topologically Nice	Definition II.4.5
WTP	Weak Thick Projections	Definition II.3.6
WTPS	Weak Two Points on each Sphere	Section II.1.4
WUSFE	Weaker Usual Square Function Estimates for the Cauchy kernel	Definition I.2.43

Table of Theorems

Let E be a d-dimensional regular set in R^n. The following is a list of necessary and sufficient conditions for E to be uniformly rectifiable. In parentheses we have included the required restrictions on the dimension, when there are any. In some cases these restrictions are needed for the conditions to make sense, while in others they represent interesting open problems. Additional open problems can be generated by attempting to remove " + WGL" in the cases where it is present.

The reader may find it convenient to consult the Table of Acronyms.

Condition	Location of Relevant Result(s)
BABI ($d = n-1$, 1)	Corollary II.4.10 ($d = n-1$), Theorem I.2.14 ($d = 1$)
BAUP	Proposition II.3.18
BALG ($d = n-1$, 1)	Corollary II.4.10 ($d = n-1$), Theorem I.2.14 ($d = 1$)
Boundedness of a few singular integral operators, + WGL	Theorem I.2.33
Boundedness of plenty of singular integral operators	Theorem I.1.57, Section I.2.3 (especially Theorem I.2.59)
BPBI	Definition I.1.65
BPLI	Theorem I.1.57
BWGL	Theorem I.2.4
CSL	Proposition II.2.100
Generalized corona decomposition	Theorem I.3.42
Geometric Lemma	Theorem I.1.47, Theorem I.1.57
Good for the Cauchy kernel, + WGL	Theorem I.2.32 (See also Theorem I.2.33.)
GWALA ($d = 1$)	Theorem III.4.94
GWEC	Theorem II.3.28
LCV	Corollary I.2.10 (See also Theorem II.2.71.)

Littlewood-Paley Conditions	Theorem I.1.66
LS	Corollary I.2.10
OUWGL	Propositions II.3.17 and II.3.18
USFE $(d = n - 1)$	Theorem I.2.41
VBPBI	Theorem I.1.57
WALA $(d = 1)$	Theorem I.2.49
WCC $(d = 1)$	Theorem I.2.14
WCD $(d = 1, 2, n - 1)$	Theorems I.2.52 and I.2.56
WCSL	Proposition II.2.102
WDD + WGL	Theorem III.5.46
Weak Littlewood-Paley Condition	Lemma I.1.86, Corollary I.2.10, and Theorem I.1.66
WEC $(d = n - 1)$	Theorem I.2.18 (See also Proposition II.3.20.)
WHIP + WTP	Theorem II.3.9
WLCV	Proposition II.2.92
WLS $(d = 1)$	Theorem II.1.21
WNB $(d = 1)$	Proposition II.3.45
WNM $(d = 1)$	Proposition II.3.41
WNR	Proposition II.3.50
WTN $(d = n - 1)$	Theorem II.4.9
WTPS $(d = 1)$	Section II.1.4
WUSFE $(d = n - 1)$	Theorem I.2.45

Index

(See also the Table of Acronyms)

Angle $(\cdot, \cdot)$, Section IV.2.1, 327
Approx $(\cdot)$, Definition I.2.21, 38
Approximate tangent, (I.1.4), 5

Big projections, Definition I.1.74, 28
Boxes, Section II.3.4, 147
BP $(\cdot)$, Section IV.2.3, 336 (See also Proposition I.1.28, 14; the rest of Part IV; and BPLG and BPBI in the Table of Acronyms, 351.)
Camembert boxes, Section II.3.4, 147
Carleson measure, (I.1.44), 19; Definition I.1.50, 20
Carleson packing condition, Definition I.3.9, 54
Carleson set, Definition I.1.69 (in $E \times \mathbf{R}_+$), 26; (I.2.44) (in $\mathbf{R}^n \setminus E$), 45
Cauchy flat, Definitions III.2.30, 227; III.3.18, 252
Cauchy integral operator, Section I.1.2, 7 (See also Theorems I.2.32, 42; I.2.33, 43.)
Center (of a cube), Lemma I.3.5, 54
Clifford analysis, Section III.3.1, 249
Closeness (of cubes), Definition I.3.23, 59
Coherent, (I.3.17), 55
Compactness arguments, Sections III.2.4, 229; III.5.1, 297
Condition B, Definitions II.4.11, 185; II.4.21, 187
Corona $(\cdot)$, Definition I.3.38, 63
Corona construction, Section IV.2.2, 328
Corona decomposition, Definition I.3.19, 57
Coronization, Definition I.3.13, 55
Cubes, Section I.3.1, 53

Densities, (I.1.2) & (I.1.3), 5; Section II.1.3, 86; Chapter III.5, 297 (See also the WCD and the WDD in the Table of Acronyms, 351.)

Garnett's counterexample, Sections I.1.1, 3; I.1.2, 7
Generalized corona decomposition, Section I.3.3, 63
Geometric lemma, Section I.1.4, 21; Definition IV.1.2, 313
(See also the rest of Chapter IV.1.)
Good for the Cauchy kernel, Section I.1.2, 7

Good for the kernels in $\mathscr{K}_d(\mathbf{R}^n)$, Definition I.1.23, 12
Good for the kernels in $\mathscr{K}_d'(\mathbf{R}^n)$, Theorem I.2.59, 49

Hausdorff measure, Section I.1.1, 3

Labellings, Definition II.4.13, 187
Lipschitz graphs, Notation and Conventions, xi
Littlewood-Paley theory, Section I.1.3, 16

Mistake, Section I.3.2, 55
Mugs, Section II.3.4, 147

Nearly ubiquitous, Definition II.4.22, 188

Projections, Theorem I.1.7, 6; Sections II.2.2, 104, and II.2.3, 110; Sections II.3.5, 154, and II.3.6, 165; Section III.3.4, 253

Radial kernels, Section I.2.3, 42
Rectifiability, Section I.1.1, 3
Reels, Section II.3.4, 147
Regular (curve), (I.1.64), 23
Regular (mappings), Section I.1.4, 21
Regular (measure), Definition I.2.50, 47
Regular (set), Definition I.1.13, 9

Second fundamental form, Section III.3.6, 261
Semicoherent, Section I.3.2, 55
Square function estimates, see USFE and WUSFE in the Table of Acronyms, 351
Stationary varifold, Sections III.2.6, 234; III.3.6, 261

Travelling salesman theorem, Theorem I.1.47, 19

Uniform rectifiability, Definition I.1.65, 24; Section I.1.4, 21
   (See also the rest of Parts I–IV.)

Weak flatness, Section III.3.6, 261
Weak Littlewood-Paley condition, (I.1.85), 30

ISBN 0-8218-1537-7